HYDROGEOLOGY
PRINCIPLES AND PRACTICE

Martin Blair

Toronto Academic Press

HYDROGEOLOGY: PRINCIPLES AND PRACTICE

Martin Blair

Toronto Academic Press
4164 Lakeshore Road
Burlington ON L7L 1A4
Canada
www.tap-books.com

Email: orders@arclereducation.com

ISBN: 978-1-77956-337-8

ABOUT THE AUTHOR

Martin Blair is a Civil Engineer with a diverse background in Geotechnical and Civil Engineering within the water industry. Currently holding the position of Design Lead within a large multinational consultancy, providing engineering consultancy services to water utility companies within the UK. Martin's academic journey includes earning his BSc degree from Glasgow Caledonian University in Scotland. Committed to continuous learning, Martin actively participates in extracurricular activities including teaching part-time for an online collage, Semester Learning & Development Limited, to share his depth of knowledge with the next generation of Engineers. Martin's passion for sharing engineering knowledge coupled with his diverse experiences, shapes him into a skilled Engineering professional.

Table of Contents

List of Figures

application in waste management. Two levels of risk assessment are recognized in the scheme. Simple risk assessment consists of quantitative calculations, typically deterministic analytical solutions using conservative (worst-case) input parameters, assumptions and methods. Complex risk assessment consists of quantitative, stochastic (probabilistic) techniques applied to analytical solutions using site-specific characterization data.

Figure 9.12. Areas of Nitrate Vulnerable Zones (nvzs) in England. The following rules apply under this mandatory, uncompensated scheme based on 'good agricultural practice': (i) closed periods for inorganic nitrogen (fertilizer) applications during the autumn and winter and for organic nitrogen (manure) applications during the autumn for both arable and grass lands; (ii) nitrogen limits applied for inorganic and organic nitrogen applied to arable and grass lands that do not exceed crop requirements; (iii) spreading controls to restrict fertilizer and manure applications on steep slopes or close to water courses; (iv) slurry storage for manure during the autumn closed period; and (v) record keeping of agricultural practices for at least 5 years.

Figure 10.1. Diagram showing the inputs, flowpaths and outputs within a river basin that comprise a catchment water balance. The total groundwater discharge, GR, is the sum of QG + QA + QU.

Figure 10.2. Map of the location of the original development area of the Great Ouse Groundwater Scheme showing the various linked components of the current Ely Ouse Essex Water Transfer Scheme.

Figure 10.3. Hydrograph records for 1970–1971 showing the net gain in flow of the River Thet from groundwater abstracted from Chalk boreholes in the pilot area of the Great Ouse Groundwater Scheme. As shown, the net gain is the difference between the actual discharge with groundwater abstraction and the estimated natural discharge without abstraction. As a result of measurements taken for identical 24-hour periods without allowance for travel times (or temporary storage effects), occasional zero flows (or even apparent losses) can arise.

Figure 10.4. Transfer of water from the Ely Ouse River to the Cut-Off Channel at Denver. Water transferred from here travels to Abberton and Hanningfield reservoirs in support of water supply and river flows in the drier south-east of England.

Figure 10.5. Schematic diagram showing the operation of an aquifer storage and recovery (ASR) scheme. ASR involves storage of available, principally surface or drinking water-quality water during the wet season through boreholes completed in a brackish-water aquifer, with subsequent retrieval from these same boreholes during dry periods. The freshwater forms a 'bubble' within the aquifer around the ASR well and can be retrieved when needed. The hydrogeological characteristics of a successful ASR storage zone include: moderate aquifer hydraulic conductivity; confinement above and below by low permeability strata; and water quality as fresh as possible to limit mixing with the native brackish water. After the United States Geological Survey.

Figure 10.6. Schematic representation of types of flow conditions at riverbank filtration sites. The majority of riverbank filtration schemes are of Type 1. Groundwater flow beneath the river (Types 3, 4 and 6) is typically neglected at most sites. The formation of unsaturated conditions beneath the river occurs if groundwater abstraction rates are not adapted to the hydraulic conductivity of the river bed or if the hydraulic conductivity of the river bed material becomes clogged due to surface water pollution inputs (Type 4). At some sites, the river bed cuts into the confining layer (Type 5). Collector wells are used with laterals at different depths, of different lengths and directions. Type 6 gives only one example with a lateral towards the river.

List of Tables

PREFACE

Hydrogeology is a scientific discipline that deals with the occurrence, distribution, and movement of water beneath the Earth's surface. This book will specifically address the occurrence and movement of groundwater below or at the edges of ancient and modern ice sheets, a field known as glaciohydrogeology. Additionally, it will explore the interaction of meltwater with the sediments and rocks within glaciated terrains. The growth and movement of ice sheets have profound and long-term effects on groundwater flow. These effects are evidenced by changes in hydraulic heads, groundwater flow patterns and velocities, the penetration depths of meltwater, recharge rates, and pore-water pressure within glaciated regions. Over the past two decades, there has been a significant increase in interest in glacial hydrogeology among glaciologists, hydrologists, and glacial geologists. This heightened interest is due to the impact of groundwater flow on ice sheet dynamics, water budgets in glacial catchments, solute fluxes, and the formation of tunnel valleys.

The implications of groundwater flow extend beyond academic interest; they are crucial for the management of potable water resources, the viability of nuclear waste repositories, and the exploitation of hydrocarbon resources. These practical concerns have greatly contributed to the development of glacial hydrogeology as a field of study.

This book will present the various factors that influence groundwater hydrology in glacial environments and will examine how the characteristics of groundwater flow affect glacier dynamics and the formation of glacial landforms. Furthermore, it aims to review the key issues and applications of glaciohydrogeology, providing a comprehensive overview of this important field.

Contents/Structure of the Book

This book is a comprehensive guide comprising ten chapters, each meticulously crafted to provide a deep understanding of hydrogeological principles and their practical applications.

Chapter 1 serves as an introduction to the field of hydrogeology, providing an overview of its history, significance, and basic concepts. It lays the groundwork for understanding the complex interactions between groundwater and geological formations.

Chapter 2 examines the physical aspects of hydrogeology, covering topics such as groundwater flow, aquifer properties, and flow dynamics. It explores the fundamental principles governing the movement and storage of groundwater in subsurface environments.

Chapter 3 explores the geological processes that influence groundwater occurrence and movement. It discusses the formation of aquifers, the effects of tectonic activity, and the role of geological structures in controlling groundwater flow.

Chapter 4 focuses on the chemical aspects of hydrogeology, including water chemistry, mineral interactions, and the processes that govern water quality. It examines how contaminants move through the subsurface and the methods used to assess and manage groundwater quality.

Chapter 5 introduces the use of environmental isotopes in hydrogeology, providing insights into groundwater age, recharge processes, and flow paths. It discusses the applications of isotopic techniques in understanding hydrological systems.

Chapter 6 explores the interactions between groundwater and surface water in catchment areas. It examines the role of groundwater in sustaining river flows, wetlands, and ecosystems, highlighting the importance of integrated water management.

Chapter 7 covers the various techniques used to investigate and characterize groundwater systems. It discusses methods such as drilling, geophysical surveys, and hydrochemical analysis, emphasizing their importance in hydrogeological assessments.

Chapter 8 focuses on groundwater contamination, discussing the sources, fate, and transport of contaminants in aquifers. It explores remediation techniques and strategies for protecting groundwater quality.

Chapter 9 examines the remediation of polluted groundwater and the protection of clean groundwater resources. It discusses technologies such as pump-and-treat systems, in-situ bioremediation, and groundwater monitoring.

Chapter 10 addresses the governance and management of groundwater resources. It discusses legal frameworks, sustainable management practices, and the role of stakeholders in ensuring the long-term sustainability of groundwater supplies.

This book offers a comprehensive exploration of hydrogeological principles and their practical applications. It is an essential resource for students, researchers, and professionals in the field of hydrogeology, providing a solid foundation for understanding and managing groundwater resources.

Audience

This book is particularly suitable for students, offering a valuable resource for undergraduate and graduate students in geology, hydrogeology, environmental science, civil engineering, and related disciplines. The book serves as an excellent textbook for hydrogeology courses, providing both theoretical foundations and practical insights.

Academics and researchers in hydrogeology and related fields will also find this book beneficial, as it presents detailed explanations and the latest research findings. It serves as a solid basis for teaching and conducting advanced research in hydrogeological processes and management.

Additionally, environmental scientists and engineers involved in environmental monitoring, remediation, and management will find the insights in this book invaluable. It covers essential topics such as groundwater quality, contaminant transport, and remediation techniques, which are crucial for addressing environmental challenges.

For educators and trainers in hydrogeology and groundwater management, this book offers a valuable resource. It provides a comprehensive overview of key concepts and practical applications in the field, making it an essential tool for training and education purposes.

Learning Features

- Start-of-chapter learning objectives
- Key terms in each chapter
- Summary Points at the end of each chapter
- Review Questions at the end of each chapter
- Case Studies and Activities in each chapter for solving real problems
- References at the end of each chapter

CHAPTER 1

Introduction to Hydrogeology

LEARNING OBJECTIVES

After studying this chapter, you will be able to:

- Understand the overview of hydrogeology
- Define hydrologic cycle and water budget
- Describe groundwater as a natural resource.

KEY TERMS FROM THIS CHAPTER

Aquatic environment	Geology
Groundwater	Groundwater contamination
Hydrological cycle	Hydrology
Managed aquifer recharge (MAR)	Soil water
Water supply management	Water treatment

INTRODUCTION

Hydrogeology, a sub-discipline of geology, focuses on the distribution and movement of groundwater within the Earth's crust. This field is crucial for understanding how water interacts with geological materials, influencing everything from the availability of fresh water for drinking and agriculture to the stability of structures built on or in the ground. Groundwater, which exists in the pores and fractures of soil and rock, is a vital resource that supports ecosystems, human consumption, and industrial processes. By studying hydrogeology, scientists can predict water flow patterns, identify potential contamination sources, and develop strategies for sustainable water management.

The science of hydrogeology integrates principles from both geology and hydrology, employing a range of techniques to analyze groundwater systems. This includes field observations, laboratory experiments, and computer modeling. Hydrogeologists assess aquifers, which are underground layers of water-bearing rock, to determine their capacity, recharge rates, and quality. They also investigate the impact of human activities, such as industrial operations, agricultural practices, and urban development, on groundwater resources. By understanding the natural processes and human influences on groundwater, hydrogeologists play a critical role in addressing environmental challenges and ensuring the long-term availability of this essential resource.

1.1. OVERVIEW OF HYDROGEOLOGY

The occurrence, distribution, movement, and geological interactions of water in the Earth's crust are highlighted in most definitions of hydrogeology. Aspects of hydrology are included in the multidisciplinary field of hydrogeology. The study of water's occurrence and movement on Earth's surface, apart from groundwater seepage and springs that maintain river flows during seasonal dry spells, is known as hydrology. But drawing too sharp a line between the two topics is not useful, especially when attempting to understand how human activity affects the aquatic environment. Our ability to approach the aquatic environment holistically will determine how well we are able to address the problems of pollution of surface and groundwater, the effects of over-exploitation of water resources, and the potential impact of climate change.

1.1.1. Early Examples of Groundwater Exploitation

It has long been known that the enormous reservoir of water below the surface of the earth is a priceless supply of water for human use and consumption. Groundwater-fed springs are valued for their life-giving or therapeutic qualities all over the world (Figure 1.1), and groundwater use existed long before knowledge of its occurrence, movement, and source.

Figure 1.1. *Lady's Well in Coquetdale, northern England (National Grid Reference NT 953 028). Groundwater seeping from glacial deposits at the foot of a gently sloping hillside is contained within an ornamental pool floored with loose gravel. The site has been used since Roman times as a roadside watering place and was walled round and given its present shape in either Roman or medieval times.*

Source: Neven Kresic; Hydrogeology 101: Introduction to Groundwater Science and Engineering; ISBN: 979-8-218-06984-1.

The wells and horizontal tunnels known as aflaj (singular: falaj) or qanats (ghanats), both Arabic terms describing a small, artificial channel excavated as part of a water distribution system, appear to have originated in Persia about 3000 years ago, are evidence that groundwater development dates back to ancient times. Examples of these systems can be seen in a strip of arid territory that stretches from Morocco to Afghanistan. The rural villages and oases supplied by aflaj are the center of Omani culture and tradition in Oman. A long-standing custom in Oman, participatory management of communal aflaj channels and distributes common-property flows to irrigation plots on a time-based basis, under the supervision of a local community.

A cross-section along a qanat is depicted in Figure 1.2, which displays its typical horizontal or gently sloping gallery that has been laboriously excavated through alluvial material. The gallery can occasionally be up to 30 km long, and access to the tunnel is provided by vertical shafts that have been dug at closely spaced intervals. Gravity flow from beneath the water table at the upper end of the qanat to a ground surface outlet and irrigation canal on the arid plain at its lower end supplies groundwater that recharges the alluvium in the foothills of the mountains. The mother well's depth (Figure 1.2) is typically less than 50 m, and discharges rarely exceed 3m^3 s^{-1} and fluctuate seasonally with changes in the water table.

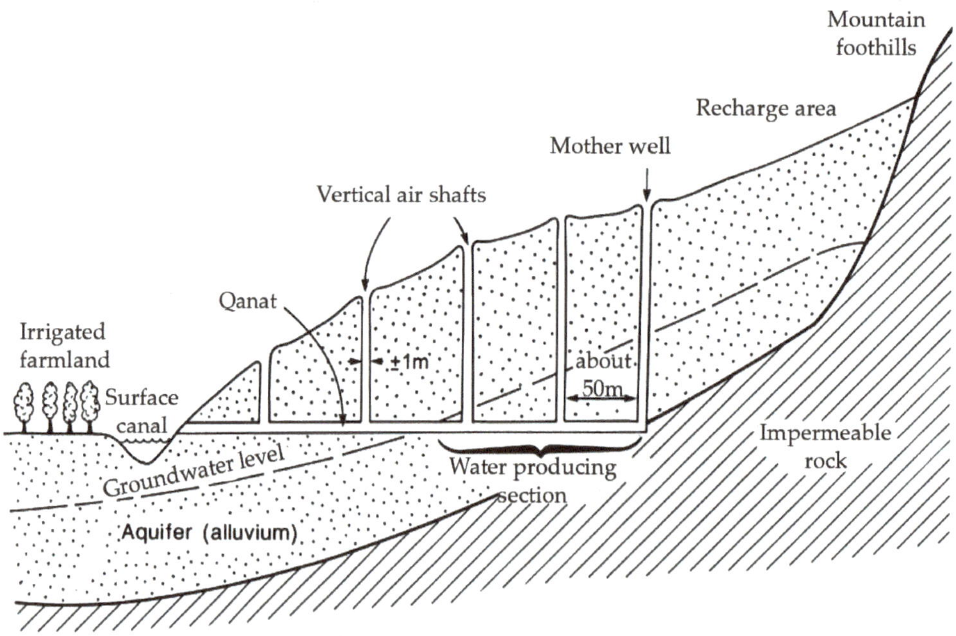

Figure 1.2. *Longitudinal section of a qanat.*

Source: Neven Kresic; Hydrogeology 101: Introduction to Groundwater Science and Engineering; ISBN: 979-8-218-06984-1.

1.1.2. The History of Hydrogeology

Hydrogeology, the study of groundwater's distribution and movement through the Earth's crust, is a field with a deep and intricate history. This discipline intersects with geology, hydrology, environmental science, and engineering, reflecting its broad applicability and essential role in sustaining life and industry. Tracing the history of hydrogeology involves exploring ancient wisdom, medieval advancements, and modern scientific breakthroughs.

1.1.2.1. Ancient Understanding of Groundwater

1. *Early Civilizations*

The knowledge of groundwater dates back to ancient civilizations such as the Egyptians, Greeks, and Chinese. These societies were among the first to recognize the importance of groundwater for irrigation, drinking, and sanitation. Ancient texts and archaeological evidence indicate that these early cultures had an intuitive understanding of groundwater flow.

- **Egyptians:** In Egypt, the annual flooding of the Nile was crucial for agriculture. The Egyptians developed early techniques for irrigation, which included wells and canals to manage water resources.
- **Greeks:** Greek philosophers like Thales of Miletus (circa 624–546 BCE) speculated about the hydrological cycle, suggesting that all matter originated from water. Later, Hippocrates (circa 460–370 BCE) wrote about the medicinal properties of water and its importance to public health.
- **Chinese:** In ancient China, water management was essential for rice cultivation. The Chinese developed complex irrigation systems and understood the necessity of managing water sources, including groundwater.

2. *Roman Engineering*

The Romans advanced hydrogeology significantly through their engineering prowess. They constructed aqueducts and wells to supply cities with fresh water. Vitruvius, a Roman architect and engineer, wrote "De Architectura" in the 1st century BCE, which included discussions on the principles of water supply, emphasizing the importance of selecting appropriate sites for wells and aqueducts based on geological conditions.

1.1.2.2. Medieval Period

During the medieval period, much of the scientific progress in Europe stagnated. However, significant advancements were made in the Islamic world. Scholars such as Al-Kindi (801–873 CE) and Al-Biruni (973–1048 CE) made considerable contributions to the understanding of geology and hydrology.

- **Al-Biruni:** He wrote extensively on geology and hydrology, proposing early theories about the origin of springs and the hydrological cycle. His works laid the groundwork for future hydrogeological studies.

1.1.2.3. Renaissance and Early Modern Period

The Renaissance era marked a resurgence in scientific inquiry in Europe, leading to renewed interest in natural sciences, including hydrogeology.

1. *Leonardo da Vinci and Georgius Agricola*

- **Leonardo da Vinci (1452–1519):** Da Vinci's notebooks reveal his keen interest in the hydrological cycle, including groundwater movement. He speculated that groundwater was replenished by precipitation, an idea ahead of its time.

- **Georgius Agricola (1494–1555):** Often considered the father of mineralogy, Agricola's work "De Re Metallica" included discussions on groundwater in the context of mining. He recognized the presence of groundwater in underground mines and suggested methods to manage it.

2. *Hydrological Cycle*

   Bernard Palissy (circa 1510–1590), a French Huguenot potter and scientist, was one of the first to articulate the concept of the hydrological cycle in detail. He described how rainfall infiltrates the ground, recharges aquifers, and emerges as springs and rivers, thus contributing to a continuous cycle.

1.1.2.4. 17th to 19th Centuries

The period from the 17th to the 19th centuries saw significant advancements in scientific understanding and technological development, which greatly influenced hydrogeology.

1. Pierre Perrault and Edme Mariotte

- **Pierre Perrault (1608–1680):** Perrault's work on the hydrology of the Seine River basin provided empirical evidence for the hydrological cycle. He demonstrated that rainfall was sufficient to account for river flow, challenging earlier theories that rivers were fed by underground sources originating from the ocean.
- **Edme Mariotte (1620–1684):** A contemporary of Perrault, Mariotte conducted experiments that further supported the concept of the hydrological cycle, particularly focusing on the measurement of precipitation and its role in recharging rivers.

2. Darcy's Law

One of the most significant milestones in hydrogeology was the formulation of Darcy's Law by Henry Darcy in 1856. While studying the public water supply of Dijon, France, Darcy developed a mathematical equation to describe the flow of groundwater through porous media. Darcy's Law remains a fundamental principle in hydrogeology, describing the relationship between hydraulic gradient and groundwater flow rate.

1.1.2.5. 20th Century: The Age of Hydrogeological Science

The 20th century marked the formal establishment of hydrogeology as a scientific discipline. This era saw the development of theories, models, and technologies that transformed our understanding and management of groundwater resources.

1. *Theis Equation*

   Charles Vernon Theis introduced a groundbreaking concept in 1935 with the Theis equation, which describes the radial flow of groundwater to a well in an unconfined aquifer. This equation has become a cornerstone of aquifer test analysis, allowing hydrogeologists to estimate aquifer properties such as transmissivity and storativity.

2. *Groundwater Modeling*

   The advent of computers in the mid-20th century revolutionized hydrogeology. Numerical modeling of groundwater flow became possible, enabling hydrogeologists to simulate complex groundwater systems and predict the impacts of human activities and natural processes.

- Finite Difference Method: Introduced by Richard E. Bellman and his colleagues, the finite difference method allowed for the discretization of groundwater flow equations, facilitating numerical solutions.
- MODFLOW: Developed by the U.S. Geological Survey in 1984, MODFLOW is a widely used groundwater modeling software that has become the industry standard for simulating groundwater flow and contaminant transport.

3. *Environmental Awareness*

   The environmental movement of the late 20th century brought increased awareness of groundwater contamination and the need for sustainable water management. Incidents such as the Love Canal disaster in the 1970s highlighted the risks of industrial pollution to groundwater resources, leading to stricter regulations and remediation efforts.

1.1.2.6. Contemporary Hydrogeology

Today, hydrogeology continues to evolve, integrating new technologies and interdisciplinary approaches to address complex challenges.

1. *Remote Sensing and GIS:* Advances in remote sensing and Geographic Information Systems (GIS) have enhanced the ability of hydrogeologists to map and analyze groundwater systems on a regional and global scale. Satellite-based measurements of soil moisture, for example, provide valuable data for monitoring groundwater recharge and depletion.
2. *Climate Change:* Climate change poses significant challenges to groundwater resources, affecting recharge rates, water demand, and the occurrence of extreme weather events. Hydrogeologists are increasingly involved in studying the impacts of climate change on groundwater and developing strategies for adaptation and mitigation.
3. *Managed Aquifer Recharge:* Managed aquifer recharge (MAR) is an innovative approach to enhancing groundwater resources by intentionally recharging aquifers with surface water. This technique is used to improve water supply reliability, especially in arid regions, and to mitigate the impacts of over-extraction.

1.1.3. The Application of Hydrogeology

Hydrogeology, the study of groundwater's distribution and movement through the Earth's crust, plays a vital role in addressing various environmental, economic, and societal challenges.

Its applications are broad and diverse, encompassing water supply management, environmental protection, agriculture, industry, and more. This comprehensive overview will delve into the multifaceted applications of hydrogeology, highlighting its critical contributions across different sectors.

1.1.3.1. Water Supply Management

1. *Groundwater Exploration and Assessment*

   One of the primary applications of hydrogeology is in the exploration and assessment of groundwater resources. Hydrogeologists use a combination of geological mapping, geophysical surveys, and drilling to locate and quantify groundwater reserves. Key activities include:

- **Aquifer Characterization:** Determining the properties of aquifers, such as porosity, permeability, and transmissivity, which influence groundwater flow and storage.
- **Water Quality Assessment:** Analyzing groundwater samples to assess water quality parameters, including pH, salinity, and the presence of contaminants like nitrates, heavy metals, and organic pollutants.
- **Recharge Estimation:** Estimating the rate at which aquifers are replenished by natural processes such as precipitation and surface water infiltration.

2. *Sustainable Groundwater Management*

   Sustainable management of groundwater resources is crucial to ensure long-term water supply for communities, agriculture, and industry. Hydrogeologists develop and implement strategies to balance groundwater extraction with recharge rates, preventing over-exploitation and depletion of aquifers. Key strategies include:

- **Groundwater Monitoring:** Regular monitoring of groundwater levels and quality to detect trends and identify potential problems.
- **Water Allocation Planning:** Developing policies and regulations to control groundwater extraction rates and allocate water resources equitably among users.
- **Artificial Recharge Projects:** Implementing techniques such as managed aquifer recharge (MAR) to augment natural recharge and maintain groundwater levels.

1.1.3.2. Environmental Protection

1. *Contaminant Transport and Remediation*

   Hydrogeologists play a crucial role in understanding and mitigating groundwater contamination. They study the movement of contaminants through soil and rock to predict their spread and design effective remediation strategies. Key activities include:

- **Contaminant Plume Mapping:** Identifying and mapping the extent of contamination in groundwater to assess the impact and plan remediation efforts.
- **Risk Assessment:** Evaluating the potential risks to human health and the environment posed by groundwater contamination.
- **Remediation Techniques:** Designing and implementing remediation methods such as pump-and-treat, bioremediation, and in-situ chemical oxidation to clean up contaminated groundwater.

1.1.3.3. Wetland and Ecosystem Protection

Groundwater plays a critical role in maintaining the health of wetlands and aquatic ecosystems. Hydrogeologists contribute to the protection and restoration of these ecosystems by:

- **Hydrological Studies:** Assessing the groundwater-surface water interactions that support wetland ecosystems.
- **Water Balance Analysis:** Evaluating the water balance of wetlands to ensure sufficient groundwater inflow and prevent degradation.
- **Conservation Planning:** Developing strategies to protect and restore wetlands, considering the impacts of land use changes, climate change, and groundwater extraction.

1.1.3.4. Agriculture

1. *Irrigation Management*

   Agriculture is a major consumer of groundwater, especially in arid and semi-arid regions. Hydrogeologists support agricultural productivity by:

- **Irrigation Scheduling:** Developing efficient irrigation schedules based on soil moisture content and crop water requirements to optimize water use.
- **Salinity Management:** Assessing and managing soil and groundwater salinity to prevent soil degradation and maintain crop yields.
- **Water Use Efficiency:** Implementing techniques such as drip irrigation and soil moisture monitoring to improve water use efficiency in agriculture.

1.1.3.5. Drought Mitigation

Hydrogeologists contribute to drought mitigation efforts by:

- **Drought Risk Assessment:** Evaluating the vulnerability of groundwater resources to drought conditions and predicting potential impacts on water supply.
- **Water Conservation Strategies:** Promoting water conservation practices and technologies to reduce groundwater demand during drought periods.
- **Emergency Water Supply Planning:** Developing contingency plans to ensure adequate water supply during prolonged droughts, including the use of alternative water sources and temporary water storage.

1.1.3.6. Industry

1. *Water Supply for Industrial Processes*

   Many industrial processes require reliable and high-quality water supply. Hydrogeologists assist industries in securing and managing their water resources by:

- **Source Identification:** Identifying suitable groundwater sources that meet the quality and quantity requirements of industrial processes.
- **Water Treatment:** Designing and implementing water treatment systems to meet specific industrial needs, such as removing contaminants or adjusting water chemistry.
- **Sustainability Planning:** Developing water management plans to ensure sustainable use of groundwater resources, considering the long-term needs of the industry and the environment.

2. *Mining and Resource Extraction*

   Mining and resource extraction activities often involve significant interactions with groundwater. Hydrogeologists contribute to the planning and management of these activities by:

- **Dewatering Systems:** Designing and managing dewatering systems to control groundwater levels in mines and prevent flooding.
- **Impact Assessment:** Evaluating the potential impacts of mining on groundwater resources, including changes in water levels and quality.
- **Remediation and Closure Planning:** Developing strategies for the remediation of contaminated groundwater and the closure of mines to ensure long-term environmental protection.

1.1.3.7. Urban Development

1. *Groundwater Management in Urban Areas*

   Urbanization can significantly impact groundwater resources through increased demand, contamination, and changes in recharge patterns. Hydrogeologists play a key role in managing these impacts by:

- **Urban Water Supply:** Ensuring a reliable groundwater supply for urban areas by identifying and managing suitable aquifers.
- **Stormwater Management:** Designing and implementing systems to manage stormwater runoff, enhancing groundwater recharge and reducing flooding risks.
- **Contamination Prevention:** Developing strategies to prevent and mitigate groundwater contamination from urban activities, such as industrial discharges, leaky sewer systems, and landfill leachate.

2. *Climate Resilience*

   Hydrogeologists contribute to building climate resilience in urban areas by:

- **Climate Impact Assessment:** Assessing the potential impacts of climate change on groundwater resources, including changes in recharge rates and water demand.
- **Adaptation Strategies:** Developing and implementing strategies to adapt urban groundwater management to climate change, such as enhancing recharge, improving water use efficiency, and diversifying water sources.
- **Integrated Water Management:** Promoting integrated water management approaches that consider the interactions between surface water, groundwater, and urban infrastructure to enhance resilience.

1.1.3.8. Groundwater Modeling and Research

1. *Numerical Modeling*

   Numerical modeling is a critical tool in hydrogeology, allowing scientists to simulate groundwater flow and transport processes under various conditions. Applications of numerical modeling include:

- **Aquifer Characterization:** Developing detailed models of aquifer systems to understand their properties and behavior.
- **Predictive Simulations:** Simulating the impacts of different scenarios, such as groundwater extraction, contamination events, and climate change, to inform management decisions.

- **Optimization Studies:** Using models to optimize groundwater management strategies, such as well placement and pumping rates, to achieve sustainable outcomes.

2. *Field Research and Data Collection*

Field research and data collection are essential components of hydrogeological studies. Hydrogeologists conduct various types of fieldwork, including:

- **Hydrogeological Surveys:** Conducting surveys to map geological formations, measure groundwater levels, and collect water quality samples.
- **Aquifer Testing:** Performing tests, such as pumping tests and slug tests, to determine aquifer properties and hydraulic parameters.
- **Environmental Monitoring:** Establishing monitoring networks to track changes in groundwater levels, quality, and flow patterns over time.

1.1.3.9. Policy and Regulation

1. *Water Rights and Allocation*

Hydrogeologists play a crucial role in developing and implementing policies and regulations related to groundwater rights and allocation. Key activities include:

- **Water Rights Assessment:** Assessing existing water rights and usage patterns to inform equitable and sustainable allocation of groundwater resources.
- **Regulatory Frameworks:** Developing regulatory frameworks to control groundwater extraction, protect water quality, and ensure compliance with environmental standards.
- **Stakeholder Engagement:** Engaging with stakeholders, including communities, industries, and policymakers, to promote informed decision-making and collaborative groundwater management.

1.1.3.10. Environmental Protection Regulations

Hydrogeologists contribute to the development and enforcement of environmental protection regulations by:

- **Contamination Prevention:** Developing guidelines and best practices to prevent groundwater contamination from industrial, agricultural, and urban activities.
- **Remediation Standards:** Establishing standards and protocols for the remediation of contaminated groundwater sites to protect public health and the environment.
- **Environmental Impact Assessments:** Conducting environmental impact assessments (EIAs) for proposed projects to evaluate potential impacts on groundwater resources and recommend mitigation measures.

1.1.3.11. Education and Public Awareness

1. *Training and Capacity Building*

Hydrogeologists are involved in training and capacity-building activities to enhance the skills and knowledge of professionals and stakeholders in groundwater management. These activities include:

- **Academic Programs:** Teaching hydrogeology courses and conducting research at universities and educational institutions to train the next generation of hydrogeologists.

- **Professional Development:** Providing training workshops and continuing education programs for practitioners in the field to keep them updated on the latest advancements and best practices.
- **Community Outreach:** Engaging with communities to raise awareness about the importance of groundwater resources and promote sustainable water use practices.

1.1.3.12. Public Awareness Campaigns

Public awareness campaigns are essential to promote the responsible use and protection of groundwater resources. Hydrogeologists contribute to these campaigns by:

- **Educational Materials:** Developing educational materials, such as brochures, posters, and online resources, to inform the public about groundwater issues.
- **Workshops and Seminars:** Organizing workshops, seminars, and public lectures to discuss groundwater management challenges and solutions with various stakeholders.
- **Media Engagement:** Collaborating with media outlets to disseminate information and raise awareness about groundwater-related topics through articles, interviews, and documentaries.

1.1.3.13. Technological Innovations

1. Remote Sensing and GIS

   Advances in remote sensing and Geographic Information Systems (GIS) have revolutionized the field of hydrogeology, providing powerful tools for groundwater exploration, monitoring, and management. Applications include:

- **Groundwater Mapping:** Using satellite imagery and aerial photography to map groundwater resources and identify potential areas for exploration.
- **Recharge Assessment:** Monitoring changes in land use and vegetation cover to assess their impact on groundwater recharge rates.
- **Data Integration:** Integrating various types of data, such as geological, hydrological, and climatic information, into comprehensive GIS databases to support decision-making.

2. Geophysical Techniques

   Geophysical techniques are widely used in hydrogeology to investigate subsurface conditions and identify groundwater resources. Common techniques include:

- **Electrical Resistivity:** Measuring the electrical resistance of subsurface materials to identify the presence of groundwater and delineate aquifer boundaries.
- **Seismic Methods:** Using seismic waves to map subsurface geological formations and detect groundwater-bearing structures.
- **Ground Penetrating Radar (GPR):** Employing radar pulses to image shallow subsurface features, useful for identifying groundwater flow paths and contamination plumes.

1.1.3.14. Future Directions and Challenges

1. *Climate Change Adaptation*

   Climate change poses significant challenges to groundwater resources, requiring adaptive management strategies. Hydrogeologists are at the forefront of developing and implementing these strategies, including:

- **Climate Impact Modeling:** Using advanced modeling techniques to predict the impacts of climate change on groundwater recharge, levels, and quality.
- **Sustainable Practices:** Promoting sustainable groundwater management practices, such as water conservation, artificial recharge, and the use of alternative water sources.
- **Policy Development:** Advocating for policies and regulations that address the impacts of climate change on groundwater resources and promote resilience.

2. *Integrated Water Resources Management (IWRM)*

   Integrated Water Resources Management (IWRM) is an approach that emphasizes the coordinated management of water resources, considering all aspects of the hydrological cycle. Hydrogeologists contribute to IWRM by:

- **Interdisciplinary Collaboration:** Working with professionals from other disciplines, such as surface water hydrologists, ecologists, and urban planners, to develop holistic water management strategies.
- **Stakeholder Involvement:** Engaging with stakeholders at all levels to ensure that diverse perspectives and needs are considered in water management decisions.
- **Sustainable Solutions:** Promoting sustainable water management practices that balance the needs of humans and ecosystems.

1.2. HYDROLOGIC CYCLE AND WATER BUDGET

The hydrologic cycle, which is the flow of water between the oceans, atmosphere, and land, depends in large part on groundwater (Figure 1.3). The primary source of groundwater is precipitation that seeps through to underlying aquifers, with surface water streams and lakes (reservoirs) providing a smaller amount of the groundwater. Merely trace amounts of groundwater originate from magmatism-related processes in the deeper crust; this water is commonly referred to as juvenile water.

With the exception of glaciers and ice caps, the amount of groundwater that is held and circulated through aquifers and aquitards in the uppermost layer of the Earth's crust is significantly greater than all other forms of movable freshwater (Figure 1.4, Left). Groundwater flows into freshwater bodies on land, such as lakes, marshes, and streams, as well as the ocean. This discharge is concentrated through seeps and springs, or it enters surface water bodies directly, where it is often invisible but occasionally quite spectacular (Figure 1.4, Right).

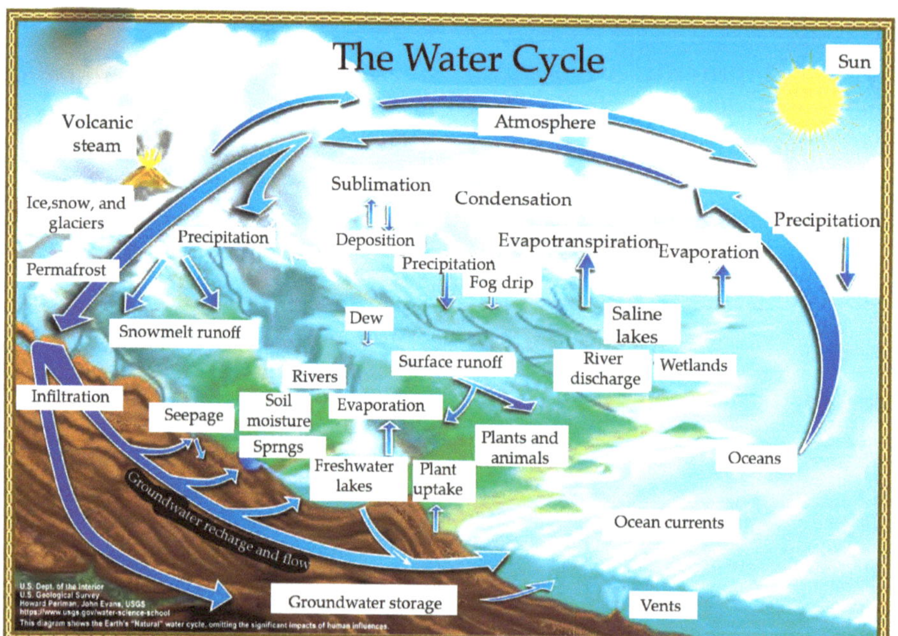

Figure 1.3. *The water (hydrologic) cycle. Modified from the United States Geological Survey (USGS).*

Source: Neven Kresic; Hydrogeology 101: Introduction to Groundwater Science and Engineering; ISBN: 979-8-218-06984-1.

From a practical hydrogeologic standpoint, subsurface water can be separated into two main zones: water stored below the water table in the saturated zone and water stored in the unsaturated zone, which is situated between the land surface and the water table. Water is stored in the pore spaces found between soil particles and the mineral grains that make up all rocks, including both consolidated and unconsolidated ("hard") rocks. Pore spaces are also partly filled with air in the unsaturated zone, also known as the vadose zone or zone of aeration. Depending on the type of soil, the climate, and the season, different amounts of water and air are present in the soil pore space in the vadose zone. This zone can be separated into the capillary fringe directly above the water table, the intermediate zone, and the uppermost zone of soil water based on the occurrence and circulation of water.

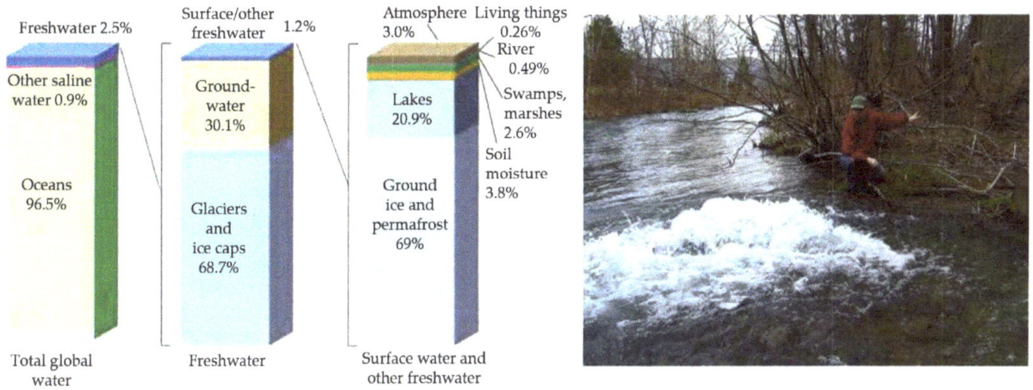

Figure 1.4. *Left: Distribution of water on Earth. Modified from USGS. Right: Boiling Spring on Potts Creek, Alleghany County, Virginia.*

Source: Neven Kresic; Hydrogeology 101: Introduction to Groundwater Science and Engineering; ISBN: 979-8-218-06984-1.

The area of the lithosphere known as the "zone of soil water" is where water is released into the atmosphere in detectable amounts by direct evaporation and plant transpiration, together known as evapotranspiration. It is only a few feet thick where the surface is covered in grass or regular field crops, but much thicker in forests and in tracts that support certain deep-rooted desert plants. Its thickness varies greatly depending on the type of soil and vegetation. Since the soil water is close enough to the surface of the ground for plant roots to access, it is the main source of interest in agriculture. Different plant species, as well as variations in soil and moisture content, have different root systems that can reach different depths for water. Most field crops and grasses take up water from depths of up to seven feet. But once well established in fine sandy loess soils, crops like alfalfa can draw groundwater from as deep as 20 to 30 feet below the surface of the earth. Substantial trees and some varieties of arid plants with deep roots extract water from significant depths. There is proof that some perennials can send their roots down to 50 or even 60 feet below the surface, and that one kind of mesquite can get water from as deep as 50 feet below the surface. Soil water is characterized by its availability to plants as either available or unavailable for plant growth. The latter cannot be absorbed by plants quickly enough to promote growth because it is adhered to so firmly by adhesion or other forces. Meinzer (1927) provides more information on the different interactions between water below the surface and vegetative cover.

Water in the soil that is in balance with atmospheric water vapor is known as hygroscopic water. In essence, it is the water that the molecular attraction of the soil grains can retain to prevent evaporation or gravity-induced downward drainage. It requires heating to between 100 and 110°C before it can be extracted from the soil (i.e., following the transformation of hygroscopic water back into vapor). Another term for it that is frequently used is residual soil moisture.

The area immediately above the water table is known as the capillary fringe. It is made up of capillary interstices, some or all of which are filled with water that is continuous with the water in the saturation zone but is kept above it by capillarity defying gravity. The combined action of two primary molecular forces, cohesion (attraction between water molecules) and adhesion (attraction between molecules of porous media), is referred to as capillarity. Aquifer recharge from percolating precipitation causes seasonal patterns in the capillary fringe to move up and down with the water table.

The soil's or rock's texture affects how thick the capillary fringe is. If the fringe is composed of materials with large capillary interstices throughout, it is comparatively thin. Sub-capillary interstices alone, as in the case of fresh crystalline rocks, are not considered indicative of a capillary fringe, nor do they constitute a functional component of one. The capillary fringe is almost nonexistent in materials with super-capillary interstices, like uniform coarse gravels. It has been commonly observed that the capillary fringe thickness in silty materials is approximately 8 feet; in very fine-grained materials (clay), it is even thicker, and in coarse sand, it is significantly thinner (Figure 1.5). When planning an irrigation system, calculating potential evapotranspiration from the water table, or researching the fate and transport (F&T) of specific groundwater contaminants that might be building up at the water table or in the unsaturated zone, it is crucial to understand the features of the capillary fringe.

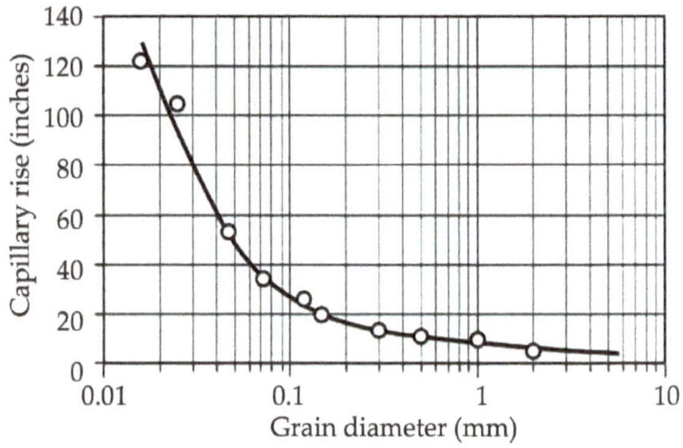

Figure 1.5. *Capillary rise in porous materials of different grain sizes.*

Source: Neven Kresic; Hydrogeology 101: Introduction to Groundwater Science and Engineering; ISBN: 979-8-218-06984-1.

The intermediate vadose zone is reached by water that seeps below the soil water zone. Here, it is either pulled into the capillary and sub-capillary pores by molecular attraction, where it may become stationary, or it is drawn further downward by gravity to the underlying zone of saturation. If the newly percolating water provides enough pressure height to overcome capillary forces, some of this "trapped" water may move downward.

The local conditions, such as the type of vegetation and the texture of the rock or soil, limit the thickness of the capillary fringe and the soil water zone, but not the intermediate zone. It is the area of the aeration zone that is left over. Depending on the hydrogeology and climate of the area, it could be completely absent or reach a thickness of several hundred feet or more. The unsaturated zone is typically thicker than 100–150 feet in arid and semi-arid regions, like the southwest of the United States, where it is isolated from rivers and groundwater discharge areas. Water that is below the water table and fully fills all voids and pore spaces in the rock within the saturated zone is known as groundwater. Gravity groundwater, or water that flows freely through rock under the influence of gravity, and groundwater that is not controlled by gravity can be distinguished based on the force that controls it. The latter is kept in place in the host rock's capillary and sub-capillary pores and fissures by resistance to gravity.

Water that is below the water table and fully fills all voids and pore spaces in the rock within the saturated zone is known as groundwater. Gravity groundwater, or water that flows freely through rock under the influence of gravity, and groundwater that is not controlled by gravity can be distinguished based on the force that controls it. The latter is kept in place in the host rock's capillary and sub-capillary pores and fissures by resistance to gravity.

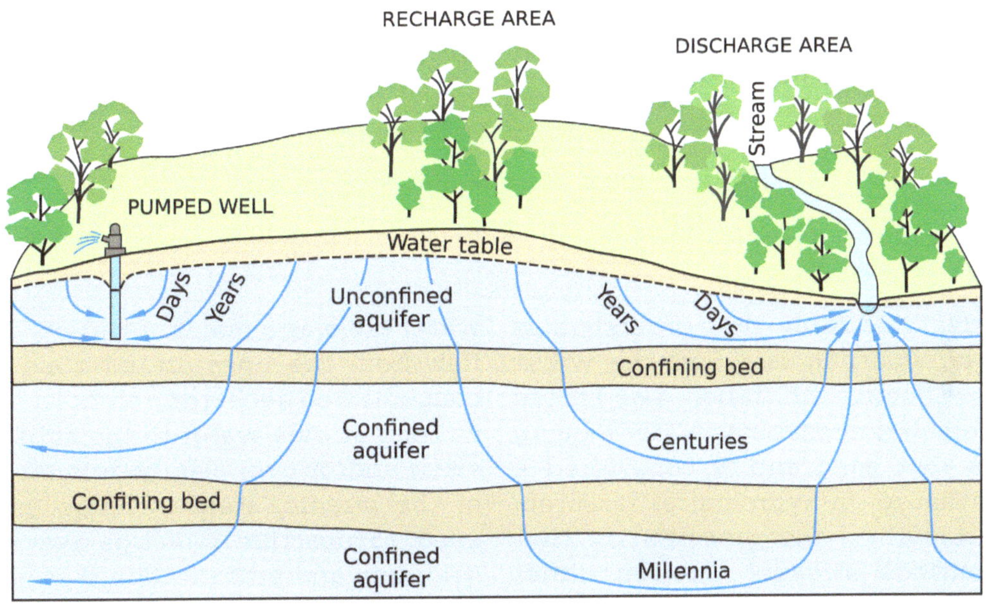

Figure 1.6. *Groundwater flow paths vary greatly in length, depth, and travel time from points of recharge to points of discharge in the groundwater system.*

Source: Neven Kresic; Hydrogeology 101: Introduction to Groundwater Science and Engineering; ISBN: 979-8-218-06984-1.

The terms "water budget" are sometimes confusing because they are frequently used interchangeably. Any water movement from the surface of the earth into the subsurface is generally referred to as infiltration. Since only a portion of this water may eventually reach the water table, also known as the saturated zone, it is sometimes referred to as potential recharge. In order to dispel any potential misunderstanding, the term "actual recharge" is being used more frequently. It refers to the amount of infiltrated water that actually reaches the aquifer and is supported by groundwater studies.

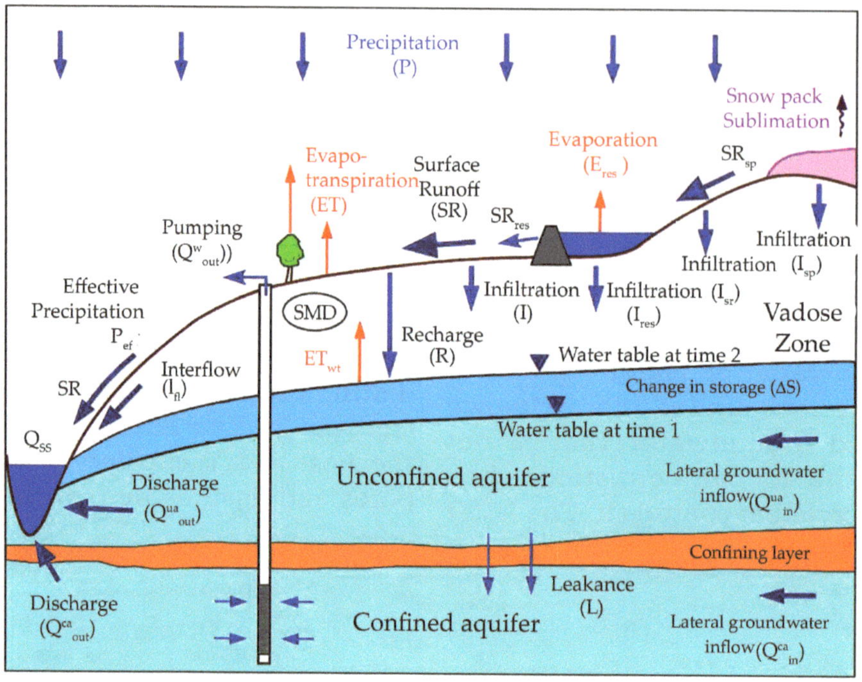

Figure 1.7. *Elements of the water budget of a groundwater system.*

Source: Neven Kresic; Hydrogeology 101: Introduction to Groundwater Science and Engineering; ISBN: 979-8-218-06984-1.

One clear indicator that groundwater recharge is occurring is when the water table rises. Effective infiltration, also known as deep percolation, involves water moving below the root zone and is considered a form of recharge. In hydrological research, effective rainfall is the precipitation that reaches surface streams through direct overland or near-surface flow, known as interflow. Rainfall excess is the portion of rainfall that generates surface runoff without seeping into the ground. Interception is the amount of rainfall that is caught by vegetation before reaching the ground, preventing it from being available for infiltration or runoff.

The phrase "net recharge" refers to the distinction between two types of water fluxes: evapotranspiration from the water table, which is an upward flux known as negative recharge, and recharge reaching the water table as a result of vertical downward flux from the unsaturated zone. Whereas concentrated recharge refers to the loss of flowing stream water to the subsurface via sinks and/or over clearly defined segments of the stream, areal recharge is the term used to describe recharge resulting from irrigation and precipitation that occurs over large areas.

The complexity of the water budget determination depends on many natural and anthropogenic factors present in the general area of interest, such as:

- Climate
- Hydrography and hydrology
- Geologic and geomorphologic characteristics
- Hydrogeologic characteristics of the surficial soils and subsurface porous media

- Land cover and land use
- Presence and operations of artificial surface water reservoirs
- Surface water and groundwater withdrawals for consumptive use and irrigation
- Wastewater management

The most general equation of water budget that can be applied to any water system has the following form:

Water Input – Water Output = Change in Storage

Equations for water budgets can be expressed in terms of volumes (for a predetermined amount of time), fluxes (volume per unit area of land surface per time, such as millimeters or inches per day), and flux densities (volume per time, such as cubic meters per day or acre-feet per year). The connections between the elements in Figure 1.7 that can be used in this instance's quantitative water budget studies of the unconfined aquifer (ua) are as follows:

$$I = P - SR - ET$$

$$I = I_{sr} + I_{res} + I_{sp}$$

$$P_{ef} = SR + I_{fl}$$

$$Q_{ss} = P_{ef} + Q_{out}^{ua} + Q_{out}^{ca}$$

$$Q_{out}^{ua} = R + Q_{in}^{ua} - L$$

$$Q_{out}^{ca} = Q_{in}^{ca} \pm L - Q_{out}^{w}$$

$$\Delta S = R + Q_{in}^{ua} \pm L - Q_{out}^{ua}$$

where I is infiltration in general, SR is surface water runoff, ET is evapotranspiration, I_{sr} is infiltration from surface runoff, I_{res} is infiltration from surface water reservoirs, I_{sp} is infiltration from snow pack and glaciers, R is groundwater recharge in general, ET_{wt} is evapotranspiration from the water table, P_{ef} is effective precipitation, I_{if} is interflow, Q_{ss} is surface stream flow, Q_{out} is direct discharge of the unconfined (ua) and confined (ca) aquifers, Qin are lateral groundwater inflows to the unconfined (ua) and confined (ca) aquifers, L is leakance from the unconfined aquifer to the underlying confined aquifer and vice-versa, Q_{out}^{w} is well pumpage, and ΔS is change in storage of the unconfined aquifer. If the area is irrigated, two more components would be added to the list: infiltration and surface runoff of the irrigation water.

To fully quantify the processes controlling the water budget, including the volumes of water stored in and flowing between the three general reservoirs—surface water, vadose zone, and saturated zone—all applicable relationships at a given site would ideally need to be established.

By default, altering one of the many parts of the water budget sets off a "chain reaction" that affects every other part. The actual physical movement of water as well as the hydraulic and hydrogeologic properties of the three main reservoirs influence how slowly these reactions occur.

1.3. GROUNDWATER AS A NATURAL RESOURCE

A valuable natural resource is groundwater. Over 2 billion people worldwide get their daily supplies from groundwater (Kemper, 2004). Along with many other industries, groundwater is essential to much of the world's irrigation and agricultural practices. Whether surface water or groundwater is used to supply water depends mainly on where aquifers are located in relation to the point of demand. While more limited but crucial water supplies may be found in poor aquifers, like weathered basement rock, in a sparsely populated rural district, a large urban population with high water demands would only be able to exploit groundwater if the aquifer, typically a sedimentary rock, has favorable storage and transmission properties. Tables 1.1 and 1.2, which provide a breakdown of water use by purpose and type (surface water and groundwater) for regions of England and Wales, can be used to infer the relationship between population and geology. The majority of freshwater abstracted for cooling purposes is returned to rivers where it can be utilized again downstream, but surface water abstraction for electricity generation is the largest category. Groundwater is particularly important in terms of abstractions for the public water supply in the Southern (73%), Anglian (41%), Midlands (39%), and Thames (33%) regions.

Table 1.1. *Estimated Abstractions from All Surface Water and Groundwater in England and Wales by Purpose and Environment Agency Region for 1996*

Region	Public water supply	Spray irrigation	Rural[1]	Electricity supply	Other industry[2]	Total
North East	2538	55	873	5142	1871	10479
Welsh	2051	10	411	6826	565	9863
North West	1595	6	174	6148	1124	9047
Southern	1451	23	1098	4210	365	7147
South West	1284	8	2046	3573	127	7038
Thames	4130	14	406	1715	255	6520
Midlands	2602	82	71	1681	406	4842
Anglian	1803	171	97	2001	497	4569
Total	17454	369	5176	31296	5210	59505

Source: Kevin M. Hiscock; Hydrogeology Principles and Practice; ISBN 0-632-05763-7.

All data are given as 103 m3 day−1. Source: Environment Agency for England & Wales.

** Category includes agricultural use (excluding spray irrigation), fish farming, public water supply (private abstractions for domestic use by individual households) and other (private domestic water supply wells and boreholes, public water supply transfer licences and frost protection use).*

† Category includes industrial and mineral washing uses.

Table 1.2. *Estimated Abstractions from Groundwaters in England and Wales by Purpose and Environment Agency Region for 1996*

Region	Public water supply	Spray irrigation	Rural[1]	Electricity supply	Other industry[2]	Total
Thames	1378 (33)[3]	8	63	0	176	1625
Southern	1056 (73)	10	202	0	155	1423
Midlands	1024 (39)	34	14	9	138	1219
Anglian	735 (41)	68	51	0	218	1072
North East	441 (17)	40	97	0	135	713
South West	407 (32)	3	185	2	31	628
North West	262 (16)	2	9	0	121	394
Welsh	113 (6)	2	9	3	28	155
Total	5416 (31)	167	630	14	1002	7229

All data are given as 10^3 m^3 day^{-1}. Source: Environment Agency for England & Wales.

Source: Kevin M. Hiscock; Hydrogeology Principles and Practice; ISBN 0-632-05763-7.

In these four regions, groundwater accounts for 42% of the total public water supply. The high-yielding Triassic and Cretaceous sandstone aquifers provide high-quality groundwater to these densely populated areas of south-east England and the English Midlands. Once more, groundwater is a vital economic resource in Europe. Table 1.3 shows that France, Germany, Italy, and Spain abstract a significant amount of groundwater (more than 5000 × 106 m^3 a^{-1}), which makes up 16% of the total water abstracted in these four nations. The average yearly water abstraction from groundwater makes up 20% of the total, with the average being over 50% in Austria, Belgium, Denmark, and Luxembourg, and only 10% and 12% in Finland and Ireland, respectively. Given the absence of a standard European method for estimating water resources and the likelihood that the data understate the contribution of groundwater to municipal water supplies, care should be taken when interpreting the data presented in Table 1.3. Roughly 75% of Europeans rely on groundwater for their water supply, according to a report written for the European Commission.

Table 1.3. *Average Annual Water Abstractions in European Union Member States by Type for the Period 1980–1995*

Country	Surface water ($\times 10^6$ m^3)	Groundwater ($\times 10^6$ m^3)	Total	% Groundwater
Denmark	9	907	916	99
Belgium	2385	4630	7015	66
Austria	1038	1322	2360	56
Luxembourg	28	29	57	51
Portugal	4233	3065	7298	42
Greece	3470	1570	5040	31
Italy	40 000	12 000	52 000	23
United Kingdom	9344	2709	12 053	22
Sweden	2121	588	2709	22
Spain	29 901	5422	35 323	15
Germany	51 151	7711	58 862	13
France	35 195	5446	40 641	13
Netherlands	10 965	1711	12 676	13
Ireland	945	125	1070	12
Finland	3011	335	3346	10
Total	193 796	47 570	241 366	20

Source: Kevin M. Hiscock; Hydrogeology Principles and Practice; ISBN 0-632-05763-7.

The data are ordered in terms of the percentage groundwater contributes to the total abstraction. Source: European Environment Agency Data Service.

A similar picture of the significance of groundwater to North America's population is presented. About 8 million people, or 26% of the total population of Canada, depend on groundwater for household needs. Of these users, five million reside in rural areas where groundwater is an affordable and dependable source of water that can be easily abstracted near the point of use. The majority of the remaining groundwater users are found in smaller municipalities. For instance, groundwater is the primary source of domestic supplies for all residents of Prince Edward Island, over 60% of those in New Brunswick, and the Yukon Territory. In Ontario, 22% of the population is dependent on groundwater, a province where groundwater is also primarily used to supply municipalities.

The abstraction of fresh and saline water in the United States from 1960 to 2000 as reported by Solley et al., (1998) and Hutson et al., (2004) is shown in Figure 1.8. The estimated total abstraction for 1995 is 1522 × 10^6 m^3 day^{-1} for all offstream uses (all uses except water used instream for hydroelectric power generation) and is 10% less than the 1980 peak estimate. This total has varied by less than 3% since 1985. In 2000, the estimated total water use in the United States is 1544 × 10^6 m^3 day^{-1}. Estimates of abstraction by source indicate that during 1995, total fresh surface water abstractions were 996 × 10^6 m^3 day^{-1} and total groundwater abstractions were 293 × 106 m^3 day^{-1} (or 23% of the combined freshwater abstractions). The respective figures for 2000 are 991 × 106 m^3 day^{-1} and 316 × 10^6 m^3 day^{-1}, with 24% of freshwater abstractions from groundwater. In 2000, the estimated total water abstraction for public water supply in the United States was 163 × 10^6 m^3 day^{-1}, representing an 8% increase from 1995. This rise contrasts with the population's 7% growth during the same time period. Although the per capita use of the public water supply increased from approximately 678 L day^{-1} in 1995 to 683 L day^{-1} in 2000, it is still less than the 696 L day^{-1} per capita recorded for 1990.

In 2000, irrigation (518 × 10^6 m^3 day^{-1} of freshwater) and cooling water for thermoelectric power generation (738 × 10^6 m^3 day^{-1}) were the two main categories of water use. The greater abstraction of freshwater between these two categories is attributed to irrigation. Between 1995 and 2000, the area under irrigation rose by almost 7%, while this water-use category saw a 2% increase in freshwater abstraction. The portion of the total irrigated by sprinkler and micro-irrigation systems has increased steadily and currently makes up more than half.

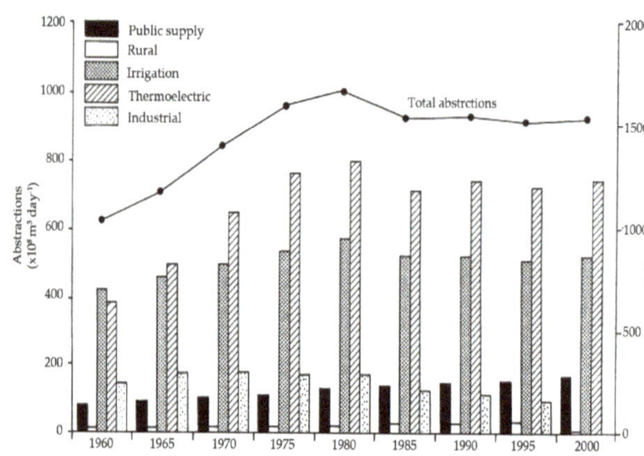

Figure 1.8. *Trends in water abstractions (fresh and saline) by water-use category and total (fresh and saline) abstractions in the United States from 1960 to 2000. After Solley et al., (1998) and Hutson et al., (2004).*

Source: Kevin M. Hiscock; Hydrogeology Principles and Practice; ISBN 0-632-05763-7.

In the arid West and the Mountain States, surface water was the main supply of irrigation water in 2000, while groundwater was the main supply in the Central States. Half of the irrigation water abstractions worldwide were accounted for by the states of California, Idaho, Colorado, and Nebraska combined. Thirty percent of groundwater abstractions were made in California and Nebraska, whereas forty percent of surface water abstractions were made in Idaho and California. Groundwater abstractions for irrigation have generally increased dramatically. Groundwater supplied 23% of the irrigation water used overall in 1950, but by 2000 that percentage had increased to 42%.

1.3.1. Management and Protection of Groundwater Resources in the United Kingdom

Our knowledge of the financial and environmental effects of groundwater exploitation has advanced alongside methods for managing and safeguarding these resources. It's interesting to see how the introduction of pertinent legislation in the UK has advanced our understanding of hydrogeology.

Prior to 1945, hydrogeological knowledge was based on a general understanding of locations that would yield good results, chemistry changes downstream from the point of recharge, and potential dangers like ground subsidence from over-exploitation of groundwater. A period of water resources assessment, including groundwater resource surveys, the creation of techniques to determine recharge amounts, and the start of groundwater studies, was sparked by the Water Act of 1945, which established legal restrictions on water abstractions. River base flow studies became necessary as a result of increased abstraction from the Chalk aquifer in the 1950s and a drought in 1959 that brought attention to the impact of groundwater abstractions on Chalk streams. Additionally, during this time, the use of quantitative pumping test analysis techniques revealed topography-related relationships and spatial variations in aquifer transmissivity.

The creation of 27 catchment-based authorities in charge of land drainage, fisheries, pollution control, and water resources was made possible by the Water Resources Act of 1963. A decade of groundwater resource management was brought about by the Act, which mandated that all abstractions in England and Wales obtain a license. The Water Resources Board was established with the responsibility of resource planning on a national level, and Section 14 of the Act mandated that every authority conduct a survey of its resources. In order to augment rivers with groundwater, perform seasonal abstraction, and create artificial recharge, regional groundwater schemes were created within the framework of river basin analysis. Future groundwater abstraction regimes can now be predicted thanks to scientific advancements in the use of numerical models to solve non-linear equations governing groundwater flow.

The Water Act of 1973 signaled the current interest in groundwater quality and acknowledged the significance of water quality aspects. Ten catchment-based regional water authorities were established as a result of the Act, and they are now in charge of all water and sewerage services as well as all aspects of the water cycle. The Control of Pollution Act of 1974 restricted certain activities that could result in polluting discharges and expanded the authority of regional water authorities to regulate effluent discharge to subsurface strata. It was during this period that the first aquifer protection laws were created.

The National Rivers Authority was established to oversee water resources planning, abstraction control, pollution prevention, and aquifer protection after the Water Act of 1989 divided the regulatory and water supply responsibilities of the regional water authorities. Following the European Communities Directive on the Protection of Groundwater Against Certain Dangerous Substances (80/68/EEC), a number of other Acts of Parliament were passed, including the Environmental Protection Act 1990 and the Water Resources Act 1991, which regulate the direct and indirect discharge of hazardous substances into groundwater. The Groundwater Regulations of 1998 brought in additional discharge controls. Furthermore, all of the Water Resources Act 1963's provisions pertaining to the regulation of groundwater abstractions were combined into the Water Resources Act 1991. The National Rivers Authority (1992) developed its practice and policy for the protection of groundwater with the aim of raising awareness of the vulnerability of groundwater to surface-derived pollution. This was done in pursuit of a strategy to protect both individual borehole sources and wider groundwater resources. The practice and policy document for groundwater protection was updated in the wake of the Environment Agency's establishment under the Environment Act 1995, which brought together the National Rivers Authority, Her Majesty's Inspectorate of Pollution, and the Waste Regulatory Authorities (Environment Agency, 1998).

In light of new European initiatives, chief among them the Water Framework Directive, the Environment Agency is currently advocating for a national framework for the protection of water resources. One piece of recent legislation that promotes the environmentally friendly and sustainable use of water resources is the Water Act of 2003. The Act replaces the licensing system based on purpose of use with one based on volume consumed, and it links licensing for water abstraction to the availability of local water resources. In order to provide flexibility in modifying abstraction rights in response to rising demand and climate change, the Act also introduces time-limited licences. Starting in 2012, licenses without a term will be cancelled in the event that an abstraction seriously harms the environment, and no compensation will be given.

1.3.2. European Union Water Framework Directive

One of the most comprehensive pieces of legislation governing the management and protection of water resources in the European Union is the Water Framework Directive (WFD), which establishes a framework for Community action in the field of water policy (Council of European Communities, 2000). Adopted in December 2000, the Directive (2000/60/EC) mandates that Member States implement suitable measures to guarantee that all water bodies have a good ecological and chemical status by the year 2015. The Directive's goal is to create a framework for the protection of groundwater, coastal waters, inland surface waters, and transitional waters (estuaries) in order to stop the further degradation of aquatic ecosystems and, in relation to their water requirements, terrestrial ecosystems and wetlands. In order to be implemented, the WFD calls for an integrated approach to river basin management and encourages the long-term preservation of the water resources that are currently available. Ensuring the progressive reduction of groundwater pollution and preventing further pollution is one of the WFD's specific goals.

In order to prevent and control groundwater pollution and achieve a good groundwater chemical status, the Commission is required by Article 17 of the WFD to submit a proposal (2003/0210(COD)) for a Groundwater Daughter Directive. (Commission of the European Communities, 2003). Furthermore, the proposal presents strategies to safeguard groundwater from indirect pollution, which refers to the release of pollutants into groundwater subsequent to their infiltration through the earth or subsurface. The proposed Directive establishes threshold values (maximum permissible concentrations) in groundwater for various pollutants by comparing monitoring data with quality standards found in EU legislation on nitrates, plant protection, and biocidal products. This comparison ensures compliance with good chemical status. The proposed Directive requires Member States to define threshold values for pollutants not covered by EU legislation at the national, river basin, or groundwater body levels. This takes into account the wide range of groundwater characteristics found throughout the EU.

In order to identify significant and persistent upward trends in pollutant concentrations and to define the starting points for when corrective action must be taken, the proposed Groundwater Daughter Directive lays out precise criteria. This definition of significance takes into account both environmental significance and time series analysis. Time series are intervals of time where regular monitoring reveals a trend. The threshold at which a pollutant's concentration begins to jeopardize groundwater quality is known as environmental significance. This cutoff point is set at 75% of the threshold value or quality standard established by the Member States. The WFD will implement a comprehensive program of actions in 2012 to prevent or reduce pollution of water, including groundwater. Measures to prevent or limit groundwater pollution will be designed using monitoring data obtained through the implementation of the Groundwater Daughter Directive.

1.3.3. Management and Protection of Groundwater Resources in the United States

In the US, there are numerous federal, state, and local programs that share responsibility for groundwater management, resulting in a highly fragmented system. Different legal authorities at different levels of government enable the management of one or more groundwater threats, such as contamination from mining, urbanization, industry, and agriculture.

Beginning with the 1972 amendments to the federal Water Pollution Control Act, and followed by the Safe Drinking Water Act 1974, the federal government's role in groundwater management has increased. The introduction of the Resource Conservation and Recovery Act (RCRA) 1976 and the Comprehensive Environmental Response, Compensation and Liability Act (CERCLA) 1980, established the federal government's current focus on groundwater remediation. With these acts, the federal government has directed billions of dollars in public and private resources towards cleaning up contaminated groundwater at 'Superfund' sites, RCRA corrective action facilities and leaking underground storage tanks. In 1994, the National Academy of Sciences estimated that over a trillion dollars, or approximately $4000 per person in the United States, will be spent in the next 30 years on remediating contaminated soil and groundwater.

The federal government's approach to groundwater protection has left state and local government authorities in charge of managing many contaminant threats, such as hazardous materials used by light industries (like dry cleaners, printers, or auto repair shops). Other groundwater threats, like over-abstraction, are typically left to the management of state and local governments rather than being covered by federal law.

The Office of Ground Water Protection was established in 1984 by the US Environmental Protection Agency (USEPA) to oversee resource protection initiatives and to start a more all-encompassing approach to groundwater resource protection. These initiatives include the Sole Source Aquifer and Wellhead Protection Programs, which were created by Safe Drinking Water Act of 1986 Amendments. Communities are encouraged to safeguard the groundwater resources that supply them with drinking water through the Wellhead Protection Program (WHPP). The Sole Source Aquifer Program restricts federal actions that may contaminate significant groundwater sources.

Programs for state groundwater management are thought to be essential to achieving effective and long-term groundwater resource protection in the future. In order to better align federal programs with state priorities for protecting groundwater resources, the USEPA established the Ground Water Strategy in 1991. This strategy aimed to promote comprehensive state management of groundwater as a resource through the development of Comprehensive State Ground Water Protection Programs (CSGWPPs) (United States Environmental Protection Agency, 1999).

ROLE MODEL

Oscar Edward Meinzer: Geologist hydrologist scientist

Background

Meinzer, Oscar Edward was born on November 28, 1876, in Davis, Illinois, United States to the family of William Meinzer and Mary Julia Meinzer, farmers. Meinzer's grandparents, who had emigrated from Prussia to escape from a culture they had found oppressive, may have directly influenced Meinzer's future religious convictions, independent thought, hatred of war, and industriousness. As a child, Meinzer exhibited an enthusiastic curiosity about his surroundings, especially about the glacial till that rested on his father's fields and the fluctuating groundwater supply that occasionally left the family well dry.

Education

Meinzer gained his early education in local schools, supplemented by study at home. In 1896 he enrolled in Beloit Academy in Wisconsin and, a year later, in Beloit College, from which he graduated magna cum laude in 1901. He was a graduate student in geology at the University of Chicago in 1906-1907 and received the Doctor of Philosophy degree, again magna cum laude, in 1922.

Career

Oscar Edward Meinzer became a junior geologist on ground-water investigations (1907), acting chief (1912), and chief, ground-water division (1913), a post which he held until retirement on 30 November 1946. In that same year, he received an honorary doctorate from Beloit College. In 1918, Meinzer has commissioned a Captain of Engineers and had his orders for overseas duty when the armistice was signed.

During his thirty-four years as the chief, groundwater division (now branch) of the United States Geological Survey, Meinzer became the main architect in the development of the modern science of ground-water hydrology. He organized and trained a large number of scientists and engineers, many of whom became recognized international authorities in this vastly expanded field. When he began, the study of underground water was an insignificant and poorly appreciated art.

During his early years as chief, Meinzer initiated the development of the science of ground-water hydrology. The need for more precise and comprehensive methods to determine the perennial yield of aquifers led him to devise a quantitative approach. In a report, Outline of Methods for Estimating Ground-Water Supplies, Meinzer described twenty-six approaches, eleven of which are applicable, though not exclusively, to aquifers and parts of aquifers under water table conditions. Five of the methods are applicable to aquifers in which water moves considerable distances from intake to discharge areas.

As part of the study of ground-water hydrology, Meinzer established a laboratory, where, along with other experiments and tests, he was able to prove that as long as the flow of water through granular material is laminar, the velocity is directly proportional to the hydraulic gradient - that is, the flow conforms to Darcy's law. For field investigations, Meinzer proposed and encouraged the development of geophysical methods and such instrumentation as automatic water-stage recorders on wells. He was in the vanguard of those pioneers who urged pumping tests and other analytical tests on wells to obtain quantitative information on the water-bearing properties of aquifers. Among these was the method of Gunter Thiem, which was tested in the field and described by L. K. Wenzel.

The quantitative methods described by Wenzel, and those developed by C. V. Theis, and later C. E. Jacobs and others under Meinzer's supervision, provided additional means for determining the perennial yield of aquifers.

Meinzer pioneered in the teaming of men of these disciplines, in particular geologists and engineers. Beginning about 1930, as the demand for ground-water investigations began to increase rapidly, Meinzer and his assistants trained and supervised dozens of geologists and engineers, many of whom, with that fundamental training, were able to develop more sophisticated tools and techniques.

Achievements

Oscar Edward Meinzer was an early or first scientist to recognize the elasticity of soils, and artesian wells, and its importance, to measuring water resources and sometimes is referred to as the "father of modern groundwater hydrology". He was awarded the Bowie Medal by the American Geophysical Union for his outstanding contributions to the science of hydrology, and in 1946 he was awarded the honorary degree of Doctor of Science by his Alma Mater on the 100th anniversary of the founding of Beloit College. The O. E. Meinzer award is named for him.

CLASS ACTIVITY

Exploring Hydrogeology

Students will investigate the principles of hydrogeology by creating models to demonstrate groundwater movement and contamination.

SUMMARY

- Hydrogeology is an interdisciplinary subject and also encompasses aspects of hydrology. Hydrology has been defined as the study of the occurrence and movement of water on and over the Earth's surface independent of the seepage of groundwater and springs which sustain river flows during seasonal dry periods.
- Advances in remote sensing and Geographic Information Systems (GIS) have enhanced the ability of hydrogeologists to map and analyze groundwater systems on a regional and global scale. Satellite-based measurements of soil moisture, for example, provide valuable data for monitoring groundwater recharge and depletion.
- Groundwater is an important part of Earth's hydrologic cycle, or the movement of water between oceans, atmosphere, and land. Groundwater is derived mostly from the percolation of precipitation and, to a lesser degree, from surface water streams and lakes (reservoirs) that lose water to underlying aquifers. Only minute quantities of groundwater have their origin from processes located in the deeper crust associated with magmatism (this water is often called juvenile water).
- Hygroscopic water is the water in the soil that is in equilibrium with atmospheric water vapor. It is essentially the water which molecular attraction of soil grains can hold against evaporation, or against downward drainage by gravity. It can be removed from the soil only after heating to about 100 to 110°C (i.e., after converting hygroscopic water back into vapor). It is also commonly referred to as residual soil moisture.
- Groundwater is an important natural resource. Worldwide, more than 2 billion people depend on groundwater for their daily supply. A large proportion of the world's agriculture and irrigation is dependent on groundwater, as are a large number of industries.

REVIEW QUESTIONS

1. What is hydrogeology and why is it important?
2. What are the main processes involved in the hydrologic cycle?
3. Describe a significant historical advancement in the understanding of groundwater flow.
4. Explain the concept of a water budget and its significance in hydrogeology.
5. What is an aquifer, and why are aquifers important for groundwater storage?
6. Discuss the various uses of groundwater in human activities.

REFERENCES

1. Appleton, J. D., Fuge, R., & McCall, G. J. H. (1996). *Environmental Geochemistry and Health with Special Reference to Developing Countries*. Geological Society, London, Special Publications, *113*.

2. Bazin, G. (2005). *Henry Bazin, L'homme, Hydraulicien, Inspecteur général des Ponts et Chaussées*. Dijon: Editions Clea.
3. Berner, E. K., & Berner, R. A. (1987). *The Global Water Cycle: Geochemistry and Environment*. Prentice-Hall.
4. Brown, G. O., Garbrecht, J. D., & Hager, W. H. (2003). *Henry P. G. Darcy and Other Pioneers in Hydraulics*. American Society of Civil Engineers.
5. Hutson, S. S., Barber, N. L., Kenny, J. F., Linsey, K. S., Lumia, D. S., & Maupin, M. A. (2004). *Estimated use of Water in the United States in 2000*. United States Geological Survey Circular 1268.
6. Jones, J. A. A. (1997). *Global Hydrology: Processes, Resources and Environmental Management*. Addison Wesley Longman.
7. Karami, G. H., & Younger, P. L. (2002). Analyzing step drawdown tests in heterogeneous aquifers. *Quarterly Journal of Engineering Geology and Hydrogeology, 35*, 295–303.
8. Kemper, K. E. (Ed.). (2004). Theme issue: Groundwater–From development to management. *Hydrogeology Journal, 12*(1).
9. Kenaga, E. E., & Goring, C. A. I. (1980). Relationship between water solubility, soil sorption, octanol-water partitioning, and concentration of chemicals in biota. In J. G. Eaton, P. R. Parrish, & A. C. Hendricks (Eds.), *Aquatic Toxicology: ASTM Special Technical Publication 707* (pp. 78–115). American Society for Testing and Materials.
10. Kirk, S., & Herbert, A. W. (2002). Assessing the impact of groundwater abstractions on river flows. In K. M. Hiscock, M. O. Rivett, & R. M. Davison (Eds.), *Sustainable Groundwater Development*. Geological Society, London, Special Publications *193*, 211–233.
11. Leeson, J., Edwards, A., Smith, J. W. N., & Potter, H. A. B. (2003). *Hydrogeological Risk Assessment for Landfills and The Derivation of Groundwater Control and Trigger Levels*. Environment Agency, Bristol.
12. Lloyd, J. W., Williams, G., Foster, S. S. D., Ashley, R. P., & Lawrence, A. R. (1991). Urban and industrial groundwater pollution. In R. A. Downing & W. B. Wilkinson (Eds.), *Applied Groundwater Hydrology: A British Perspective*. Clarendon Press, 134–148.
13. Meinzer, O. E. (1927a). *Plants as Indicators of Ground Water*. U.S. Geological Survey Water-Supply Paper 577. Washington, DC.
14. Meinzer, O. E. (1927b). *Large Springs in the United States*. U.S. Geological Survey Water-Supply Paper 557. Washington, DC.
15. Solley, W. B., Pierce, R. R., & Perlman, H. A. (1998). *Estimated use of Water in the United States in 1995*. United States Geological Survey Circular 1200.

CHAPTER 2

Physical Hydrogeology

LEARNING OBJECTIVES

After studying this chapter, you will be able to:

- Define porosity
- Explain Darcy's law and hydraulic conductivity
- Describes the terms aquifers, aquitards and aquicludes
- Discuss the hydraulic properties of fractured rocks
- Define karst aquifers
- Elaborate groundwater potential and hydraulic head
- Illustrate interpretation of hydraulic head and groundwater conditions
- Explain groundwater flow theory

KEY TERMS FROM THIS CHAPTER

Aquiclude	Aquifer	Aquitard
Atmospheric pressure	Darcy velocity	Epikarstic aquifer
Hydraulic conductivity	Hydraulic gradient	Hydraulic head
Isotropy	Mass density	Montmorillonite clay
Porosity	Potentiometric surface	Storativity
Transmissive capability	Transmissivity	

INTRODUCTION

The geomorphology of landforms, the lithology of geological materials, the presence of recharge sources, and the occurrence of groundwater within the Earth's crust all influence the emergence of springs at the ground surface. Physical laws describing changes in groundwater energy govern the infiltration of rainfall into the water table and the flow of groundwater in an aquifer towards a discharge area. This chapter discusses the various rock types and hydrogeological conditions in relation to the physical characteristics of aquifer storage and permeability. The equations of steady-state and transient groundwater flow are then derived for the hydraulic conditions encountered above and below the water table, beginning with Darcy's law, the fundamental law of groundwater flow. The presentation of analytical solutions to basic one-dimensional groundwater flow problems is followed by an explanation of how topography affects groundwater movement at different scales, such as local, intermediate, and regional flow patterns.

2.1. POROSITY

Porosity refers to the fraction of void space or interstices within a given volume of soil or rock. It is represented by the symbol n and is typically calculated as the ratio of the volume of voids (V_v) to the total unit volume (V_t) of the material, $n = V_v/V_t$. The porosity of a soil or rock can be determined in the laboratory by knowing the bulk mass density (ρ_b) and particle mass density (ρ_s) of the porous material, using the relationship:

$$n = 1 - \frac{\rho_b}{\rho_s} \quad (1)$$

The relation $n_f = Fa$, where F is the number of joints per unit distance intersecting a straight scan line across a rock outcrop and a is the mean aperture of the fractures, can be used to estimate secondary or fracture porosity in fractured rocks.

Porosity and void ratio are closely related; the relationship between the two can be written as follows: $e = V_v / V_s$., which is the ratio of the volume of voids to the volume of the solid material, V_s.

$$n = \frac{e}{(1+e)} \quad (2)$$

or

$$e = \frac{n}{(1-n)} \quad (3)$$

The void ratio shows a large range of numbers. The range of the void ratio is 0.001 to 2.3 in soils and rocks with a total porosity between 0.001 and 0.7.

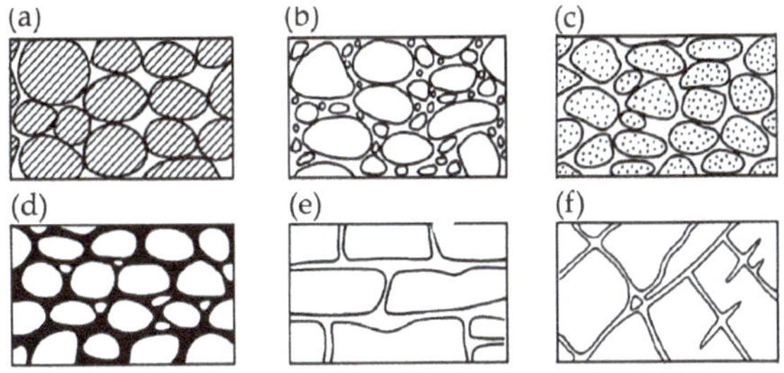

Figure 2.1. *Types of porosity with relation to rock texture: (a) well-sorted sedimentary deposit having high porosity; (b) poorly sorted sedimentary deposit having low porosity; (c) well-sorted sedimentary deposit consisting of pebbles that are themselves porous, so that the whole deposit has a very high porosity; (d) well-sorted sedimentary deposit whose porosity has been reduced by the deposition of mineral matter (cementation) in the interstices; (e) soluble rock made porous by solution; (f) crystalline rock made porous by fracturing. After Meinzer (1923).*

Source. https://www.geokniga.org/bookfiles/geokniga-hydrogeologyprinciplesandpractice.pdf.

Generally speaking, indurated, consolidated sediments like sandstone and limestone have smaller porosities than unconsolidated sediments like gravels, sands, silts, and clays, which are made up of angular and rounded particles. The porosity of crystalline igneous and metamorphic rocks is particularly low because the pores are limited to the surfaces between the crystals. On the other hand, high porosity values can be attained in formations that are rich in very fine-grained platy clay minerals.

As demonstrated in Figure 2.1, the sorting, compaction, cementation, fracturing, and solutional weathering of the constituent grains, as well as their shape and arrangement, all affect porosity. When it comes to weathered crystalline rocks, porosity values range from minuscule (0%), to 0–0.7% (40–70%) for unconsolidated clay deposits, and from 0–1% (10%) for unfractured rocks.

Primary porosity, the intrinsic properties of a soil or rock matrix that emerged during formation, is different from secondary porosity.

Secondary porosity can arise due to controlled regional fracturing and nearsurface weathering in hard rocks like igneous and metamorphic rocks, or from secondary physical and chemical weathering along the bedding planes and joints of indurated sediments like sandstones and limestones. A dual-porosity system is identified in cases where both primary and secondary porosities are present, such as in porous limestone or sandstone that has fractured and fissured.

It is not possible to consider all of the water present in a soil or rock's pore space to be accessible to groundwater flow, especially in aquifers that are fractured or have fine grains. In an aquifer with a water table, the specific yield, or S_y, is the amount of water released from groundwater storage per unit aquifer surface area per unit drop in the water table.

Table 2.1. *Range of Values of Hydraulic Conductivity and Porosity for Different Geological Materials*

Geological material	Hydraulic conductivity, K (m s^{-1})	Porosity, n
Fluvial deposits (alluvium)	10^{-5}–10^{-2}	0.05–0.35
Glacial deposits		
Basal till	10^{-11}–10^{-6}	0.30–0.35
Lacustrine silt and clay	10^{-13}–10^{-9}	0.35–0.70
Outwash sand and gravel	10^{-7}–10^{-3}	0.25–0.50
Loess	10^{-11}–10^{-5}	0.35–0.50
Sandstone	10^{-10}–10^{-5}	0.05–0.35
Shales		
Unfractured	10^{-13}–10^{-9}	0–0.10
Fractured	10^{-9}–10^{-5}	0.05–0.50
Mudstone	10^{-12}–10^{-10}	0.35–0.45
Dolomite	10^{-9}–10^{-5}	0.001–0.20
Oolitic limestone	10^{-7}–10^{-6}	0.01–0.25
Chalk		
Primary	10^{-8}–10^{-5}	0.15–0.45
Secondary	10^{-5}–10^{-3}	0.005–0.02
Coral limestones	10^{-3}–10^{-1}	0.30–0.50
Karstified limestones	10^{-6}–10^{0}	0.05–0.50
Marble, fractured	10^{-8}–10^{-5}	0.001–0.02
Volcanic tuff	10^{-7}–10^{-5}	0.15–0.40
Basaltic lava	10^{-13}–10^{-2}	0–0.25
Igneous and metamorphic rocks: unfractured and fractured	10^{-13}–10^{-5}	0–0.10

Source. https://www.geokniga.org/bookfiles/geokniga-hydrogeologyprinciplesandpractice.pdf.

Based on data contained in Freeze and Cherry (1979) and Back et al., (1988). The specific retention, or S_r, is the percentage of water that is held in the soil or rock by gravity. As depicted in Figure 2.2: The total porosity, n, is equal to the sum of the specific yield and specific retention ($S_y + S_r$). Differentiating between a porous material's effective porosity, or n_e, and total porosity is useful. While the effective porosity is related to the material's transmissive ability, the total porosity is related to the material's storage capacity.

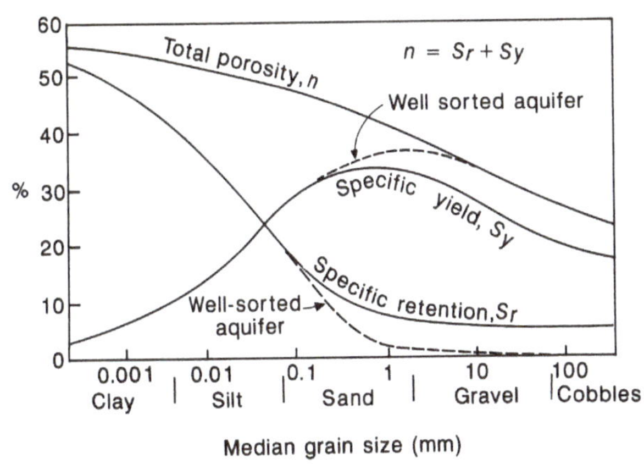

Figure 2.2. *Relation between median grain size and water storage properties of typical alluvial sediments. After Davis and De Wiest (1966).*

Source. https://www.geokniga.org/bookfiles/geokniga-hydrogeologyprinciplesandpractice.pdf.

The capillary films that encircle the solid particles in coarse-grained rocks with large pores occupy a small percentage of the pore space, so S_y and ne will nearly equal n. Capillary forces predominate in fine-grained rocks and clay, causing S_r to nearly equal n while ne will be significantly less than n. The phrase "specific surface area," or S_{sp}, which is defined as the ratio of the interstitial voids' total surface area to the porous material's total volume, can be used to explain these variations. S_{sp} will be approximately 1.5×10^4 m^{-1} in sands, but approximately 1.5×10^9 m^{-1} in montmorillonite clay (Marsily, 1986). These characteristics play a significant role in the adsorption of dissolved ions and water molecules on mineral surfaces, particularly clay.

The only effective porosity in an aquifer that is fissured or fractured, like weathered limestone and crystalline rocks, is linked to the mobile water that is present in the fissures and fractures. The water in the solid matrix is usually immobile. The effective porosity of these formations significantly decreases as depth increases because the frequency of fissures and fractures decreases and the overburden pressure increases, closing any remaining openings.

2.2. DARCY'S LAW AND HYDRAULIC CONDUCTIVITY

The fluid that is present in the networked spaces between rocks and soils can move, and a rock's hydraulic properties are determined by its capacity to both store and transfer water. The principles governing the behavior of groundwater flow in saturated material were developed empirically in 1856 by Henry Darcy, the French municipal engineer for Dijon. With the kind of test equipment depicted in Figure 2.3 Using manometer tubes at both ends of the column, Darcy measured the difference in water level, $h_1 - h_2$, as well as the cross-sectional area of flow, A. He discovered that the total flow, Q, is inversely proportional to the column length, L. This information is presented in Figure 2.3. After adding the proportionality constant, K, Darcy got:

$$Q = KA\frac{(h_1 - h_2)}{L} \tag{4}$$

In general terms, Darcy's law, as it is known, can be written as:

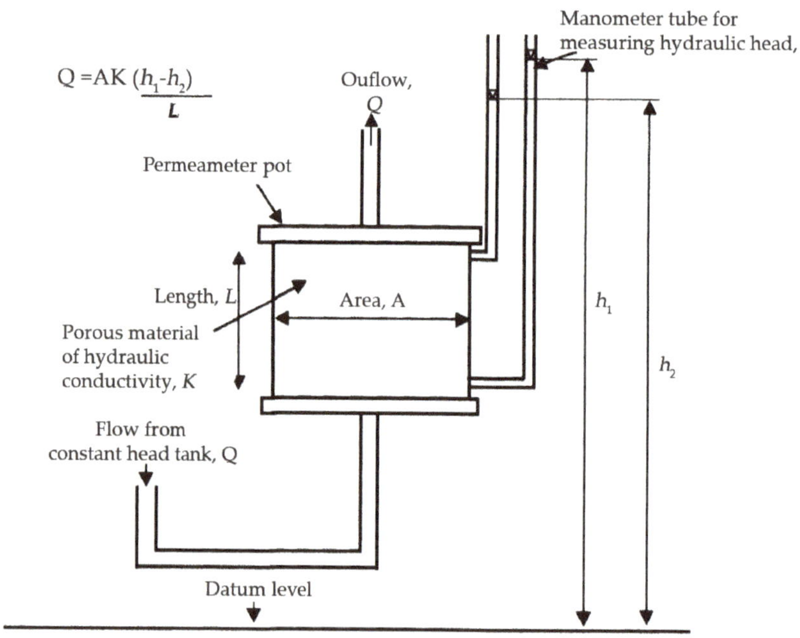

Figure 2.3. *Permeameter apparatus for determining the hydraulic conductivity of saturated porous material using Darcy's law.*

Source. https://www.geokniga.org/bookfiles/geokniga-hydrogeologyprinciplesandpractice.pdf.

$$Q = -KA\frac{dh}{dl} \qquad (5)$$

where the hydraulic gradient is represented by dh/dl, and the negative sign denotes flow in the direction of decreasing hydraulic head. The hydraulic conductivity of the porous material is denoted by K. Equation 5 can be expressed as follows if the shorthand of dh/dl equal to i is used.

$$Q = -AiK \qquad (6)$$

Hydraulic conductivity, also known as the coefficient of permeability in older publications, is a measurement of the ease with which water can pass through a porous material. Its dimensions are [L T^{-1}]. Hydraulic conductivity values exhibit a broad range of 13 orders of magnitude (Table 2.1). Hydraulic conductivity is generally high for coarse-grained and fractured materials and low for fine-grained silts and clays. Figure 2.4 provides an example of the correlation between hydraulic conductivity and grain size for two alluvial aquifers. In these aquifers, the sediment grain size typically increases with depth, resulting in the deepest part of the aquifer typically having the highest hydraulic conductivity (Sharp, 1988). The isotropy and homogeneity of the hydraulic conductivity distribution will be strongly influenced by the characteristics of the geological material.

The hydraulic conductivity of geological materials depends on the specific weight, γ (= ρg, where ρ is the fluid's density and g is its gravitational acceleration), and viscosity, μ, of the migrating fluid as well as the physical characteristics of the porous material, such that:

$$K = k_i\frac{\gamma}{\mu} \qquad (7)$$

where the physical property intrinsic to the porous material alone is the constant of proportionality, k_i, which is called the intrinsic permeability.

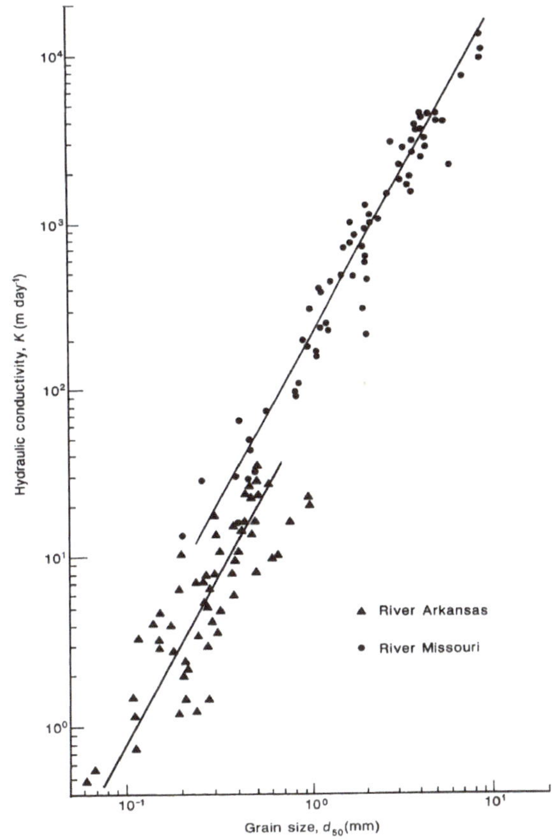

Figure 2.4. *Laboratory-determined values of hydraulic conductivity as a function of grain size for alluvial aquifers in the Rivers Missouri and Arkansas. Note the log–log scales. After Sharp (1988).*

Source. https://www.geokniga.org/bookfiles/geokniga-hydrogeologyprinciplesandpractice.pdf.

The density and viscosity of water are functions of temperature and pressure but these effects are not great for the ranges of temperature and pressure encountered in most groundwater situations. A one-third increase in hydraulic conductivity is calculated using Equation 7 for a temperature increase from 5°C (ρ = 999.965

kg m−3 , μ = 1.5188 × 10−3 Nsm−2) to 15°C (ρ = 999.099 kg m−3 , μ = 1.1404 × 10−3 Nsm−2), although a groundwater flow system exhibiting such a temperature change would be considered unusual. An example is groundwater that penetrates deep in the Earth's crust, becomes heated and returns rapidly to the surface as highly mineralized hot springs. Equally, in coastal areas, saline intrusion into fresh groundwater will cause variations in fluid density such that information about both ki and K is required in any investigation.

The size of the apertures through which the fluid passes determines the intrinsic permeability, which is a measure of the porous material's characteristics only. The following empirical relationship was derived for unconsolidated sand by Krumbein and Monk (1943), where ρ is the standard deviation of the grain size in phi units (−log2(grain diameter in mm)) and GMd is the geometric mean of the grain diameter (mm):

$$k_i = 760(GM_d)^2 e^{-1.3\sigma} = Cd^2 \quad (8)$$

As shown, Equation 8 is more generally expressed as $k_i = Cd^2$ where d is equal to the mean pore diameter and C represents a dimensionless 'shape factor' assessing the contribution made by the shape of the pore openings.

As influenced by the relationship between the pore and grain sizes and their effect on the tortuosity of fluid flow. Intrinsic permeability has the dimensions of [L^2] and, using nomenclature common in the petroleum industry, the unit of k_i is the darcy where 1 darcy is equivalent to 9.87 × 10^{-13} m^2 .Now, combining Equations 5, 7 and 8 gives a full expression of the flow through a porous material as:

$$\frac{Q}{A} = q = -K\frac{dh}{dl} = -\frac{Cd^2\rho g}{\mu}\frac{dh}{dl} \quad (9)$$

The quotient Q/A, or q, indicates the discharge per unit cross-sectional area of saturated porous material. The term q, referred to as the specific discharge, has the dimensions of velocity [L T^{-1}] and is also known as the darcy velocity or darcy flux.

It is important to remember that the darcy velocity is not the true, microscopic velocity of the water moving along winding flowpaths within the soil or rock. Instead, by dividing the specific discharge by the fraction of open space (in other words, effective porosity, n_e) through which groundwater flows across a given sectional area, this provides an average measure of groundwater velocity such that:

$$\frac{Q}{An_e} = \frac{q}{n_e} = \bar{v} \quad (10)$$

where $\bar{v}$ is the average linear velocity.

To obtain first estimates of groundwater flow and velocity, simple hydrogeological situations can be applied to Equations 5 and 10. The application of more sophisticated methods, like flow net analysis and groundwater modeling, is necessary for more accurate calculations. (Figure 2.5).

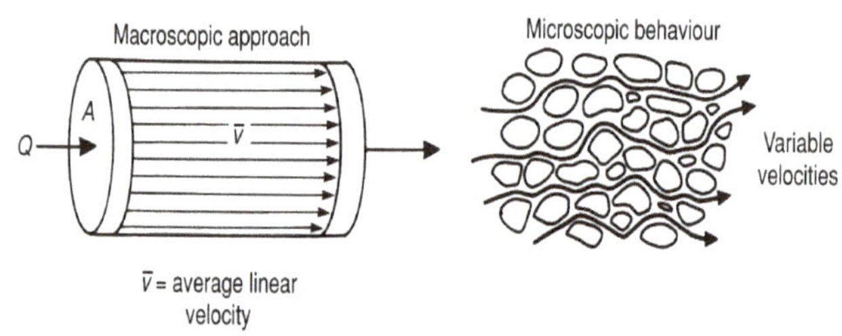

Figure 2.5. *Macroscopic (Darcian) approach to the analysis of groundwater flow contrasted with the true, microscopic behaviour of tortuous flowpaths.*

Source. https://www.geokniga.org/bookfiles/geokniga-hydrogeologyprinciplesandpractice.pdf.

2.2.1. Isotropy and Homogeneity

Whether viewed at the microscopic or regional scale, aquifer properties like hydraulic conductivity are unlikely to match the idealized, uniform porous material. The spatial variance and directional trends in aquifer property values are referred to as isotropy, anisotropy, homogeneity, and heterogeneity.

An area of geology is homogeneous if its hydraulic conductivity, K, is constant across the formation. The formation is heterogeneous if the hydraulic conductivity varies from location to location. The geological setting that gave rise to the deposit or rock type will determine the type of heterogeneity. As depicted in Figure 2.6, where each bed in the formation has a different hydraulic conductivity value, is a common example of layered heterogeneity in sedimentary rocks. Interbedded clay and sand deposits will exhibit strong, layered heterogeneity. Large contrasts can also occur when there is discontinuous heterogeneity brought on by extensive stratigraphic features or faults. Where there is a grading and sorting of the material deposits, such as in glacial outwash plains, deltas, and alluvial fans, there is trending heterogeneity. Consolidated rocks exhibit vertical trends in hydraulic conductivity, as permeability is contingent upon joint and fracture density.

It is commonly acknowledged that a log-normal probability density function can accurately describe the statistical distribution of hydraulic conductivity for a given geological formation, and that the geometric mean can be used to determine the average hydraulic conductivity. A trend in the mean hydraulic conductivity value can be interpreted as trending heterogeneity within a geological formation.

When the hydraulic conductivity at a particular location within a geological formation is constant regardless of the direction of measurement, the formation is said to be isotropic. At a point, the formation is anisotropic if the hydraulic conductivity changes with the direction of measurement. The maximum and minimum values of hydraulic conductivity correspond to the principal directions of anisotropy, which are typically perpendicular to one another. In sedimentary rocks and unconsolidated sediments, the orientation of clay minerals is the main source of anisotropy on a small scale. Strong anisotropy can be induced at different scales, ranging from the local to the regional, in consolidated rocks depending on the direction of jointing or fracturing.

Combining the aforementioned definitions, as illustrated in Figure 2.7. When characterizing a formation's hydraulic conductivity, four possible combinations of heterogeneity and anisotropy can be identified.

It is also important to understand that, due to the introduction of anisotropy, in a three-dimensional flow system the particular discharge or darcy velocity (Eq. 9) is a vector quantity with components q_x, q_y, and q_z given by: according to Darcy's law.

$$q_x = -K_x \frac{\partial h}{\partial x},\; q_y = -K_y \frac{\partial h}{\partial y},\; q_z = -K_z \frac{\partial h}{\partial z}$$

where K_x, K_y and K_z are the hydraulic conductivity values in the x, y and z directions.

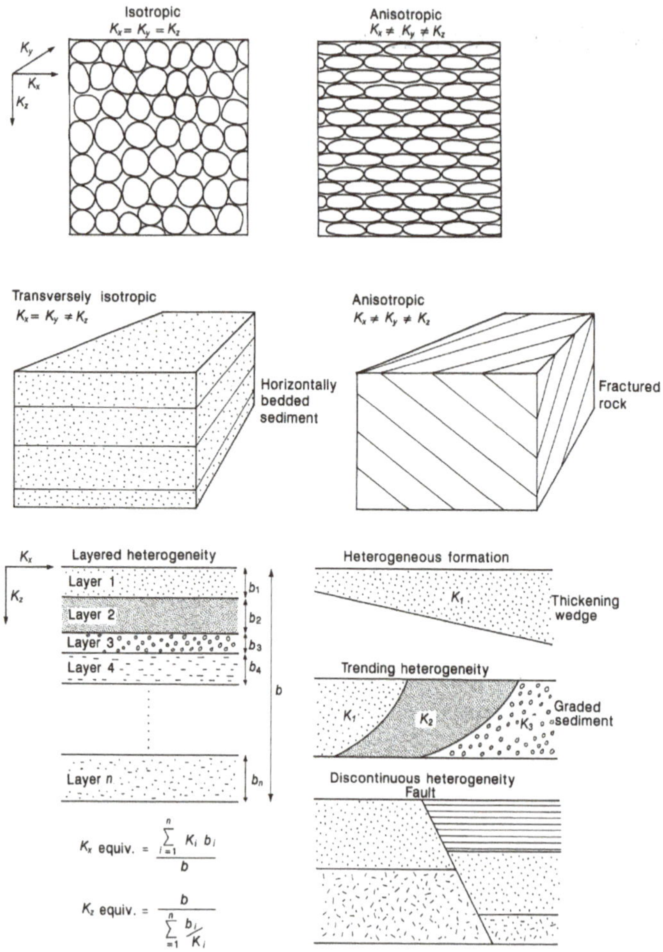

Figure 2.6. *Examples of isotropy, anisotropy and heterogeneity showing the influence of grain shape and orientation, sedimentary environment and geological structure on hydraulic conductivity. After Fetter (2001).*

Source. https://www.geokniga.org/bookfiles/geokniga-hydrogeologyprinciplesandpractice.pdf.

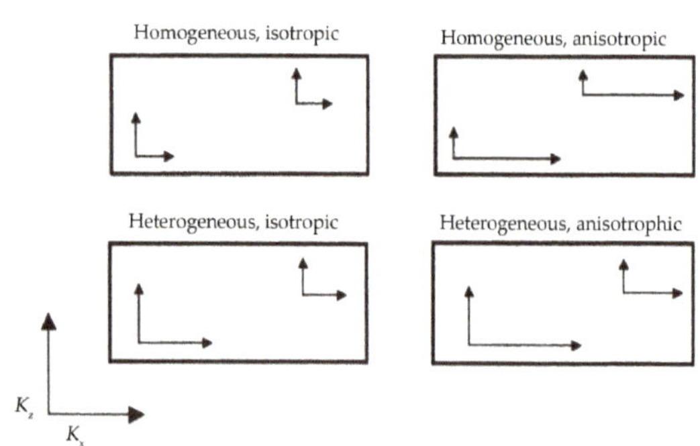

Figure 2.7. *Four possible combinations of heterogeneity and anisotropy describing the hydraulic conductivity of a porous material. After Freeze and Cherry (1979).*

Source. https://www.geokniga.org/bookfiles/geokniga-hydrogeologyprinciplesandpractice.pdf.

2.3. AQUIFERS, AQUITARDS, AND AQUICLUDES

Aquifers, aquitards, and aquicludes are identified by natural differences in the permeability and ease of transmission of groundwater in various geological materials. A layer or layered sequence of rock or sediment that consists of one or more geological formations and contains water and has the capacity to transfer sizable amounts of water under typical hydraulic gradients is known as an aquifer. Because of this, aquifers have enough permeability to transfer groundwater from wells or springs that can be profitably used. Sands, gravels, solution-weathered limestones, and fractured sandstones are the typical environments in which good aquifers develop.

A formation with lower permeability that can transfer substantial amounts of water for regional groundwater flow but from which only small amounts of groundwater can be extracted is referred to as an aquitard. Aquitards include, among other things, fractured crystalline rock, sandy clays, glacio-fluvial and riverine silts, and sedimentary rocks with few fractures.

An aquiclude is a saturated geological unit that has such low permeability that it can obstruct local groundwater flow and is unable to transfer appreciable amounts of water under typical hydraulic gradients. Clays, shales, and metamorphic rocks are examples of aquiclude rocks.

2.4. HYDRAULIC PROPERTIES OF FRACTURED ROCKS

It is implied that the physical arrangement of the grains that make up the porous material is thought of as a representative continuum and that macroscopic laws, like Darcy's law, provide macroscopically averaged descriptions of the microscopic behavior by using the Darcian approach to analyze groundwater flow. Put differently, Darcy's law characterizes groundwater flow as a flow via a porous material whose properties are thought to be continuous and smoothly varying. When analyzed at a scale comparable to the spacing of the dominant pore size, intergranular and fractured porous materials are actually very heterogeneous. As a result, Darcy's law can be effectively applied, but only on a scale big enough to include a typical assemblage of pores. The continuum scale is shown here. The local pore network geometry has a significant impact on flow and pollutant transport at subcontinuum scales. This is especially true for fractured rocks, where the fracture spacing's dimension can give rise to a continuum scale larger than many real-world issues.

The conceptual model of groundwater flow in fractured material, like crystalline and carbonate rocks, and fissured clay sediments, like glacial tills, can be either overly simplified or a detailed description of the aquifer properties can be attempted, as shown in Figure 2.8a points. Fracture flow models typically assume that both fracture apertures and flow velocities are small enough to apply Darcy's law and result in laminar flow, with the exception of conduit flow in karst aquifers. In the corresponding porous material example displayed in Figure In 2.8b, the equivalent or effective hydraulic properties of a continuous porous material are represented by the distributions of primary and secondary porosity and hydraulic conductivity. The inability to accurately estimate the size of the representative elemental volume of material from which to derive the effective hydraulic property values is a disadvantage of this method. As a result, while the equivalent porous material approach might accurately depict the behavior of a regional flow system, it is probably not going to accurately replicate local conditions.

More complex methods, like the dual-porosity and discrete fracture models in Figures 2.8c and 2.8d, depict the flow of groundwater through the fracture network.

An idealized scenario for flow through a single fracture would be between two parallel plates that have a uniform fracture aperture, or separation, 2b. Under laminar flow conditions, the relationship between flow and hydraulic gradient for individual fractures is typically thought to be governed by the "cubic law," which was first proposed by Snow (1969) and later confirmed by Witherspoon et al., Gale (1982) and (1980). The flow rate through a fracture in this treatment, Q_f, can be written as follows:

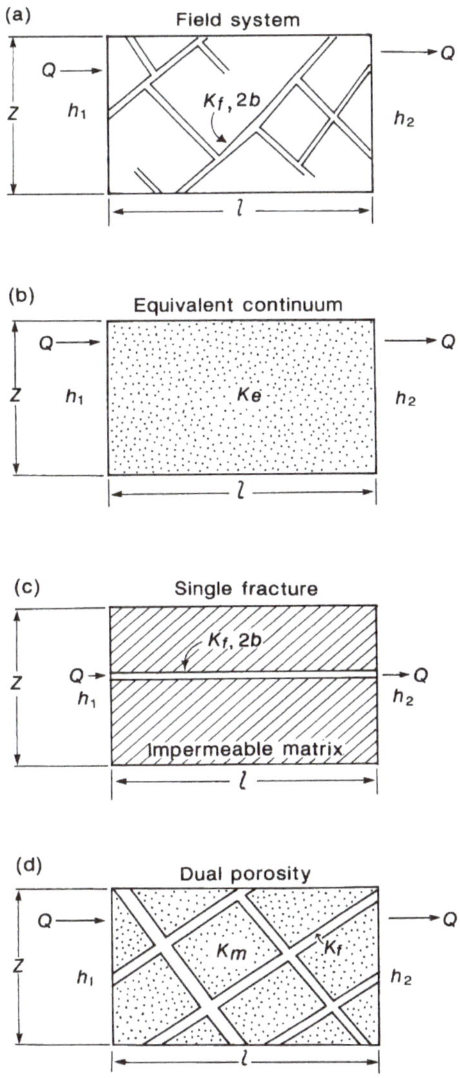

Figure 2.8. *Conceptual models to represent a fractured rock system. The fracture network of aperture 2b and with groundwater flow from left to right is shown in (a). The equivalent porous material, discrete fracture and dual porosity models representative of (a) are shown in (b), (c) and (d), respectively. After Gale (1982).*

Source. https://www.geokniga.org/bookfiles/geokniga-hydrogeologyprinciplesandpractice.pdf.

$$Q_f = -2bwK_f \frac{dh}{dl} \quad (11)$$

where w is the width of the fracture, K_f the hydraulic conductivity of the fracture and l the length over which the hydraulic gradient is measured. The hydraulic conductivity, K_f, is calculated from:

$$K_f = \frac{\rho g (2b)^2}{12\mu} \quad (12)$$

where ρ is fluid density, μ is fluid viscosity and g is the gravitational acceleration.

If the expression for K_f (Equation 12) is substituted in Equation 11, then:

$$Q_f = -\frac{2}{3}\frac{w\rho g b^3}{\mu}\frac{dh}{dl} \quad (13)$$

Equation 13 shows that as the fracture aperture's cube grows, so does the flow rate. The description of the fracture network, including the mapping of fracture apertures and geometry, is necessary for using a model based on these equations and can only be ascertained through meticulous fieldwork.

When it comes to the dual-porosity model, water and solute are exchanged between the fractures and the surrounding porous rock matrix.

A term that characterizes the rate of mass transfer may be used to represent exchange between the fracture network and the porous blocks. Field mapping and hydraulic testing are required in this model because it evaluates the hydraulic properties of the porous rock matrix as well as the fracture network.

2.5. KARST AQUIFERS

The characteristic landforms that arise on easily dissolved rock types like halite, gypsum, and limestones are commonly referred to as karst. The word "karst," which originally meant "stony ground," was used to characterize the Kras region, which is now a part of Slovenia and Croatia and is known for its unusually well-developed karst landforms. Swallow holes and enclosed depressions are common features of karst areas, along with the absence of permanent surface streams. Rainfall runoff typically happens underground in channels that have been solutionally enlarged; some of these channels are big enough to create caves. (Figure 2.9).

Figure 2.9. *The disappearance of the upper River Fergus at An Clab, south-east Burren, County Clare, Ireland, where surface runoff from Namurian shales disappears into a swallow hole at the contact with Carboniferous limestone.*

Source. https://www.geokniga.org/bookfiles/geokniga-hydrogeologyprinciplesandpractice.pdf.

A few well-known karst regions are the Greek Islands; the Dordogne, Vercors, and Tarn regions of France; the Postojna Caves in Slovenia; the Burren, County Clare, Ireland; and the pinnacle karst of the Guilin area, southern China. The Karst Working Group (2000) describes the karst in Ireland, and Ford and Williams (1989) give a thorough treatment of karst hydrology and geomorphology in general.

The Mendip Hills karst aquifer in western England is described here, with reference to Figure 2.10, which depicts a typical model of groundwater flow. The hills stretch 10 km north-south and 50 km east-west. Developed in Carboniferous limestone, they rise above the surrounding lowlands in the west to form a broad karst plateau at around 260 meters above sea level. The Mendips are made up structurally of four en echelon periclines with Devonian sandstone cores. On the northern limbs of the folds, the limestone dip is typically steep (60–90°), but it is more gentle (20–40°) to the south. In the east, the periclines are covered in Mesozoic strata that are being worn away by erosion, but in the west, they are visible through younger Triassic rocks. Because of this, the karstic features in the west are more developed and most likely older than those in the east. The main source of water supply is the Carboniferous limestone aquifer, which has been heavily mined, mostly through spring abstraction. Because spring discharges typically happen quickly after a storm and are visually spectacular, abstracted water is typically kept in surface reservoirs.

As shown in Figure 2.10, groundwater discharge is via springs located at the lowest limestone outcrop, often where the limestones dip below Triassic mudstones.

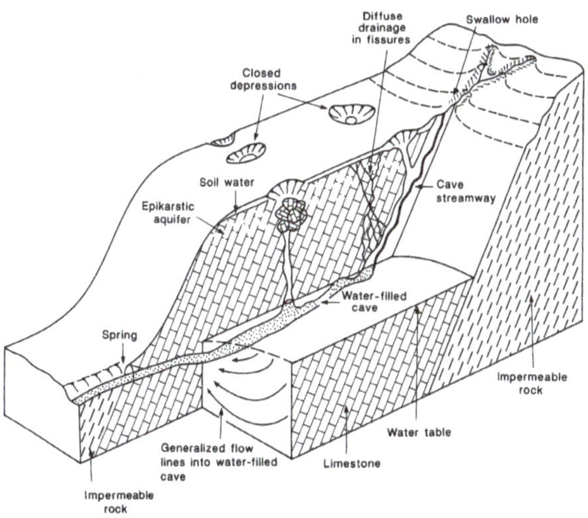

Figure 2.10. *Block diagram showing the occurrence of groundwater in karst aquifers. After an original by T.C. Atkinson.*

Source. https://www.geokniga.org/bookfiles/geokniga-hydrogeologyprinciplesandpractice.pdf.

Conduits or flooded cave systems provide water to the larger springs. Groundwater in fissures and fractures flows toward the conduits, which serve as drains in the saturated zone of the aquifer. Conduit flow experiences turbulence within the saturated zone, whereas diffuse fissure flow follows Darcy's law. Both allogenic and autogenic recharges can occur in the aquifer. Sinking streams that gather on exposed sandstone and shale in the center of the periclines are the source of allogenic recharge. Through swallow holes, these streams enter the conduit system directly. Autogenic recharge takes place as diffuse infiltration through the soil or is concentrated by closed depressions (dolines). The conduit system's first-order tributaries, known as closed depressions, direct concentrated recharge into shafts and caverns. Dense fissuring created by weathering in the top few meters of bedrock stores water in the unsaturated zone of what is sometimes called the epikarstic aquifer. The epikarstic aquifer receives replenishment from infiltration and empties into the saturated zone through fissures and fractures. However, drainage frequently concentrates into shafts that serve as tributaries to cave systems.

A thorough examination of hydrographs, baseflow recession curves, water balances, and tracer tests reveals that the Mendip Hills' diffuse flow component of the saturated zone has a hydraulic conductivity of 10^{-4} to 10^{-3} m s^{-1} and a storativity of approximately 1% (Atkinson, 1977). Conduits carry about 70% of the flow through the saturated zone, but they only make up less than 30% of the active storage. Direct investigation has revealed that the conduit circulation depth below the water table is greater than 60 m, which suggests that at least 600 mm of precipitation—roughly the equivalent of a year's worth of runoff—have been stored in the saturated zone. The epikarstic aquifer also has significant storage, though it's unclear how much in total.

As illustrated in Figure2.11, Karst aquifers can be categorized based on the relative importance of diffuse flow and conduit flow, the level of recharge concentration, and the volume of storage in the aquifer. The Mendip Hills karst aquifer has a high storage capacity, with 70% of conduit flow occurring in the saturated zone and roughly 50% of recharge concentrated into streams and closed depressions.

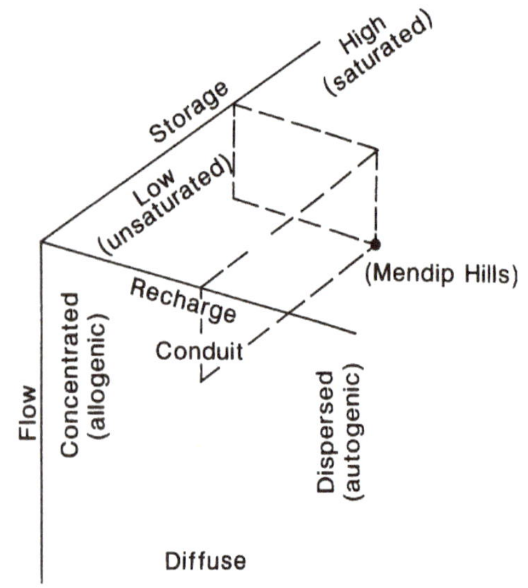

Figure 2.11. Conceptual classification of karst aquifers from a consideration of recharge and groundwater flow mechanisms and the degree of saturated aquifer storage. After an original by T.C. Atkinson.

Source. https://www.geokniga.org/bookfiles/geokniga-hydrogeologyprinciplesandpractice.pdf.

The challenge of accurately characterizing the hydraulic properties makes it difficult to represent the hydraulic behavior of the system in karst aquifers, where turbulent flow conditions may arise in conduits that have developed through solution. The behavior of karst aquifers is frequently modeled using a variety of techniques. The first is to use one of the models depicted in Fig 2.8. and assume that Darcy's law governs groundwater flow. Additional methods for simulating flow in karst conduits include using the Darcy-Weisbach pipe flow equation or, in the case of mature karst landscapes, a "black box" model that replicates input and output responses, particularly recharge and spring flow, by developing empirical functions based on field observations of flow. The typical aquifer characteristics, such as hydraulic conductivity, storativity, and porosity, may or may not be included in these functions. Utilizing the aquifer response functions created for the "black box" approach and applying them to an equivalent porous material model is a third hybrid approach. However, it must be acknowledged that significant uncertainties still exist, necessitating rigorous field validation.

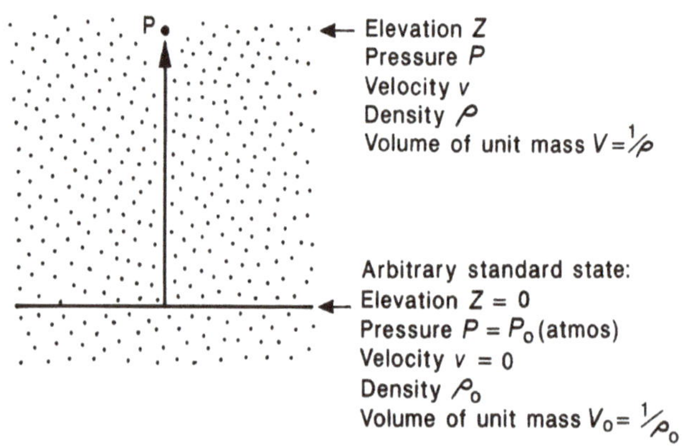

Figure 2.12. *Work done in moving a unit mass of fluid from the standard state to a point P in a groundwater flow system.*

Source. https://www.geokniga.org/bookfiles/geokniga-hydrogeologyprinciplesandpractice.pdf.

2.6. GROUNDWATER POTENTIAL AND HYDRAULIC HEAD

Groundwater distribution and mobility in geological formations are characterized by the porosity and hydraulic conductivity of porous material. Analogies between the flow of heat through solids at higher temperatures and the flow of electrical current at lower voltages are made when analyzing the physical process of groundwater flow. Similar to how groundwater flow is controlled by potential gradients, the rates of heat and electrical flow are proportionate to potential gradients. The flow of groundwater is governed by the laws of physics and thermodynamics and contains energy in mechanical, thermal, and chemical forms. In relation to Figure 2.12, the following three elements make up the work required to move a unit mass of fluid from the standard state to a point, P, in the flow system:

1. Potential energy (mgz) required to lift the mass to elevation, z.
2. Kinetic energy ($mv^2/2$) required to accelerate the fluid from zero velocity to velocity, v.
3. Elastic energy required to raise the fluid from pressure P_o to pressure P.

The latter quantity can be thought of as the change in potential energy per unit volume of fluid and is found from:

$$m\int_{P_0}^{P}\frac{v}{m}dp = m\int_{P_0}^{P}\frac{dP}{\rho} \qquad (14)$$

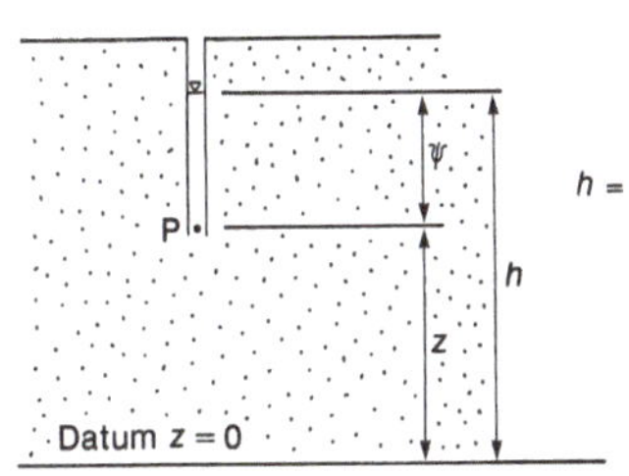

Figure 2.13. *Relation between hydraulic head, h, pressure head, ψ, and elevation head, z, at a point P in a column of porous material.*

Source. https://www.geokniga.org/bookfiles/geokniga-hydrogeologyprinciplesandpractice.pdf.

Given that groundwater velocities in porous material are very small, the kinetic energy term can be ignored such that at the new position, P, the fluid potential, Φ, or mechanical energy per unit mass (m = 1) is:

$$\Phi = gz + \int_{P_o}^{P} \frac{dP}{\rho} \tag{15}$$

For incompressible fluids that have a constant density, and therefore are not affected by a change in pressure, then:

$$\Phi = gz + \frac{(P - P_o)}{\rho} \tag{16}$$

To relate the fluid potential to the hydraulic head measured by Darcy in his experiment, Figure 2.13 demonstrates that the fluid pressure at position P in a column containing porous material is found as follows:

$$P = \rho g \psi + P_o \tag{17}$$

where ψ is the height of the water column above P and P_o is atmospheric pressure (the pressure at the standard state).

It can be seen that ψ = h − z and so, substituting in Equation 17:

$$P = \rho g(h - z) + P_o \tag{18}$$

By substituting this expression for pressure into the equation for fluid potential, Equation 16, then:

$$\Phi = gz + \frac{[\rho g(h - z) + P_o - P_o]}{\rho} \tag{19}$$

and, thus:

$$\Phi = [gz + gh - gz] = gh \tag{20}$$

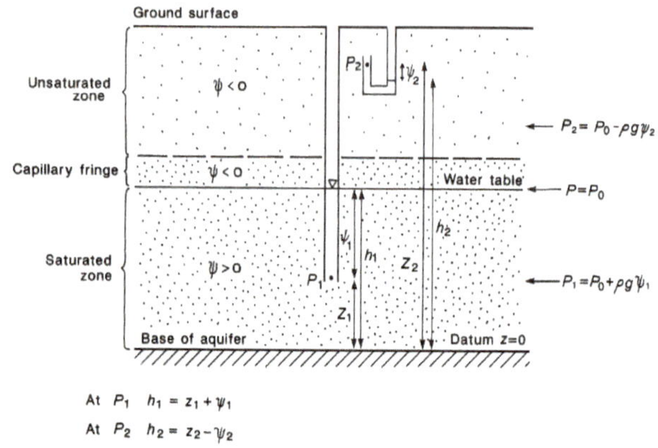

Figure 2.14. *Condition of pressure head, ψ, for the unsaturated and saturated zones of an aquifer. At the water table, fluid pressure is equal to atmospheric (Po) and by convention is set equal to zero. Note also that the unsaturated or vadose zone is the region of a geological formation containing solid, water and air phases while in the saturated or phreatic zone, pore spaces of the solid material are all water-filled.*

Source. https://www.geokniga.org/bookfiles/geokniga-hydrogeologyprinciplesandpractice.pdf.

An important relationship in hydrogeology is revealed by Equation 20: the fluid potential, Φ, at any point in a porous material can be easily determined by multiplying the hydraulic head by the acceleration caused by gravity. Near the Earth's surface, gravity is practically constant, so there is an almost perfect correlation between Φ and h. The reason this matters is that hydraulic head is a physical quantity that can be measured, making it an appropriate way to calculate fluid potential, or Φ.

It is now demonstrated that groundwater flow is driven by a fluid potential gradient that is comparable to a hydraulic head gradient, using the analogy with heat and electricity where rates of flow are determined by potential gradients. To put it succinctly, groundwater moves from higher hydraulic head regions to lower ones.

With reference to Equations 16 and 20, and, by convention, setting the atmospheric pressure, P_o, to zero, then:

$$gh = gz + \frac{P}{\rho} \quad (21)$$

The pressure at point P in Figure 2.13 is equal to $\rho g \psi$, and so it can be shown by substitution in Equation 21 and by dividing through by g that:

$$h = z + \psi \quad (22)$$

Equation (22) provides a fundamental relationship for comprehending groundwater flow by confirming that the hydraulic head at a point in a saturated porous material is equal to the sum of the elevation head (z) and pressure head (ψ). As illustrated in Figure 2.14, this expression holds true for both the saturated and unsaturated zones of porous material, but recognition is required. Due to the convention of working in gauge pressures and setting atmospheric pressure to zero, the pressure head term, ψ, is a negative quantity in the unsaturated zone. This implies that the water pressure is equal to zero (i.e., at the water table level) (i.e., atmospheric pressure).

The aquifer material is fully saturated in the capillary fringe above the water table, but the porewater pressure is negative—that is, lower than atmospheric pressure—due to capillary suction pulling water up from the water table. The thickness of the capillary fringe varies with the pore space diameter and can be as little as a few centimeters for deposits with coarse grains and as much as several meters for deposits with fine grains.

2.7. INTERPRETATION OF HYDRAULIC HEAD AND GROUNDWATER CONDITIONS

2.7.1. Groundwater Flow Direction

The directions of groundwater flow in an aquifer system can be ascertained by measuring the hydraulic head, which is typically accomplished by installing a piezometer or well point.

Within Figure In 2.15a, the direction of groundwater flow can be ascertained by installing three piezometers to the same depth. Using Darcy's law (eq. 5) the computation of the flow's horizontal component. Within Figure In 2.15b, two illustrations of piezometer nests that enable the determination of groundwater flow direction and hydraulic head at different levels within the same aquifer formation or in separate formations are provided.

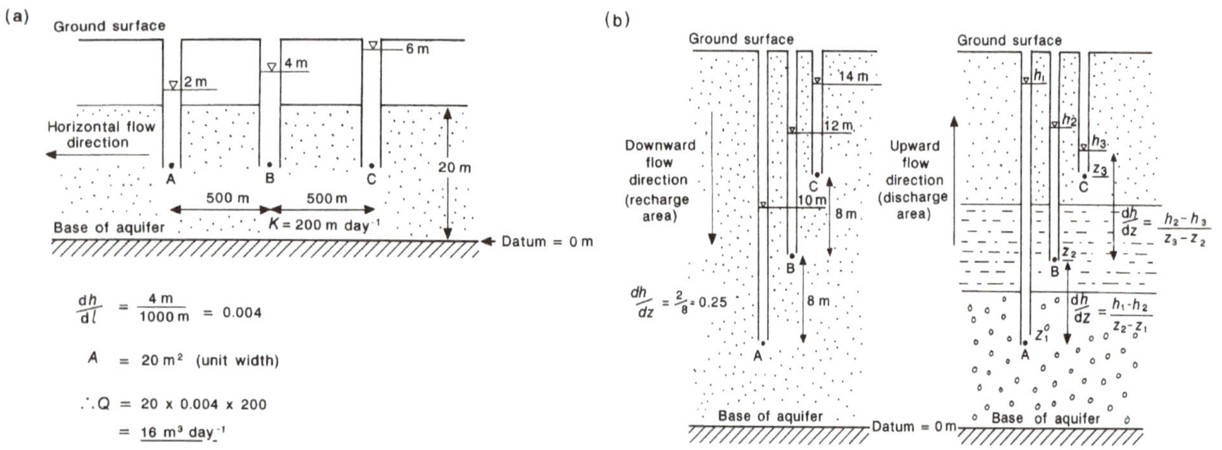

Figure 2.15. *Determination of groundwater flow direction and hydraulic head gradient from piezometer measurements for (a) horizontal flow and (b) vertical flow. The elevation of the water level indicating the hydraulic head at each of the points A, B and C is noted adjacent to each piezometer.*

Source. https://www.geokniga.org/bookfiles/geokniga-hydrogeologyprinciplesandpractice.pdf.

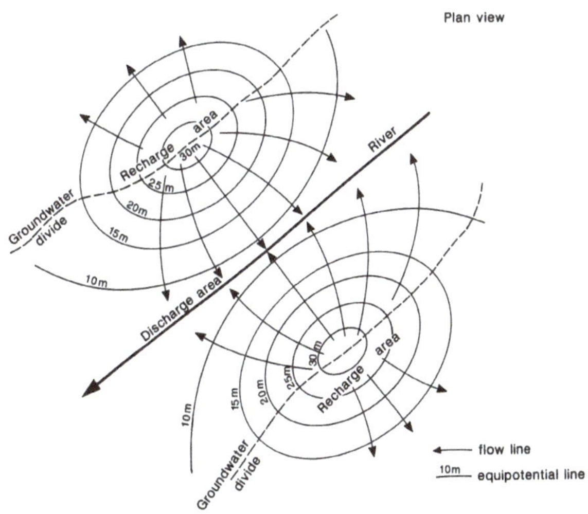

Figure 2.16. *Sketch map of the surface of the water table in an unconfined aquifer showing recharge and discharge areas and the position of groundwater divides.*

Source. https://www.geokniga.org/bookfiles/geokniga-hydrogeologyprinciplesandpractice.pdf.

2.7.2. Water Table and Potentiometric Surface Maps

A district's observation boreholes and piezometers paint a picture of the hydraulic head's three-dimensional distribution throughout the aquifer system. Equipotential lines are lines that connect locations with equal groundwater potential, or groundwater head. A flow net can be built using lines that are perpendicular to the equipotential lines since they are flow lines. The process of creating equipotential contours in plan view yields a potentiometric surface map. The potentiometric surface in an unconfined aquifer is a map of the water table, where groundwater is naturally at atmospheric pressure. The potentiometric surface in a confined aquifer forecasts the location of the water level rise in a borehole that penetrates the buried aquifer. As depicted in Figure 2.16: Low hydraulic head areas are usually in groundwater discharge zones, whereas high hydraulic head areas can be understood as groundwater recharge zones.

2.7.3. Types of Groundwater Conditions

The interaction of lithological units and geological structure has a significant impact on groundwater conditions. Aquifer geometry can lead to four main types of groundwater conditions, as illustrated in Figure 2.17, and also plays a role in the formation of springs. An unconfined aquifer is present when a water table separates the unsaturated zone from the saturated zone. Sometimes, an unconfined aquifer can form below an aquitard layer, resulting in a concealed unconfined aquifer.

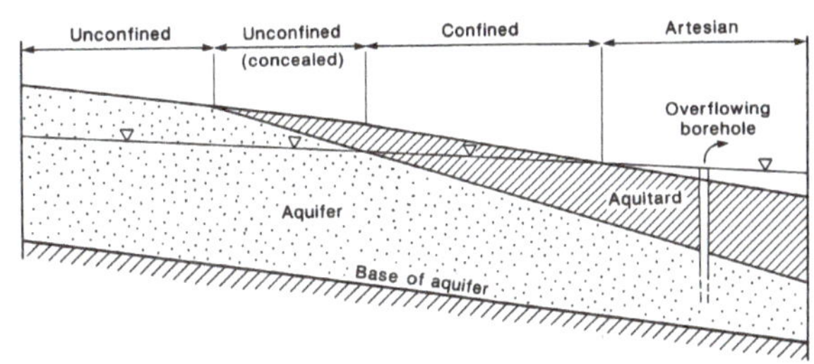

Figure 2.17. *Cross-section showing four types of groundwater conditions. A water table is developed where the aquifer is unconfined or concealed and a potentiometric surface is present where the aquifer experiences confined or artesian conditions.*

Source. https://www.geokniga.org/bookfiles/geokniga-hydrogeologyprinciplesandpractice.pdf.

Perched water table conditions can arise in heterogeneous material, such as sedimentary units with intercalated lenses or clay layers. As depicted in Figure 2.18: A clay layer within a sand matrix causes water to be held above the lower permeability material, creating a perched water table above the regional water table and within the unsaturated zone. Inverted water tables can be found at the base of the perched lens of water in the clay layer and below the ground surface after a rainfall event that causes water to seep into the soil zone. This is because water table conditions arise where groundwater is at atmospheric pressure.

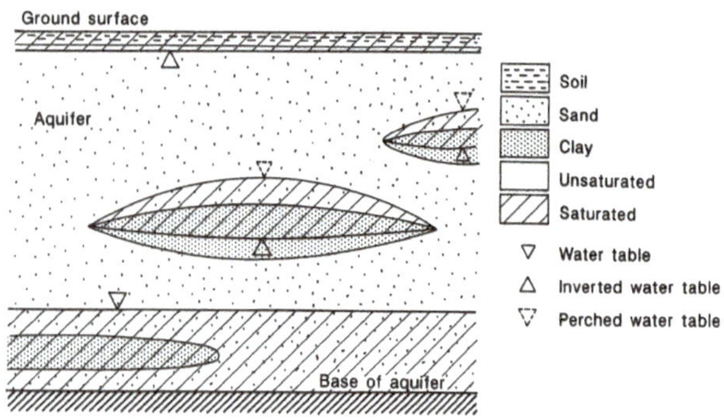

Figure 2.18. *Perched and inverted watertable conditions developed within a sand aquifer containing clay lenses. An inverted water table is also shown below the wetted ground surface in the soil zone.*

Source. https://www.geokniga.org/bookfiles/geokniga-hydrogeologyprinciplesandpractice.pdf.

Between two aquitards or aquicludes is a confined aquifer. Because confined aquifers are under pressure, groundwater in a borehole that penetrates one will rise to a level above the aquifer's top. An overflowing artesian groundwater condition occurs when the groundwater level rises to the top of the borehole above ground level and overflows. It is possible for the groundwater level to be drawn down below the aquifer's top and become unconfined if overabstraction of groundwater from boreholes that exploit a confined aquifer happens. Between the early 1800s and the early 1960s, excessive groundwater extraction was necessary to support the expansion of London's industry and population, which led to the development of this condition in the Chalk aquifer of the London Basin.

Water leakage to the underlying aquifer may occur if a vertical hydraulic gradient develops across the aquitard–aquifer boundary and the overlying geological unit behaves like an aquitard. This condition, known as semiconfined aquifer condition, is frequently found where sand clays and fluvial or glacio-fluvial silts cover an aquifer. Surveyed groundwater-level data, mapped geological and geomorphological information, and hydrogeological field observations must all be combined and analyzed for a more thorough regional hydrogeological interpretation. In this extensive glacial meltwater channel feature, the relationship between groundwater and spring-fed lakes is demonstrated through the interpretation of the relationship between geology, groundwater occurrence, and potentiometric head distribution.

2.7.4. Classification of Springs and Intermittent Streams

Since Bryan (1919), who identified the following types of springs: volcanic, fissure, depression, contact, artesian, and springs in impermeable rock, the classification of springs has been discussed. In a nutshell, springs indicate the end of subterranean flow systems and the transition from fluvial to dominant processes.

At the point of aquifer discharge, the vertical position of the spring indicates the elevation of the water table or a minimum elevation of the potentiometric surface. Springs' impact on the aquifers they drain is primarily determined by their topography and structural characteristics. Discussing hydrogeological controls on springs, Ford and Williams (1989) identify three main types of springs: free draining, dammed, and confined. They primarily discuss karst aquifers, which are home to some of the largest springs in the world.

Considering Figure 2.19 Free draining springs are fully or mostly in the unsaturated (vadose) zone and undergo groundwater discharge due to gravity. Commonly found in areas where there is a significant obstruction in the underground drainage system's path are dammed springs. The barrier may result from aggradation of the valley, for example, by the deposition of glacial deposits, or it may result from another lithology, either faulted or in conformable contact. Freshwater discharge from submerged groundwater is likewise prevented by the sea's denser salt water. In all situations, there's a chance that intermittent overflow springs will develop when the water table rises. Confined springs emerge in areas where an impermeable formation on top creates artesian conditions. Sometimes, as with the Bath thermal springs and the Permian Magnesian Limestones in South Yorkshire, fault planes offer a way for the confined groundwater to escape. In other places, where erosion has removed the underlying strata, groundwater escapes. (Table 2.2).

Table 2.2. *Discharges of 10 of the World's Largest Karst Springs*

Spring	Discharge ($m^3 s^{-1}$)			Basin area (km^2)
	Mean	Max.	Min.	
Matali, Papua New Guinea	90	>240	20	350
Bussento, Italy	–	117	76	–
Dumanli*, Turkey	50	–	25	2800
Trebišnijca, Bosnia-Herzegovina	50	250	3	–
Chingshui, China	33	390	4	1040
Vaucluse, France	29	200	4.5	2100
Frió, Mexico	28	515	6	>1000?
Silver, USA	23.25	36.5	15.3	1900
Waikoropupu, New Zealand	15	21	5.3	450
Maligne, Canada	13.5	45	1	730

Source. https://www.geokniga.org/bookfiles/geokniga-hydrogeologyprinciplesandpractice.pdf.

After Ford and Williams (1989).

** Dumanli spring is the largest of a group of springs that collectively yield a mean flow of 125–130 m3 s –1 at the surface of the Manavgat River.*

The spring pool may appear to be "boiling" due to dissolved gases since the emerging water typically acclimates to atmospheric pressure quickly.

Depending on how high the river bed is relative to the water table, rivers that flow over an aquifer outcrop may experience influent and effluent conditions. The portions of rivers that receive both surface runoff and groundwater discharge will fluctuate according to the seasonal fluctuations in the water table. The sporadic streams that emerge over portions of the Chalk outcrop in southern England are an excellent illustration of this kind of river. The low specific yield of the Chalk aquifer in these regions results in significant variations in the summer-winter water table position and, consequently, in the intermittent streams' length. The seasonal streams, also referred to as winterbournes or chalk bournes, run intermittently for a portion of the year, usually right before or right after the season with the highest precipitation. One such is the River Bourne, which is depicted in Figure 2.21a and is situated northeast of Salisbury Plain. The intermittent portion of the River Bourne in this region of rolling Chalk downland in central southern England is 10 km long, or until the point where the perennial stream head is met, below which the river bed and Chalk water table are permanently intersected. As seen in Figure (2.21b), the intermittent section may stay dry during drought years while becoming wet in wet years. The upper reaches maintain a bank full discharge.

2.8. GROUNDWATER FLOW THEORY

Following the presentation of the steady-state and transient groundwater flow equations mathematically, basic analytical solutions to groundwater flow issues will be shown. This leads to the conclusion that regional aquifer systems have a variety of flow system scales. For both confined and unconfined aquifers, it is essential to define the two aquifer properties, transmissivity and storativity, before proceeding.

2.8.1. Transmissivity and Storativity of Confined Aquifers

For a confined aquifer of thickness, b, the transmissivity, T, is defined as:

$$T = Kb \qquad (23)$$

and represents the rate at which water of a given density and viscosity is transmitted through a unit width of aquifer or aquitard under a unit hydraulic gradient. Transmissivity has the units of $L^2\ T^{-1}$.

The storativity (or storage coefficient), S, of a confined aquifer is defined as:

$$S = S_s b \qquad (24)$$

where S_s is the specific storage term, and represents the volume of water that an aquifer releases from storage per unit surface area of aquifer per unit decline in the component of hydraulic head normal to that surface. Storativity values are dimensionless and range in value from 0.005 to 0.00005, such that large head changes over extensive areas are required to produce significant yields from confined aquifers.

2.8.2. Release of Water from Confined Aquifers

Meinzer and Hard (1925) noted in their study of the Dakota sandstone at the turn of the century that the water-bearing formation was exhibiting elastic behavior in releasing water from storage because more water was being pumped out of the area than could be explained. As water was pumped, a cone of depression developed and the rate of abstraction decreased, but this had no apparent effect on groundwater levels in the recharge zone.

Later, Jacob (1940) formally described the elastic behavior of porous rocks by deriving the general partial differential equation describing transient groundwater flow. Two processes account for the water produced by confined aquifers: first, compaction reduces the aquifer's porosity, releasing groundwater; second, because water is slightly compressible, the water itself expands.

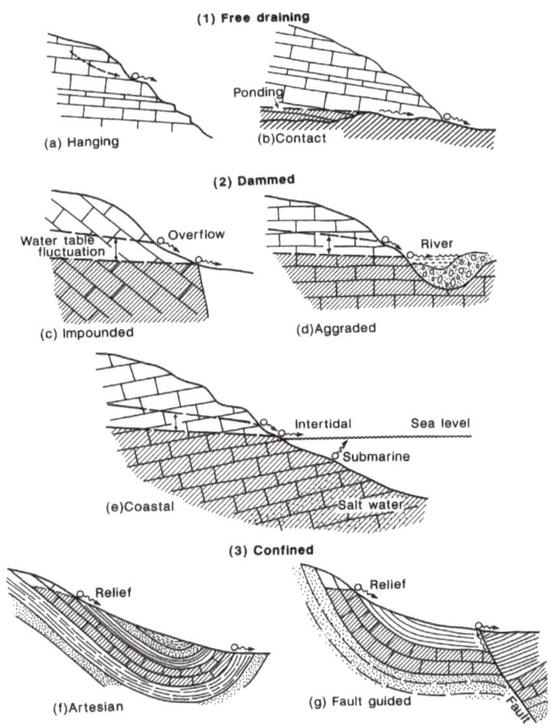

Figure 2.19. *Types of springs. After Ford and Williams (1989).*

Source. https://www.geokniga.org/bookfiles/geokniga-hydrogeologyprinciplesandpractice.pdf.

As shown in Figure 2.23, the total downward stress, σ_T, applied at the top of a confined aquifer is supported by an upward effective stress, σ_e , on the aquifer material and the water pressure contained in the pore space, P_w, such that:

$$\sigma_T = \sigma_e + P_w \tag{25}$$

The aquifer's material will experience compression if the porewater pressure is reduced by groundwater pumping or by naturally occurring groundwater outflow.

The inverse of the modulus of elasticity, or compressibility, is a property of a material that expresses the change in volume (strain) induced in a material under an applied stress (equal to a change in stress divided by a change in strain). β, the water's compressibility, is defined as:

$$\beta = \frac{-dV_w/V_w}{dP_w} \tag{26}$$

where P_w is porewater pressure, V_w is volume of a given mass of water and dV_w/V_w is volumetric strain for an induced stress dP_w. For practical purposes, β can be taken as a constant equal to 4.4×10^{-10} m2 N^{-1} (or Pa^{-1}).

The compressibility of aquifer material, α, is defined as:

$$\alpha = \frac{-dV_T/V_T}{d\sigma_e} \tag{27}$$

where V_T is total volume of aquifer material and dV_T/V_T is volumetric strain for an induced change in effective stress $d\sigma_e$.

Now, with reference to Equation 27, for a reduction in the total volume of aquifer material, dV_T, the amount of water produced by compaction of the aquifer, dV_w, is:

$$dV_w = -dV_T = \alpha V_T d\sigma_e \tag{28}$$

If the total stress does not change (dσT = 0), then from a knowledge that Pw = ρgψ and ψ = h − z (eq. 22), with z remaining constant, then, using Equation 25:

$$d\sigma_e = 0 - \rho g d\psi = -\rho g dh \tag{29}$$

For a unit decline in head, dh = −1, and if unit volume is assumed (V_T = 1), then Equation 28 becomes:

$$dV_w = \alpha(1)(-\rho g)(-1) = \alpha\rho g \qquad (30)$$

The water produced by the expansion of water is found from Equation 26 thus:

$$dV_w = -\beta V_w dP_w \qquad (31)$$

Recognizing that the volume of water, Vw, in the total unit volume of aquifer material, VT, is nVT where n is porosity, and that dP = ρgdψ or −ρg for a unit decline in hydraulic head (where ψ = h − z (eq. 22), with z remaining constant), then for unit volume, VT = 1, Equation 31 gives:

$$dV_w = -\beta n(1)(-\rho g) = \beta n\rho g \qquad (32)$$

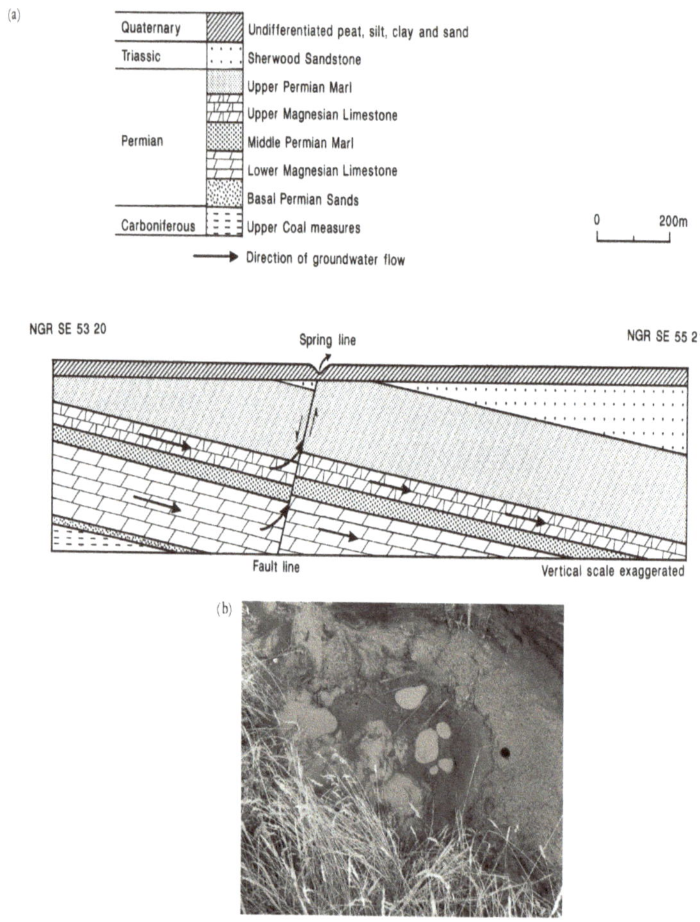

Figure 2.20. *(a) Geological section at Gale Common, South Yorkshire showing a spring line at the position of a normal fault that acts as a barrier to groundwater movement. The existence of the fault causes groundwater flow through the overlying Upper Permian Marl via a zone of enhanced permeability. (b) Springs from Permian Magnesian Limestones appearing through Quaternary deposits at Gale Common, South Yorkshire. For scale, a camera lens cap is shown to the right of the spring pool.*

Source. https://www.geokniga.org/bookfiles/geokniga-hydrogeologyprinciplesandpractice.pdf.

Finally, the volume of water that a unit volume of aquifer releases from storage under a unit decline in hydraulic head (the specific storage, S_s) is the sum of the volumes of water produced by the two mechanisms of compaction of the aquifer (eq. 30) and expansion of the water (eq. 32) thus:

$$S_s = \alpha\rho g + \beta n\rho g = \rho g(\alpha + n\beta) \qquad (33)$$

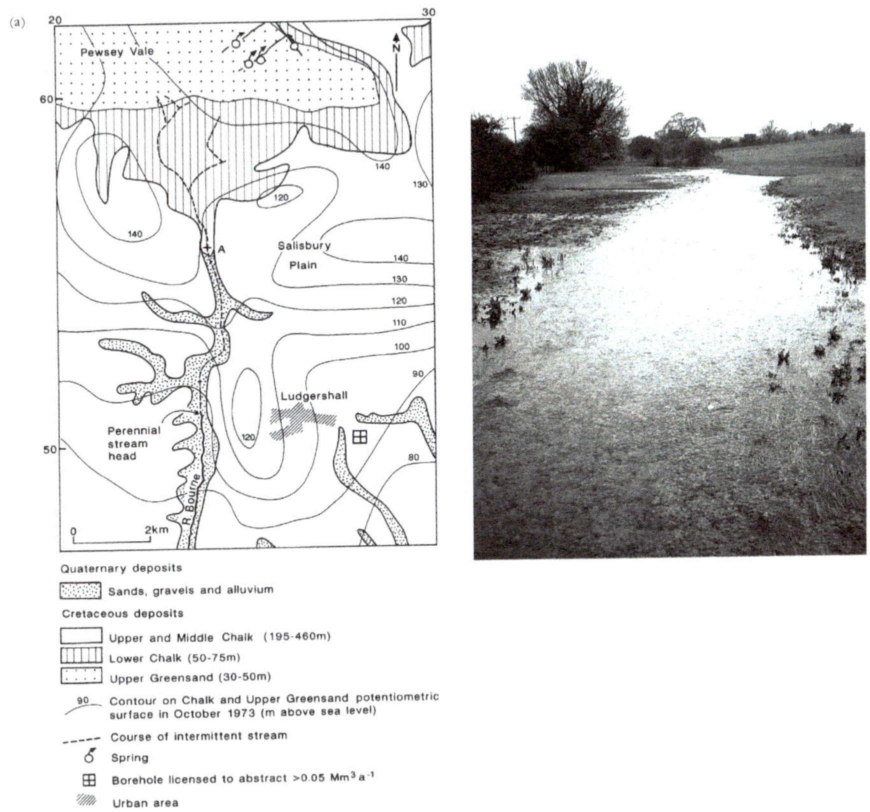

Figure 2.21. *(a) Hydrogeological map of the north-east of Salisbury Plain showing the intermittent and perennial sections of the River Bourne. Position A is the site of the photograph shown in (b) located at Collingbourne Kingston and looking north in December 2002. The river bed is covered by lesser water parsnip Berula erecta, a plant that proliferates in still to medium flows of base-rich water. Hydrogeological information based on the British Geological Survey (1978); plant species identification using Haslam et al., (1975).*

Source. https://www.geokniga.org/bookfiles/geokniga-hydrogeologyprinciplesandpractice.pdf.

Stated differently, groundwater pumped from a confined aquifer arises from the secondary effects of aquifer compaction and water expansion rather than representing a dewatering of the aquifer's physical pore space. Thus, yields from confined aquifers are significantly lower than yields from unconfined aquifers for an equivalent unit decline in hydraulic head. As a result, confined aquifer storage coefficient values are substantially lower than those of unconfined aquifers. For clay and sand, the values of material compressibility, or α, range from 10^{-6} to 10^{-9} m^2 N^{-1}, while for gravel and jointed rock, they range from 10^{-8} to 10^{-10} m^2 N^{-1}. These values suggest that in a previously unconsolidated clay aquitard, a larger, mostly irreversible compaction is anticipated, whereas in gravel or indurated sedimentary aquifers, smaller, elastic deformations are probably more likely. Ground subsidence after aquifer compaction is a potential effect of groundwater abstraction from confined aquifers, particularly in sand–clay aquifer–aquitard systems.

2.8.3. Transmissivity and Specific Yield of Unconfined Aquifers

The transmissivity of an unconfined aquifer is not as well defined as it is for a confined aquifer; however, Equation 23 can still be used, with b standing for the aquifer's saturated thickness or the water table's height above a lower aquitard boundary. (Table 2.3).

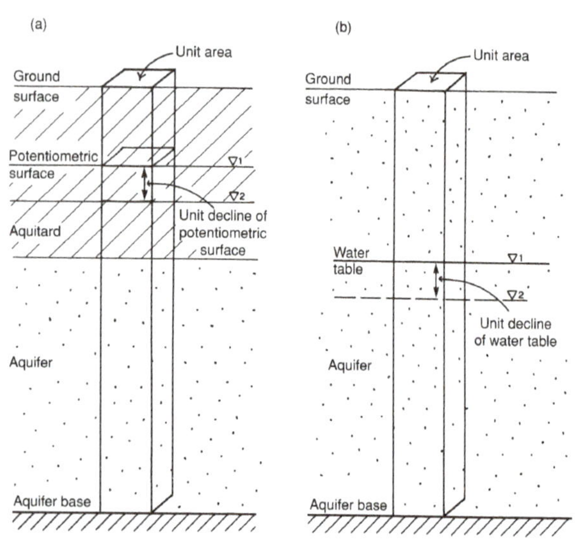

Figure 2.22. *Schematic representation of storativity in (a) a confined aquifer (the storage coefficient) and (b) an unconfined aquifer (the specific yield).*

Source. https://www.geokniga.org/bookfiles/geokniga-hydrogeologyprinciplesandpractice.pdf.

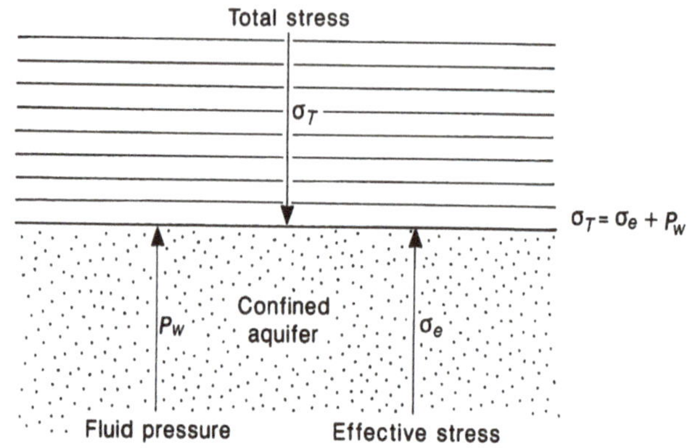

Figure 2.23. *Total stress, effective stress and fluid pressure at the top of a confined aquifer.*

Source. https://www.geokniga.org/bookfiles/geokniga-hydrogeologyprinciplesandpractice.pdf.

Table 2.3. *Range of Values of Compressibility*

	Compressibility, α ($m^2\ N^{-1}$ or Pa^{-1})
Clay	10^{-6}–10^{-8}
Sand	10^{-7}–10^{-9}
Gravel	10^{-8}–10^{-10}
Jointed rock	10^{-8}–10^{-10}
Sound rock	10^{-9}–10^{-11}
Water (β)	4.4×10^{-10}

Source. https://www.geokniga.org/bookfiles/geokniga-hydrogeologyprinciplesandpractice.pdf.

After Freeze and Cherry (1979).

Therefore, if there are significant seasonal variations in the water table's elevation or if the aquifer's saturated thickness exhibits lateral variation due to an uneven lower aquitard boundary or variations between its recharge and discharge areas, the transmissivity will change.

The volume of water that an unconfined aquifer releases from storage per unit surface area of aquifer per unit decline in the water table is known as the specific yield, or the unconfined storativity. It is roughly equivalent to the total porosity of a soil or rock. The normal range of the dimensionless term "specific yield" is 0.01 to 0.30. The higher values represent the real dewatering of pore space as the water table drops in comparison to confined aquifers. As a result, an unconfined aquifer can yield the same amount of water over smaller areas than a confined aquifer, with smaller head changes. Though it is uncommon, it is possible to define a single formation parameter, the hydraulic diffusivity, D, defined as either of the following, by combining the aquifer properties of transmissivity (T or K) and storativity (S or S_s).

$$D = \frac{T}{S} \text{ or } \frac{K}{S_s} \tag{34}$$

When groundwater abstraction occurs, for example, aquifer formations with a high hydraulic diffusivity react fast to transfer altered hydraulic conditions from one area of the aquifer to other regions.

CASE STUDY

PHYSICAL HYDROGEOLOGY OF THE OGALLALA AQUIFER

The Ogallala Aquifer, part of the High Plains Aquifer system, is one of the largest and most important aquifers in the United States. Spanning approximately 174,000 square miles across eight states (South Dakota, Nebraska, Wyoming, Colorado, Kansas, Oklahoma, New Mexico, and Texas), it plays a crucial role in supporting agricultural activities, providing about 30% of the groundwater used for irrigation in the country.

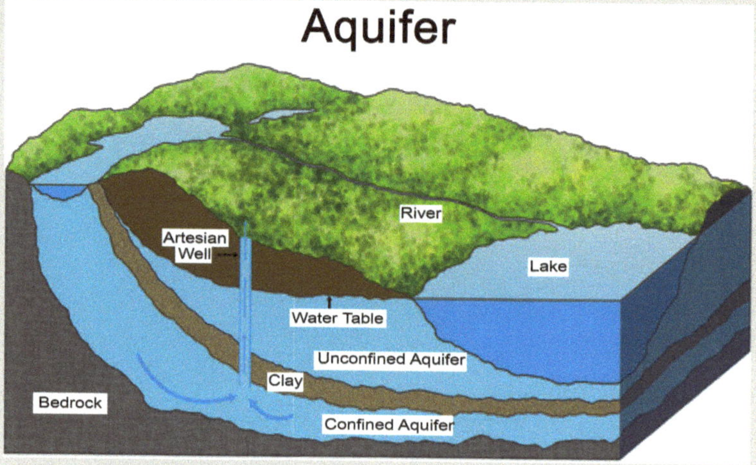

The Ogallala Aquifer consists primarily of unconsolidated sand, gravel, clay, and silt deposited by ancient rivers. The aquifer's thickness varies significantly, from a few feet to over 1,000 feet, and its permeability is largely dependent on the proportion of coarse-grained materials.

Water Balance and Recharge

Recharge to the Ogallala Aquifer is relatively low, averaging only about 0.5 inches per year due to low precipitation, high evaporation rates, and significant agricultural water use. The primary sources of recharge are local precipitation, return flow from irrigation, and seepage from streams and rivers. However, in many areas, the recharge is insufficient to balance the withdrawals, leading to significant declines in water levels.

Groundwater Flow and Storage

Groundwater flow within the Ogallala Aquifer is generally from west to east, following the natural slope of the land. The hydraulic conductivity of the aquifer varies widely, with higher values in areas dominated by sand and gravel. The specific yield of the aquifer, representing the volume of water that can be drained by gravity, ranges from

10% to 30%. These variations in hydraulic properties influence the aquifer's capacity to store and transmit water.

Challenges and Impacts

The primary challenge facing the Ogallala Aquifer is the unsustainable rate of groundwater extraction for irrigation. Since large-scale irrigation began in the 1940s, water levels have declined by more than 100 feet in some areas, particularly in Texas and Kansas. This depletion has led to several significant impacts:

1. **Reduced Water Availability**: Decreased groundwater levels have resulted in reduced well yields, increased pumping costs, and the need for deeper wells.
2. **Decline in Agricultural Productivity**: Water scarcity threatens the long-term viability of agriculture in the region, which relies heavily on irrigation.
3. **Environmental Consequences**: Lower water levels affect surface water bodies connected to the aquifer, impacting ecosystems and wildlife habitats.

Management and Sustainability

Efforts to manage and sustain the Ogallala Aquifer involve a combination of regulatory measures, technological innovations, and conservation practices:

1. **Regulation and Policy**: States have implemented various policies to regulate groundwater withdrawals, such as well permitting, metering, and allocation limits.
2. **Water Conservation**: Farmers are adopting water-saving technologies and practices, such as drip irrigation, no-till farming, and crop rotation.
3. **Recharge Enhancement**: Projects to enhance artificial recharge, such as the construction of recharge basins and the use of treated wastewater for irrigation, are being explored.

Conclusion

The Ogallala Aquifer case study illustrates the complex interplay between hydrogeology, human activity, and sustainability. Addressing the challenges of groundwater depletion requires an integrated approach that combines scientific understanding with effective management practices. Ensuring the long-term sustainability of this vital water resource is crucial for the economic and environmental health of the region.

CLASS ACTIVITY

Does spring water really come from a natural spring? Some water labeled as spring water comes from underground aquifers, layers of rock and sediment underground that are saturated with water. In the How Dirt Cleans Water activity, students explore how surface water and surface run-off seep into the ground where they are stored in underground aquifers. With this hands-on activity, students use plastic bottles to model different reservoirs and investigate to see how clean the groundwater collected from these reservoirs is. Students learn about infiltration and experiment to see how the size of the particles in soil relates to how permeable the soil is and how water moves through the soil.

Questions:

1. What happens when water reaches an impermeable layer?
2. What happens when underground areas are saturated with water?
3. What is an aquifer?
4. Why is groundwater often clean (and drinkable) compared to surface water collected in containers like lakes and rivers?

SUMMARY

- -The occurrence of groundwater within the Earth's crust and the emergence of springs at the ground surface are determined by the lithology of geological materials, regional geological structure, geomorphology of landforms and the availability of recharge sources.
- The porosity of a soil or rock is that fraction of a given volume of material that is occupied by void space, or interstices.
- Water contained within the interconnected voids of soils and rocks is capable of moving, and the ability of a rock to store and transmit water constitutes its hydraulic properties.
- The density and viscosity of water are functions of temperature and pressure but these effects are not great for the ranges of temperature and pressure encountered in most groundwater situations.
- Aquifer properties, such as hydraulic conductivity, are unlikely to conform to the idealized, uniform porous material whether viewed at the microscopic or regional scale.
- An aquifer is a layer or layered sequence of rock or sediment comprising one or more geological formations that contains water and is able to transmit significant quantities of water under an ordinary hydraulic gradient.

- An aquiclude is a saturated geological unit of such low permeability that it is incapable of transmitting significant quantities of water under ordinary hydraulic gradients and can act as a barrier to regional groundwater flow.
- The term karst is used widely to describe the distinctive landforms that develop on rock types such as limestones, gypsum and halite that are readily dissolved by water.
- Measurements of hydraulic head, normally achieved by the installation of a piezometer or well point, are useful for determining the directions of groundwater flow in an aquifer system.

REVIEW QUESTIONS

1. What do you understand by Darcy's law?
2. Explain about karst aquifers.
3. Discuss about groundwater flow direction.
4. What is groundwater flow theory?
5. Write short notes on transmissivity and storativity of confined aquifers.

REFERENCES

1. Albu, M., Banks, D., & Nash, H. (1997). *Mineral and Thermal Groundwater Resources*. Chapman & Hall.
2. Amery, H. A., & Wolf, A. T. (2004). *Water in the Middle East: A Geography of Peace*. University of Texas Press.
3. Andrews, J. N., Burgess, W. G., Edmunds, W. M., et al., (1982). The thermal springs of Bath. *Nature, 298*, 339–343.
4. Araújo, L. M., Franca, A. B., & Potter, P. E. (1999). Hydrogeology of the Mercosul aquifer system.
5. Atkinson, T. C. (1977). Diffuse flow and conduit flow in limestone terrain in the Mendip Hills, Somerset (Great Britain). *Journal of Hydrology, 35*, 93–110.
6. Atkinson, T. C., & Davison, R. M. (2002). Is the water still hot? Sustainability and the thermal springs at Bath.
7. Back, W., Rosenshein, J. S., & Seaber, P. R. (1988). *Hydrogeology. The Geology of North America, Vol. O-2*. The Geological Society of North America.
8. Bentley, H. W., Phillips, F. M., Davis, S. N., et al., (1986). Chlorine 36 dating of very old groundwater: The Great Artesian Basin, Australia. *Water Resources Research, 22*, 1991–2001.
9. Bergkamp, G., & Cross, K. (2006). Groundwater and ecosystem services: Towards their sustainable use. In *Proceedings on the International Symposium on Groundwater Sustainability (ISGWAS)*, Alicante, Spain, 177–193.

10. British Geological Survey. (1978). *1:100000 Hydrogeological Map of the South West Chilterns and the Berkshire and Marlborough Downs*. Natural Environment Research Council.
11. British Geological Survey. (1988). *1:625000 Hydrogeological Map of Scotland*. Natural Environment Research Council.
12. British Geological Survey. (1994). *1:250000 Hydrogeological Map of Northern Ireland*. Natural Environment Research Council.
13. Bryan, K. (1919). Classification of springs. *Journal of Geology, 27*, 552–561.
14. Burgess, W. G., Edmunds, W. M., Andrews, J. N., et al., (1980). Investigation of the geothermal potential of the UK. *The Hydrogeology and Hydrochemistry of The Thermal Water in the Bath-Bristol Basin*. Institute of Geological Sciences.
15. Burnett, W. C., Aggarwal, P. K., Aureli, A., et al., (2006). Quantifying submarine groundwater discharge in the coastal zone via multiple methods. *Science of the Total Environment, 367*, 498–543.
16. Burnett, W. C., Kim, G., & Lane-Smith, D. (2001). A continuous radon monitors for assessment of radon in coastal ocean waters. *Journal of Radioanalytical and Nuclear Chemistry, 249*, 167–172.
17. Campos, H. (2000). Mapa Hidrogeológico del Acuífero Guaraní. In *Proceedings of the First Joint World Congress on Groundwater*, Fortaleza, Brazil.
18. Cave, R. R., & Henry, T. (2011). Intertidal and submarine groundwater discharge on the west coast of Ireland. *Estuarine, Coastal and Shelf Science, 92*, 415–423.
19. Cedergren, H. R. (1967). *Seepage, Drainage and Flow Nets*. J. Wiley and Sons.
20. Christiansen, E. A., Acton, D. F., Long, R. J., et al., (1981). *Fort Qu'Appelle Geolog. The valleys–Past and Present*. Interpretive Report No. 2. The Saskatchewan Research Council.
21. Cripps, J. C., Bell, F. G., & Culshaw, M. G. (1986). *Groundwater in Engineering Geology*. In *Proceedings of the 21st Annual Conference of the Engineering Group of the Geological Society*. The Geological Society of London.
22. Davis, S. N., & De Wiest, R. J. M. (1966). *Hydrogeology*. John Wiley & Sons.
23. Döll, P., & Fiedler, K. (2008). Global-scale modelling of groundwater recharge. *Hydrology and Earth System Sciences, 12*, 863–885.
24. Döll, P., Kaspar, F., & Lehner, B. (2003). A global hydrological model for deriving water availability indicators: Model tuning and validation. *Journal of Hydrology, 270*, 105–134.
25. Domenico, P. A., & Schwartz, F. W. (1998). *Physical and Chemical Hydrogeology* (2nd ed.). John Wiley.
26. Downing, R. A. (1993). Groundwater resources, their development and management in the UK: An historical perspective. *Quarterly Journal of Engineering Geology, 26*, 335–358.
27. Downing, R. A., Price, M., & Jones, G. P. (1993). *The Hydrogeology of the Chalk of North-West Europe*. Clarendon Press.

28. Downing, R. A., Price, M., & Jones, G. P. (1993). The making of an aquifer. In R. A. Downing, M. Price, & G. P. Jones (Eds.), *The Hydrogeology of the Chalk of North-West Europe* (pp. 1–13). Clarendon Press.

29. Famiglietti, J. S., Lo, M., Ho, S. L., et al., (2011). Satellites measure recent rates of groundwater depletion in California's central valley. *Geophysical Research Letters, 38*, L03403. https://doi.org/10.1029/2010GL046442.

30. Faunt, C. C. (2009). *Groundwater Availability of the Central Valley Aquifer, California*. United States Geological Survey Professional Paper 1766.

31. Fetter, C. W. (2001). *Applied Hydrogeology* (4th ed.). Pearson Higher Education.

32. Fili, M., Da Rosa Filho, E. F., Auge, M., et al., (1998). El acuífero Guaraní. Un recurso compartido por Argentina, Brasil, Paraguay y Uruguay (América del Sur). *Hidrología Subterránea. Boletin Geologico y Minero, 109*, 389–394.

33. Ford, D. C., & Williams, P. W. (1989). *Karst Geomorphology and Hydrology*. Unwin Hyman.

34. Foster, S., Hirata, R., Vidal, A., et al., (2009). The Guarani Aquifer Initiative–Towards realistic groundwater management in a transboundary context. *GW Mate Case Profile Collection No. 9*. World Bank.

35. Freeze, R. A., & Cherry, J. A. (1979). *Groundwater*. Prentice-Hall.

36. Freeze, R. A., & Witherspoon, P. A. (1967). Theoretical analysis of regional groundwater flow. Effect of water table configuration and subsurface permeability variation. *Water Resources Research, 3*, 623–634.

37. Gale, J. E. (1982). Assessing permeability characteristics of fractured rock. In *Geological Society of America, Special Paper 189* (pp. 163–181). Geological Society of America.

38. Geological Museum. (1978). *Britain Before Man*. HMSO for the Institute of Geological Sciences.

39. Gleeson, T., & Manning, A. H. (2008). Regional groundwater flow in mountainous terrain: Three-dimensional simulations of topographic and hydrologic controls. *Water Resources Research, 44*, W10403. https://doi.org/10.1029/2008WR006848.

40. Gleeson, T., Marklund, L., Smith, L., & Manning, A. H. (2011). Classifying the water table at regional to continental scales. *Geophysical Research Letters, 38*, L05401. https://doi.org/10.1029/2010GL046427.

41. Habermehl, M. A. (1980). The great artesian basin, Australia. *BMR Journal of Australian Geology and Geophysics, 5*, 9–38.

42. Habermehl, M. A., & Lau, J. E. (1997). *1:2500000 Map of the Hydrogeology of the Great Artesian Basin, Australia*. Australian Geological Survey Organization.

43. Haitjema, H. M., & Mitchell-Bruker, S. (2005). Are water tables a subdued replica of the topography? *Ground Water, 43*, 781–786.

44. Harris, C. R. (1981). Oases in the desert: The mound springs of northern South Australia. *Proceedings of the Royal Geographical Society of Australasia, South Australian Branch, 81*, 26–39.

45. Haslam, S., Sinker, C., & Wolseley, P. (1975). British water plants. *Field Studies, 4*, 243–351.
46. Hiscock, K. M., & Paci, A. (2000). Groundwater resources in the Quaternary deposits and Lower Paleozoic bedrock of the Rheidol catchment, West Wales. In N. S. Robins & B. D. R. Missteer (Eds.), *Groundwater in the Celtic Regions: Studies in Hard Rock and Quaternary Hydrogeology* (pp. 141–155). Geological Society, London, Special Publications 182.
47. Institute of Geological Sciences. (1977). *1:62500 Hydrogeological map of England and Wales*. Natural Environment Research Council.
48. Jacob, C. E. (1940). On the flow of water in an elastic artesian aquifer. *Transactions of the American Geophysical Union, 22*, 574–586.
49. Kandel, R. (2003). *Water From Heaven: The Story from the Big Bang to the Rise of Civilization, and Beyond*. Columbia University Press.
50. Karst Working Group. (2000). *The Karst of Ireland*. Geological Survey of Ireland.
51. Khan, S., & Rushton, K. R. (1996). Reappraisal of flow to tile drains I. Steady state response. *Journal of Hydrology, 183*, 351–366.
52. Kirkaldy, J. F. (1954). General principles of geology. In *Hutchinson's Scientific and Technical Publications* (pp. 284–286). Hutchinson.
53. Kohout, F. A., & Kolipinski, M. C. (1967). Biological zonation related to groundwater discharge along the shore of Biscayne Bay, Miami, Florida. *Estuaries, 83*, 488–499.
54. Krumbein, W. C., & Monk, G. D. (1943). Permeability as a function of the size parameters of unconsolidated sand. *Transactions of the American Institution of Mining and Metallurgy Engineers, 151*, 153–163.
55. Langevin, C. D. (2003). Simulation of submarine ground water discharge to a marine estuary: Biscayne Bay, Florida. *Ground Water, 41*, 758–771.
56. Love, A. J., Herczeg, A. L., Sampson, L., et al., (2000). Sources of chloride and implications for 36Cl dating of old groundwater, southwestern Great Artesian Basin, Australia. *Water Resources Research, 36*(6), 1561–1574.
57. Lucas, H. C., & Robinson, V. K. (1995). Modelling of rising groundwater levels in the Chalk aquifer of the London Basin. *Quarterly Journal of Engineering Geology, 28*(S1), S51–S62.
58. Margat, J., & van der Gun, J. (2013). *Groundwater Around the World: A Geographic Synopsis*. CRC Press.
59. Marsh, T. J., & Davies, P. A. (1983). The decline and partial recovery of groundwater levels below London. *Proceedings of the Institution of Civil Engineers, Part 1, 74*(3), 263–276.
60. Marsily, G. (1986). *Quantitative Hydrogeology: Groundwater Hydrology for Engineers*. Academic Press.
61. Meinzer, O. E. (1923). The occurrence of groundwater in the United States with a discussion of principles. *United States Geological Survey Water Supply Paper 489.*
62. Meinzer, O. E., & Hard, H. H. (1925). The artesian water supply of the Dakota

sandstone in North Dakota, with special reference to the Edgeley Quadrangle. *United States Geological Survey Water Supply Paper 520–E,* 73–95.

63. Moore, W. S. (1999). The subterranean estuary: A reaction zone of ground water and sea water. *Marine Chemistry, 65*(1–2), 111–125.
64. Moore, W. S. (2010). The effect of submarine groundwater discharge on the ocean. *Annual Review of Marine Science, 2*(1), 59–88.
65. Moore, W. S., Sarmiento, J. L., & Key, R. M. (2008). Submarine groundwater discharge revealed by 228Ra distribution in the upper Atlantic Ocean. *Nature Geoscience, 1*(5), 309–311.
66. Office of Public Works (OPW). (1997). An investigation of the flooding problems in the gort ardrahan area of south Galway. *Final Report,* vols 1–12. Office of Public Works, Ireland.
67. Paraná and Chaco-Paraná Basins, South America, and comparison with the Navajo-Nugget aquifer system, USA. (1999). *Hydrogeology Journal, 7*(4), 317–336.
68. Prescott, J. R., & Habermehl, M. A. (2008). Luminescence dating of spring mound deposits in the southwestern great artesian basin, northern South Australia. *Australian Journal of Earth Sciences, 55*(2), 167–181.
69. Puri, S., Appelgren, B., Arnold, G., et al., (2001). *Internationally Shared (Transboundary) Aquifer Resources Management: Their Significance and Sustainable Management.* UNESCO.
70. Robins, N. S. (1990a). *Hydrogeology of Scotland.* HMSO for the British Geological Survey.
71. Robins, N. S. (1990b). Groundwater chemistry of the main aquifers in Scotland. *British Geological Survey Report 18/2.*
72. Robins, N. S., & Misstear, B. D. R. (2000). *Groundwater in the Celtic Regions: Studies in Hard Rock and Quaternary Hydrogeology.* Geological Society, London, Special Publications 182.
73. Rushton, K. R. (2003). *Groundwater Hydrology: Conceptual and Computational Models.* John Wiley & Sons.
74. Rushton, K. R., & Tomlinson, L. M. (1999). Total catchment conditions in relation to the Lincolnshire Limestone in South Lincolnshire. *Quarterly Journal of Engineering Geology, 32*(3), 233–246.
75. Sharp, J. M. (1988). Alluvial aquifers along major rivers. In W. Back, J. S. Rosenshein, & P. R. Seaber (Eds.), *Hydrogeology. The Geology of North America, Vol. O-2* (pp. 273–282). The Geological Society of North America.
76. Shepherd, R. G. (1989). Correlations of permeability and grain size. *Ground Water, 27*(5), 633–638.
77. Sherlock, R. L. (1962). *British Regional Geology: London and Thames Valley* (3rd ed.). HMSO.
78. Simpson, B., Blower, T., Craig, R. N., & Wilkinson, W. B. (1989). *The Engineering Implications of Rising Groundwater Levels in the Deep Aquifer Beneath London.* Construction Industry Research & Information Association, Special Publication 69.

79. Snow, D. T. (1969). Anisotropic permeability of fractured media. *Water Resources Research, 5*(6), 1273–1289.

80. Sracek, O., & Hirata, R. (2002). Geochemical and stable isotopic evolution of the Guarani Aquifer System in the state of São Paulo, Brazil. *Hydrogeology Journal, 10*(6), 643–655.

81. Taniguchi, M., Burnett, W. C., Cable, J. E., & Turner, J. V. (2002). Investigation of submarine groundwater discharge. *Hydrological Processes, 16*(11), 2115–2129.

82. Tóth, J. (1963). A theoretical analysis of groundwater flow in small drainage basins. *Journal of Geophysical Research, 68*(16), 4795–4812.

83. Tynan, S., Gill, M., & Johnston, P. (2007). Water Framework Directive: Development of a methodology for the characterization of a karst groundwater body with particular emphasis on the linkage with associated ecosystems such as turlough ecosystems (2002-W-DS8-M1). Final report to the environmental protection agency, Ireland. Department of civil, structural and environmental engineering, Trinity College, Dublin.

84. UNESCO. (1976). *1:1500000 International Hydrogeological Map of Europe. Map B4 London*. Bundesanstalt für Geowissenschaften und Rohstoffe.

85. UNESCO. (1980). *1:1500000 International Hydrogeological Map of Europe. Map B3 Edinburgh*. Bundesanstalt für Geowissenschaften und Rohstoffe.

86. UNESCO. (2006). *Groundwater Resources of the World–Transboundary Aquifer Systems*. Special Edition for the 4th World Water Forum, Mexico City.

87. Walton, N. R. G. (1981). A detailed hydrogeochemical study of groundwaters from the Triassic Sandstone Aquifer of South-West England. *Institute of Geological Sciences, Report No. 81/5*. HMSO.

88. Williamson, A. K., Prudic, D. E., & Swain, L. A. (1989). Ground-water flow in the Central Valley, California. *United States Geological Survey Professional Paper 1401-D*.

89. Wilson, G., & Grace, H. (1942). The settlement of London due to the under drainage of the London Clay. *Journal of the Institution of Civil Engineers, 12*(1), 100–127.

90. Witherspoon, P. A., Wang, J. S. Y., Iwal, K., & Gale, J. E. (1980). Validity of cubic law for fluid flow in a deformable rock fracture. *Water Resources Research, 16*(6), 1016–1024.

91. Zektser, I. S., & Everett, L. G. (2004). *Groundwater Resources of the World and Their Use*. International hydrological programme-VI, series on groundwater No. 6. UNESCO.

CHAPTER 3

Groundwater and Geological Processes

LEARNING OBJECTIVES

After studying this chapter, you will be able to:

- Describe the geological processes driving fluid flow
- Understand the topography-driven flow in the context of geological processes
- Learn about compaction-driven fluid flow and variable-density driven fluid flow
- Explain the regional groundwater flow systems driven predominantly by variable-density flow
- Describe the regional groundwater flow systems driven predominantly by shifting topography and stress changes
- Understand the coupling and relative importance of processes driving fluid flow

KEY TERMS FROM THIS CHAPTER

Earth's crust
Fluid flow
Geothermal systems
Hydraulic gradients
Rainfall
Topography
Earthquakes
Geofluids
Groundwater
Hydrocarbons
Thermodynamics
Erosion
Geologic systems
Groundwater recharge
Potential energy

INTRODUCTION

A significant part of many geologic processes involves groundwater. For instance, it is now known that the fluid pressures that accumulate on faults control fault movement and the production of earthquakes. On the other hand, groundwater plays a crucial role in processes like the formation of geothermal systems, the thermodynamics of pluton emplacement, and the genesis of profitable mineral deposits because subsurface flow systems are in charge of transferring heat and chemical components throughout geologic systems. Groundwater flow systems regulate the movement and buildup of petroleum at depths. They are involved in geomorphologic processes closer to the surface, like stream bed erosion, natural slope development, and karst formation.

The reaction of groundwater systems to geological phenomena like plate tectonics, glaciation, and variations in sea level is covered in this chapter. Even though topography-driven groundwater flow dominates groundwater circulation in many parts of the world, understanding the additional influence of geological processes on groundwater flow is frequently crucial to understanding the dynamics of groundwater systems over comparatively long timescales (millennia). For example, the existence of fresh water beneath the seafloor along a large portion of the continental shelf can only be understood by taking into account the hydrogeological effects of glaciation and the associated sea-level low stand. Regional groundwater flow systems are characterized by flow driven by the variable density of groundwater and flow induced by stresses imposed on pore fluids. These flow drivers often constitute the only source of fluid flow in the absence of topographically driven flow, and they typically reach circulation depths of several kilometers. Numerous geological processes can be the cause of anomalous pore pressures and variable groundwater density. Therefore, an understanding of how the relevant geological processes cause fluid flow is necessary to understand fluid flow patterns in relatively deep hydrogeological systems. The hydrogeologist's job is to assess which subsurface fluid flow-promoting processes, given a particular geological setting, can be anticipated to be dominant.

The response of hydrocarbons and other geofluids, like groundwater, to geological forcing frequently has a direct impact on the progression of geological processes. For example, this feedback mechanism is crucial when taking into account how fluids affect seismicity. However, as a more focused discussion on feedback mechanisms would be beyond the scope of this chapter, it will be restricted to a more or less one-way description of the primary impact of geological processes on fluid movement in the Earth's crust.

This chapter illustrates the importance of comprehending how geological processes influence groundwater flow for society. Groundwater and geological processes interact, and this knowledge is crucial for understanding current issues like the safe subsurface sequestration of CO_2 or the design of nuclear waste repositories in deep aquifers.

3.1. GEOLOGICAL PROCESSES DRIVING FLUID FLOW

The hydrogeological process of groundwater recharge results in groundwater flow that is determined by topography. In addition to changing climatic conditions, a number of geological processes, including ice sheet build-up and retreat, mountain building, and erosion, alter water table gradients over geological timescales. These processes also affect the rates of groundwater recharge. Hydraulic head gradients in groundwater flow systems can also result from two other categories of processes unrelated to topography-driven groundwater flow, which are frequently important when considering flow on a regional scale and on geological timescales. First, hydraulic gradients may be caused by differences in fluid densities as a result of geological processes. Developing temperature gradients and/or contrasts in the total dissolved solids content (salinity) will result in density contrasts strong enough to generate significant hydraulic head gradients. Density-driven flow plays an important role in hydrothermal systems, deep sedimentary basins, and coastal areas where highly mineralized fluids are common and dense. Second, fluid movement may arise from stresses that geological processes place on pore fluids. Examples of these processes covered in this chapter include erosion and mountain building, as well as sediment compaction beneath developing ice sheets and in subsiding sedimentary basins.

3.2. TOPOGRAPHY-DRIVEN FLOW IN THE CONTEXT OF GEOLOGICAL PROCESSES

The topography of the water table close to the surface creates hydraulic gradients in aquifers that are necessary for groundwater flow in topography-driven groundwater flow systems. Water table gradients will decrease as a result of the dissipation of potential energy reflected by hydraulic gradients caused by fluid movement. Therefore, energy must be added to the system in order to maintain fluid flow. In topography-driven flow systems, this energy is added through groundwater recharge. Therefore, groundwater will only continue to flow from recharge areas to groundwater discharge areas if there is enough groundwater recharge to sustain the driving water table gradient. If not, groundwater movement will eventually stop as the water table gradient decreases. Arid regions with deep water tables and little groundwater recharge serve as examples of this. As a result, the rate of topography-driven flow will fluctuate depending on geological time and potential topographical changes in addition to climate factors influencing recharge. The formation of mountains, erosion, and ice sheets can all alter the topography. These modifications in the topographic gradients that drive them are typically accompanied by other geological processes that result in density gradients and/or variations in the pressure of the pore water, which offer more ways to create fluid flow. Prior to going through some examples of how shifting topographic configurations affect groundwater systems, it is important to briefly introduce density-driven flow and pressure anomaly-driven flow in groundwater.

3.3. COMPACTION-DRIVEN FLUID FLOW

Sediment piles in sedimentary basins (e.g., the Gulf of Mexico and North Sea Basins) have a thickness of several kilometers. Sediments in these basins will be progressively buried beneath younger sediments as they develop. Porosity is lost as a result of mechanical compaction that occurs during this process. Since water is not very compressible, the fluids inside the pores will be forced out when the pore volume is reduced. A rise in pore fluid pressure will complete this process. As a result, the increasing weight of the overburden provides the energy needed for the compaction process to propel fluid flow. A closer look at porosity versus depth measurements for various kinds of sediments provides more information about the compaction process.

Figure 3.1 illustrates an example of such data that makes two observations. First, the upper few hundred meters of the subsurface are where porosity is lost the fastest. Second, the porosity (water content) of clay-rich sediments is higher at the surface than that of sand-rich sediments. On the other hand, porosity in clay-rich sediments decreases more quickly with depth than in sand-rich sediments. Since platy clay minerals can retain water molecules between them, the compressibility of clay differs from that of sands.

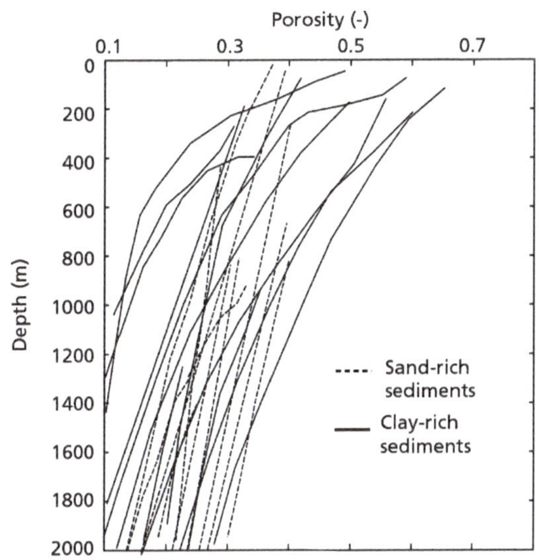

Figure 3.1. *Variation of porosity with depth in predominantly sand-rich sediments and clay-rich sediments. Data compiled from various data sources as listed in Bense and Person (2006).*

Source: Kevin M. Hiscock and Victor F. Bense. Hydrogeology Principles and Practice. Wiley-Blackwell. ISBN: 978-1-119-56951-0.

As outlined by Terzaghi in his classic work from the 1920s (Terzaghi, 1925), mechanical compaction of a pore network requires an increase of the effective stress (σ_e) which is equal to:

$$\sigma_e = \sigma_t - p \tag{3.1}$$

Either the fluid pressure (p) must drop or the overall weight of the overburden (σ_t) must rise in order to achieve an increase in effective stress. While the former happens as sediment burial increases, the latter will be achieved by the drainage of pore fluids. The sediment framework must support a greater share of the overburden's weight (total stress) when effective stress increases. Compaction and a decrease in porosity will arise from an increase in effective stress if the sediment is mechanically compressible. When subjected to a mechanical load, relatively porous sediments like clays and shales compress more readily than lower porosity sands, which are less easily compressible.

Porosity (n) variations with depth as shown in Figure 3.1 are often described (e.g., Bethke and Corbet, 1988) using a simple relationship taking into account effective stress, compressibility (β, Pa^{-1}) and assuming a porosity at the surface (n_o), of the form:

$$n = n_o \cdot e^{-\sigma_e \beta} \tag{3.2}$$

Porosity, n, in Equation 3.2 decreases exponentially as effective stress increases. Equation 3.2, assuming a linear relationship between effective stress and depth, can be used to compute porosity-depth curves, which are represented by the dashed lines in Figure 3.1, for a range of compressibility and n_o values.

Hydrostatic conditions will be satisfied and increases in pore pressure caused by porosity loss will swiftly level off when the pore network's vertical connectivity is effective. The pore pressure in the subsurface increases linearly with depth under hydrostatic conditions, indicating only the weight of the fluids that are hydraulically connected to the surface through the pore network. The hydrostatic pressure (p_h) with increasing depth (z) is thus simply given by:

$$p_h = z \rho_f g \tag{3.3}$$

in which g is the acceleration due to gravity and ρ_f is the water density. For purely fresh water (ρ_f = 1000 kg m^{-3}), the hydrostatic gradient is about 10^4 Pa m^{-1}. One kilometer of water thus exerts a pressure of about 10 MPa. Likewise, the total vertical stress, σ_T, often called the lithostatic pressure, is equal to the pressure exerted by the overburden of sediment only, which is equal to:

$$\sigma_T = z \rho_s g \tag{3.4}$$

in which ρ_s is the density of the rock matrix. A typical density for sediments is 2700 kg m^{-3}. Hence, the lithostatic gradient under such conditions is 2.7 times larger than the hydrostatic gradient (Figure 3.2a).

Pore pressures in deeper aquifers (>1 km) are higher in many areas of active sedimentation than hydrostatic conditions would predict (Figure 3.2b). In the literature, terms like "overpressure," "anomalous pressure," "excess pressure," or "geopressure" are frequently used to describe elevated pressure above the hydrostatic gradient. Drilling in strata that are overly pressurized can be dangerous and lead to a "blow-out" when pressurized fluids begin to leak out of the borehole. Overpressure is a sign that the high pressures created during compaction are not dissipating effectively. In environments with low permeability, this is usually the case. The rate of dissipation is in competition with the rate of overpressure generation resulting from continuous mechanical loading (sedimentation rate), even though any out-of-equilibrium pressures eventually vanish. A one-dimensional version of the groundwater flow equation can incorporate the net effect of sedimentary loading-generated hydraulic gradients (and thus fluid fluxes) by expressing flow in the depth (z) direction via a source term, Q_c, as follows.

$$K\frac{\partial^2 h}{\partial z^2} = S_s\frac{\partial h}{\partial t} + Q_c \quad (3.5)$$

Kooi (1999) showed that the magnitude of the source term (Qc) relates to the subsidence rate (vz), the rate of sedimentation (vsed), the density of the sediment settling at the sea bed (ρgr) and its porosity (nsed), and the compressibilities of sediment (α) and the fluid residing in the pores (β), as given by:

$$Q_c = \alpha(\rho_{gr} - \rho_w)(1 - n_{sed})gv_{sed} - n\beta\rho_w g v_z \quad (3.6)$$

Equation 3.6 demonstrates that the hydrodynamic effects of compaction will be greater in basins that are subsiding more quickly and in sediments that are more easily compressible.

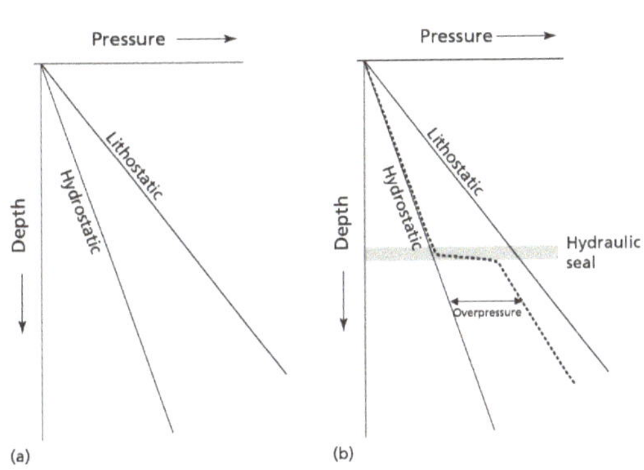

Figure 3.2. *Fluid pressure in a pore network will follow a hydrostatic trend when the vertical connectivity of the pore network is efficient. In (a) lithostatic pressures reflect the pressure exerted by the rock matrix. In (b) when the vertical equilibration of fluid pressures is hampered during basin subsidence, for example by the presence of a hydraulic seal such as a low-permeability shale, so called 'overpressures' can develop at depth.*

Source: Kevin M. Hiscock and Victor F. Bense. Hydrogeology Principles and Practice. Wiley-Blackwell. ISBN: 978-1-119-56951-0.

Although the example of flow driven by an increasing overburden was used to introduce the idea of compaction-driven flow above, other geological processes can also cause pore space to compress and cause fluid movement. For instance, fluid flow in the aquifers beneath an ice sheet can be induced by the increasing overburden weight exerted by a developing glacier or ice sheet. Moreover, fluid pressure may increase as a result of tectonic stresses (Ge and Garven, 1994). The load's application rates and magnitude (e.g., tectonic movement, sedimentation rate, or ice-sheet buildup rate) will largely determine how these various processes affect fluid pressure.

Understanding that compaction-driven flow is a transitory phenomenon associated with the geological process of basin formation—which results in the burial of sediments bearing fluid—is crucial. It is expected that the effects of compaction-driven flow will be less significant in more developed basins.

Flow driven by compaction is a frequent occurrence in sedimentary basins that are actively subsiding. While topography-driven flow can be relatively less dominant in these basins during their early development, it is more dominant in more mature sedimentary basins where compaction has stopped due to the sediments' increased strength from diagenesis and decreased compressibility (Figure 3.3).

The Bath Basin is an example of a basin that falls into the latter category, whereas the Gulf of Mexico Basin is a prime example of an active basin where flow is influenced by both compaction and topography.

A two-dimensional numerical model is used in the case study that describes overpressure generation in the Gulf of Mexico to further analyze the interaction between compaction-driven and topography-driven flow.

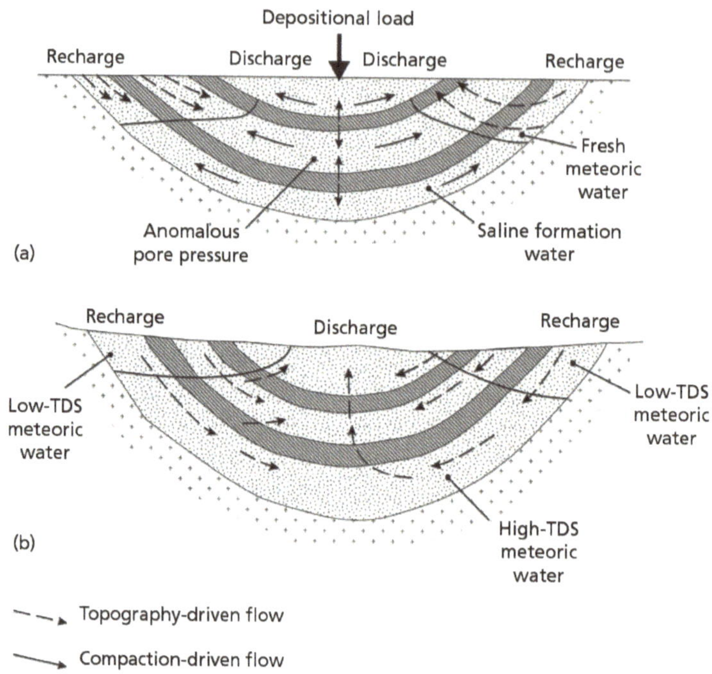

Figure 3.3. *(a) Groundwater circulation pattern in an actively subsiding sedimentary basin where topography-driven flow from the flanks competes with compaction-driven flow originating mainly from the depositional center of the basin. TDS = total dissolved solids. (b) A more geologically mature basin in which deposition has halted and topography-driven flow is dominant.*

Source: Kevin M. Hiscock and Victor F. Bense. Hydrogeology Principles and Practice. Wiley-Blackwell. ISBN: 978-1-119-56951-0.

3.3.1. Determining Porosity and Sediment Compressibility from Drill Core and Well Data

In order to demonstrate the fundamental principles of porosity and the relationship between porosity loss and sediment compressibility, let's look at some dry, water-free drill core that was extracted from a 324-meter-deep borehole that penetrated sandstone. The circular drill-core measures 50 mm in diameter and 10 cm in length. This core is immersed in a water-filled tank with measurements of 0.1 × 0.1 × 0.2 meters (l × w × h). This tank's water level rises by an initial, undefined amount. The water level has dropped by 3.1 mm and is not dropping any further four hours after the core was submerged. It is now reasonable to assume that the rock's pore space was initially empty of water and only held air. Therefore, the volume of water filling the pores can be calculated as: 0.0031m (the depth of water used to fill the pores in meters) × 0.1 × 0.1 (the surface area over which the drop in water level occurred in the tank) = 3.1×10^{-5} m^3.

Now, to calculate the porosity of the rock in the drill-core, the volume of the drill-core is found from: $\pi \times \text{radius}^2 \times h = \pi \times 0.025^2 \times 0.1\text{m}^3 = 1.96 \times 10^{-4}\text{m}^3$; and the porosity of the rock at a depth of 324m follows from: $n = V_{pores}/V_{drill\text{-}core} = 3.1 \times 10^{-5}/1.96 \times 10^{-4} = 0.158$.

A comparable analysis from a rock outcrop of the same geological formation that the drill core was taken from produced a porosity of 0.27. From this additional information, the compressibility of the sandstone can be derived using Equation (3.2):

$$n = n_0 \cdot e^{-\sigma_e \beta}$$

and by assuming that the effective stress (Equation 3.1), σ_e, follows a simple linear relationship with depth. Therefore, σ_e can be replaced with depth, z, and as a result the compressibility, β, takes on units of length rather than pressure:

$$n = n_o \cdot e^{-z\beta}$$

β is calculated using the data obtained above using n_o = 0.27 (porosity of 27% at the surface), and n = 0.158 at z = 324 m. This calculation yields:

$$n(324\text{ m}) = n_0(0\text{ m}) \times \exp(-\beta \times 324)$$

$$n(324\text{ m})/n_0(0\text{ m}) = \exp(-\beta \times 324)$$

$$\ln(0.158/0.27) = -\beta \times 324$$

$$\beta = -\ln(0.158/0.27))/324$$

$$\beta = \sim 1.7 \times 10^{-3}\text{ m}^{-1}$$

With the value of β discovered in this manner, a porosity-depth profile for the entire sandstone formation can now be computed. These determined values for Well #1 are displayed in Table 3.1.

Figure 3.4 displays a plot of Well #1's porosity values. Two more wells' worth of data (Well #2 and Well #3 in Table 3.1) are also plotted. It is possible to verify the significantly higher compressibility of 4.04 × 10^{-3}m^{-1} of the sediments found in Well #2 by following the above described procedure.

The porosity-depth profile for Well #3 indicates a porosity increase with depth over a distance of approximately 100–300 m that cannot be attributed to compaction.

At this depth, more processes have to be at work, like the interaction between the water and the rock, which can enhance porosity and cause dissolution. If a zone of fracturing intersects the borehole during this depth interval, it could also account for an interval of enhanced porosity in borehole data like these.

Table 3.1. *Example Porosity-Depth Data for Three Wells*

Depth (m)	Well #1	Well #2	Well #3
0	27.0	55.0	15.0
50	24.9	45.2	14.2
100	22.9	37.1	13.5
150	21.1	30.5	13.8
200	19.4	25.0	14.1
250	17.8	20.5	19.5
300	16.4	16.9	19.9
350	15.1	13.8	18.3
400	13.9	11.4	13.8
450	12.8	9.3	11.3
500	11.8	7.7	8.8
550	10.9	6.3	8.3
600	10.0	5.2	7.9
650	9.2	4.2	7.5
700	8.5	3.5	7.1
750	7.8	2.9	6.7

Source: Kevin M. Hiscock and Victor F. Bense. Hydrogeology Principles and Practice. Wiley-Blackwell. ISBN: 978-1-119-56951-0.

Porosity values for Well #1 were calculated from field data as outlined in the text. Note that values of porosity, n, are given in %.

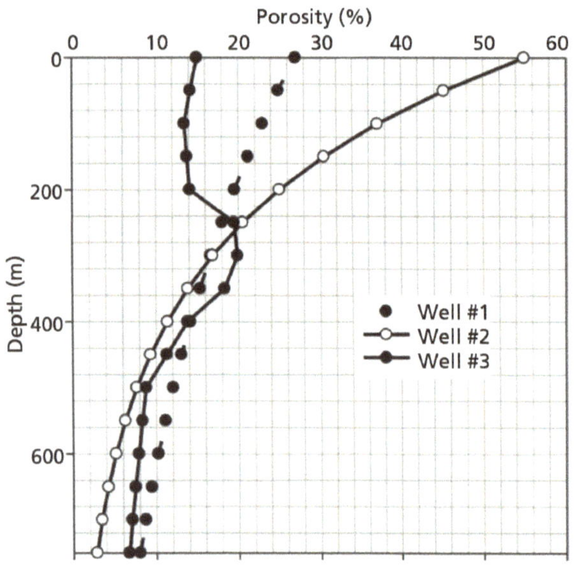

Figure 3.4. *A plot of the porosity-depth data presented in Table 3.1.*

Source: Kevin M. Hiscock and Victor F. Bense. Hydrogeology Principles and Practice. Wiley-Blackwell. ISBN: 978-1-119-56951-0.

3.3.1.1. Compaction- and Topography-Driven Fluid Flow in the Gulf of Mexico Basin

A well-known illustration of a sedimentary basin that is actively subsiding is the Gulf of Mexico Basin (Figure 3.5). In addition to economically significant hydrocarbon reserves, the Gulf of Mexico Basin has groundwater aquifers that are used as a source of water. The Gulf of Mexico's hydrodynamic conditions are comparable to those of other global basins, including the Persian Gulf, Amazon Delta, Niger Delta, and North Sea Basin.

Sand and clay sediments build up offshore in the Gulf of Mexico, and it is anticipated that the growing overburden will cause compaction and compaction-driven flow.

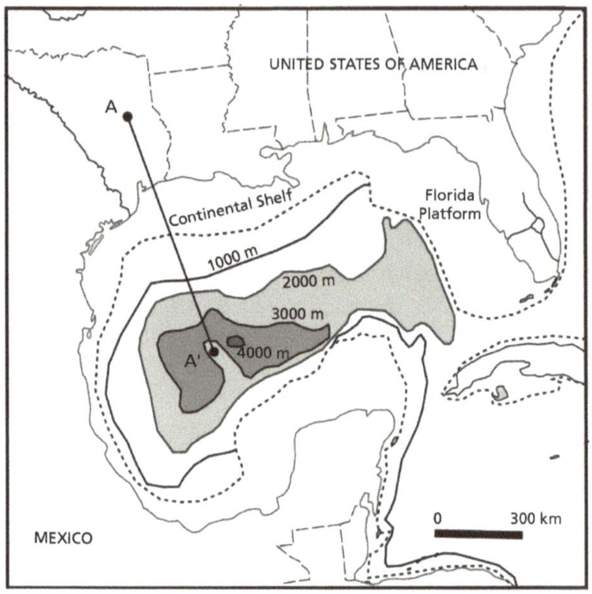

Figure 3.5. *Outline of the shape and size of the Gulf of Mexico sedimentary basin and its approximate bathymetry. The location of cross-section A-A′ is indicated, for which Harrison and Summa (1991) carried out a numerical model to simulate the hydrodynamic history of the basin.*

Source: Kevin M. Hiscock and Victor F. Bense. Hydrogeology Principles and Practice. Wiley-Blackwell. ISBN: 978-1-119-56951-0.

On the other hand, meteoric recharge takes place onshore along the basin's flanks, where topography-driven flow will be predominant.

Where compaction drives flow and topography drives flow, interesting hydrodynamic systems arise. Examining pressure-depth profiles taken within the basin reveals the interplay between compaction-driven and topography-driven flow (Figure 3.6). In the Gulf of Mexico Basin, profiles five and six are captured both onshore and nearshore. Below a depth of approximately 4 km, these profiles show hydrostatic conditions; below that, pressures are significantly higher than hydrostatic (overpressure). Further offshore, at much shallower depths, the conditions change from hydrostatic to over-pressured (Profiles 1-4). The relationship between the occurrence of overpressure and the geological build-up of the basin is demonstrated by model simulations that are constrained by pressure and geological data from the Gulf of Mexico (Figure 3.7).

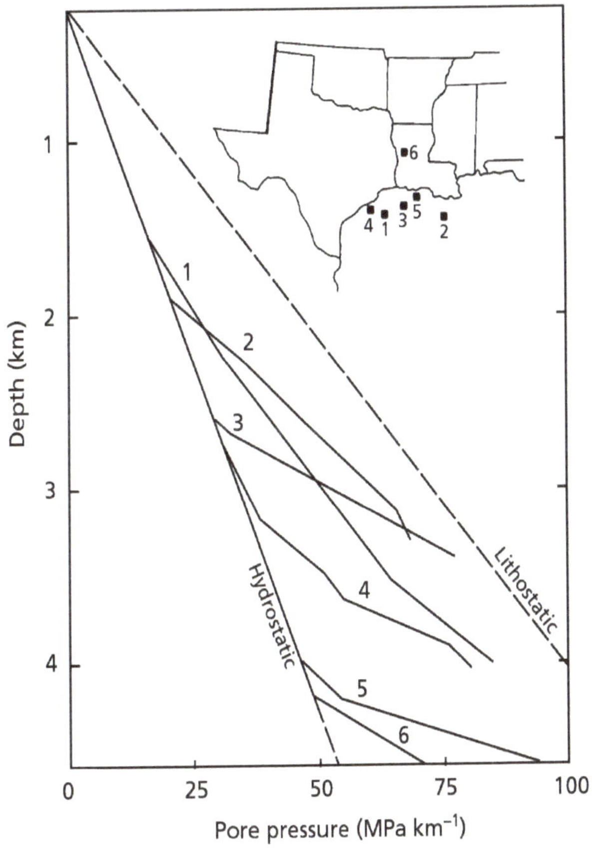

Figure 3.6. *Pore pressure variation with depth as observed in six wells in the Gulf of Mexico sedimentary basin. Locations of these wells are indicated on the accompanying map inset. The occurrence of overpressure at depth is observed in all wells but nearer to the center of the basin overpressure occurs at increasingly shallower depth.*

Source: Kevin M. Hiscock and Victor F. Bense. Hydrogeology Principles and Practice. Wiley-Blackwell. ISBN: 978-1-119-56951-0.

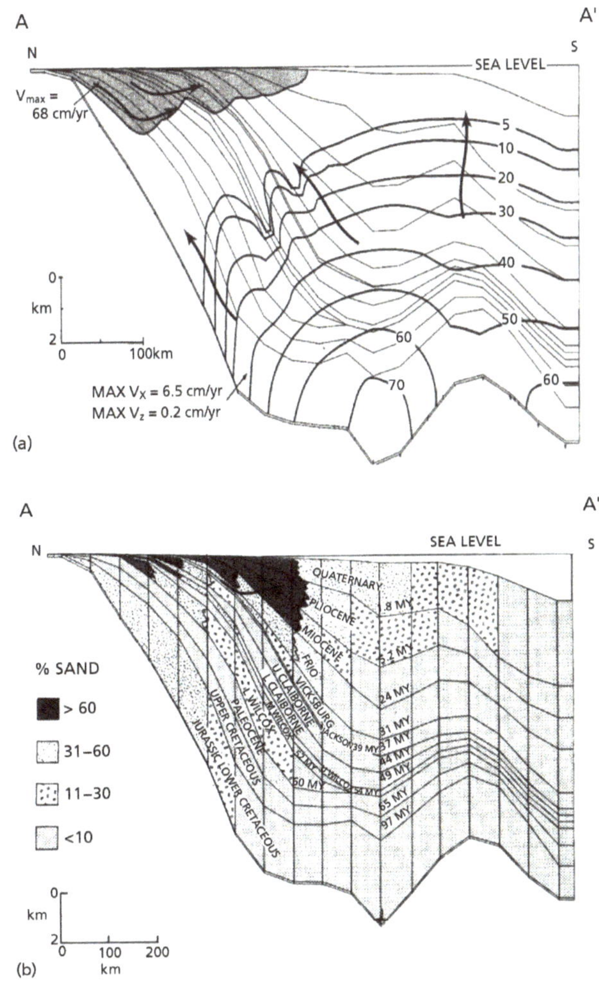

Figure 3.7. *(a) The current distribution of overpressure in the Gulf of Mexico Basin as simulated by Harrison and Summa (1991) with approximate direction of fluid flow directions indicated by arrows. The shaded area represents that part of the basin where fluids are driven by topographic gradients, while in the rest of the basin compaction-driven flow predominates. (b) The geological build-up of the basin shows that the largest overpressures develop in the more easily compressible lithologies which are more clay-rich, such as shale, and in the deepest parts of the basin.*

Source: Kevin M. Hiscock and Victor F. Bense. Hydrogeology Principles and Practice. Wiley-Blackwell. ISBN: 978-1-119-56951-0.

The two-dimensional distribution of overpressure in the basin's lower depths is shown in Figure 3.7. Fluids rise out of the basin center due to the overpressures at depth. According to Figure 3.7a, the model results indicate that the fluid expulsion rates are approximately 6.5cma^{-1}, accounting for pressure gradients and sediment permeabilities. In the onshore and near-shore portion of the basin, overpressure is absent at depths of up to about 2 km, and topography-driven groundwater flow is directed toward the center of the basin. Shallow groundwater flow rates can reach up to ~70 cma^{-1}, which is an order of magnitude higher than deeper basin rates. When elements of the basin stratigraphy are taken into account, the observed patterns can be understood, at least partially (Figure 3.7b). The deeper basin sediments are more clay-rich than the shallower sediments, which are primarily composed of sand, due to the basin's geological past. Relatively high clay content makes sediments more compressible and less permeable. Therefore, the presence of clay-rich sediments can help to explain, at least in part, the dominance of compaction-driven flow and low fluid flow rates in the basin center. Similarly, the presence of permeable sediments speeds up the topography-driven groundwater flow circulation rates at shallow depths. Detailed analysis reveals that, in addition to sediment compaction, other geological processes may also be responsible for the overpressures currently seen in the Gulf of Mexico Basin. These include the dehydration of clay minerals like smectite during compaction, which releases interlayer water into pore space and increases pore pressure, and the role that thermal expansion of pore waters plays in producing overpressure. Furthermore, current pore pressure distributions may be significantly impacted by the palaeohydrology of the Gulf of Mexico system, where sea level and subsidence rates have varied over the course of millions of years. A series of two-dimensional numerical models that took into account the palaeohydrology of the Gulf of Mexico Basin during the Tertiary and Quaternary were presented by Harrison and Summa (1991). These models show the historical variations in the depth of meteoric groundwater circulation and the force of fluid expulsion from the basin center. Nonetheless, the results of the Harrison and Summa (1991) model provide strong evidence in favor of the theory that deeper fluid flow regimes in the Gulf of Mexico Basin are primarily driven by compaction-driven flow.

3.4. VARIABLE-DENSITY DRIVEN FLUID FLOW

In groundwater hydrology, where fluid density varies in space and time as a function of pressure, temperature, and concentrations of multiple species, variable density flow is becoming more and more important.

3.4.1. Salinity Gradients Leading to Variable-Density Flow

Groundwater flow is driven by the hydraulic gradients that result from buoyancy effects caused by density gradients in the water. Coastal regions, where fresh water from meteoric recharge over the land surface meets salty seawater, are the classic settings in which to study variable-density flow. Furthermore, when examining the behavior of certain organic contaminants in groundwater flow systems, it is necessary to consider their density because some of them may possess characteristics that make them denser or less dense than water.

Mathematically, variable-density flow can be expressed by adding buoyancy terms to the groundwater flow equation. Variations in temperature, pressure, and solute concentration can result in contrasts in fluid density. Variable-density driven flow is often described as happening through free-convection when it forms in the absence of topographic gradients. Thus, forced-convection—a term that is rarely used in practice—would be used to describe groundwater convection brought on by the water table's topography. Only when the aquifer's permeability, thickness, and the density difference between its groundwater and seawater are sufficiently high will free-convection take place. These parameters can be combined to calculate a theoretical stability criterion, a Rayleigh number, Ra, for the onset of free-convection due to solute concentration effects (e.g., Schincariol and Schwartz, 1990) and is given as:

$$Ra = \frac{gkL\left(\rho_s - \rho_f\right)}{\mu D} \qquad (3.7)$$

in which μ is the dynamic viscosity, ρ_s is seawater density, ρ_f is density of water in the aquifer, k is intrinsic permeability, D is solute diffusivity coefficient, and L is thickness of the aquifer. If Ra is larger than ~10, free-convection of saline groundwater into fresh groundwater underneath is likely to occur.

3.4.2. Hydrothermal Systems Driven by Variable-Density Flow

Thermal convection, also known as convective fluid flow patterns, can occur when temperature gradients are high enough to produce adequate fluid density gradients. There are zones of upwelling and downwelling in freely convecting systems. Hot fluids with a low density rise to the surface and cool in a hydrothermal system powered by free convection. Subsequently, as their density increases, it offsets the buoyancy and causes them to sink.

Again, a Rayleigh number can be formulated to determine under which conditions thermal convection will start to occur. In a simplified system, in which the upper and the lower boundaries of a horizontal aquifer with thickness, L, are kept at a fixed temperature (denoted as T_u and T_l, respectively), the Rayleigh number for the onset of thermal convection is given as:

$$Ra = \frac{\alpha_f \rho_w^2 c_w g k L (T_l - T_u)}{\mu \kappa} \tag{3.8}$$

in which α_f is the thermal expansivity, k is aquifer permeability, κ is thermal conductivity, ρ_w is fluid density, μ is fluid dynamic viscosity, and c_w is the fluid volumetric heat capacity. The reference temperature chosen to evaluate temperature-dependent properties in Equation 3.8 is typically taken as $(T_u + T_l)/2$.

When Ra rises above a critical value, free convection takes place. An often-cited critical value is approximately $4\pi^2$. Figure 3.8 displays the critical thermal gradients as a function of aquifer thickness and for three different permeability values, using Ra = $4\pi^2$ and standard values for the other parameters in Eq (3.8). This implies that thermal gradients must primarily be greater than those found under typical geothermal conditions (e.g., 25°C km^{-1}) significantly for density differences brought on by thermal effects to result in free-convection. Free-convection is more likely to occur in permeable rock if there is good vertical continuity in the direction of the geothermal gradient. Thus, it should come as no surprise that numerous permeable fault zones have been found to be hotspots for groundwater thermal convection that cause hot springs to surface. To some extent, the high crustal heat flow necessary to sustain the necessary temperature gradients also controls many of these systems.

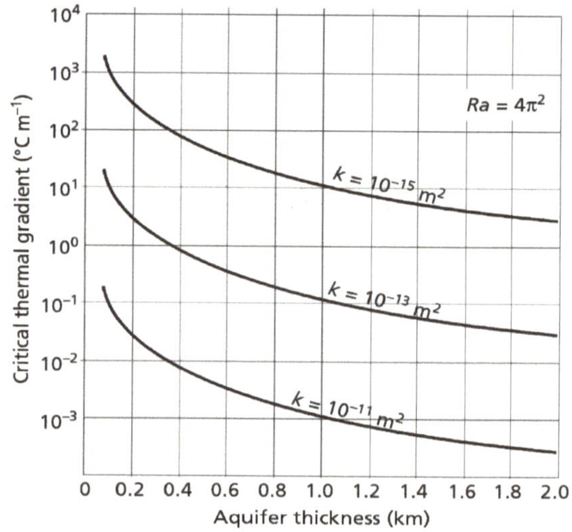

Figure 3.8. *In this diagram, by assuming a Rayleigh number (Ra) for the onset of hydrothermal convection (eq. 3.8) the critical thermal gradient can be calculated which, if exceeded, leads to groundwater convection. However, as shown, this critical thermal gradient is also a function of aquifer thickness and permeability (k).*

Source: Kevin M. Hiscock and Victor F. Bense. Hydrogeology Principles and Practice. Wiley-Blackwell. ISBN: 978-1-119-56951-0.

3.4.2.1. Hot Springs along Permeable Fault Zones in the Great Basin, Western USA

Several extensional basins enclosed by typical fault systems combine to form the Great Basin in the Western United States of America. Because of its relatively small crustal thickness (~20 km), which lies halfway between oceanic and continental crust, the Great Basin is reminiscent of its past as a location where oceanic crust formed before it was incorporated into the American continent during the Mesozoic era. New local basin formation and extensive normal-fault systems were produced by Cenozoic uplift and related crustal extension. Numerous hydro-geothermal systems are found in these fault systems due to the high geothermal heat flow in the region.

The historical movement of hydrothermal water along faults in this region is linked to world-class ore deposits created by gold-silver accumulations along fault zones (e.g., Person et al., 2007a).

The spring system situated in northern Nevada, north of Borax Lake, is one of the most extensively researched spring systems linked to active faulting in the Great Basin (Figure 3.9). There are no boreholes in this region that can be used to gather geological and hydraulic head data or to extract samples from the deeper groundwater. Nonetheless, it is possible to measure the spring waters' temperature and geochemistry using comparatively easy methods. In addition to measuring the temperature of the near-surface materials near these springs, Fairley and Hinds (2004) also measured the water temperature of 175 springs using a handheld digital thermometer. After correctly interpolating these data using geostatistical techniques, an image of the near-surface temperature distribution near the springs was created (Figure 3.10).

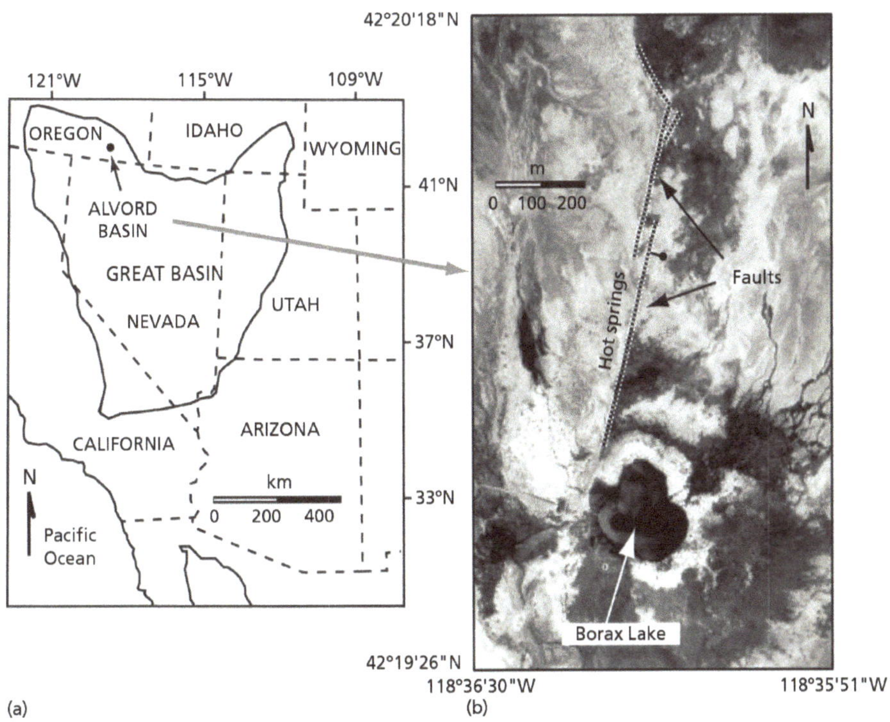

Figure 3.9. *Locations of: (a) the Great Basin; and (b) Borax Lake.*

Source: Kevin M. Hiscock and Victor F. Bense. Hydrogeology Principles and Practice. Wiley-Blackwell. ISBN: 978-1-119-56951-0.

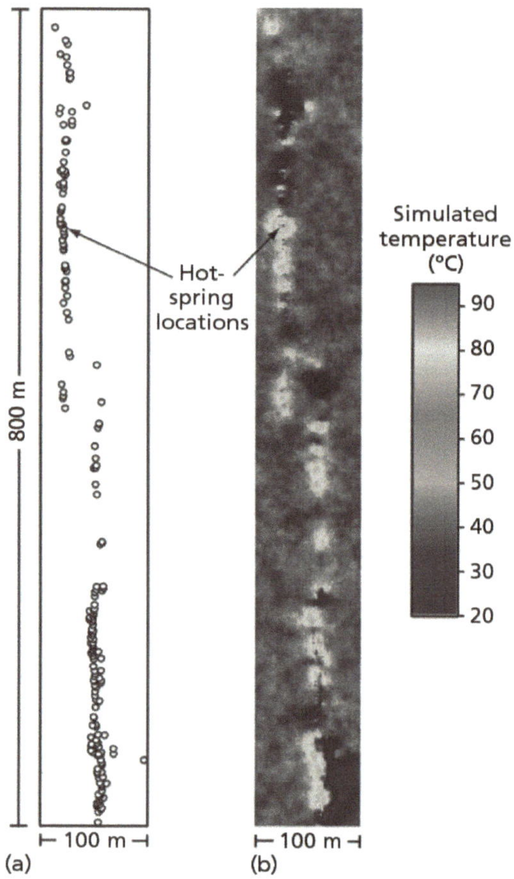

Figure 3.10. *Location and interpolated (simulated) near-surface temperatures of springs emerging in the Borax Lake area. Clearly, the springs align along the trace of a set of linear fault segments.*

Source: Kevin M. Hiscock and Victor F. Bense. Hydrogeology Principles and Practice. Wiley-Blackwell. ISBN: 978-1-119-56951-0.

The thermal image (Figure 3.10b) displays several noteworthy patterns. Two fault traces are thought to be the alignment of the hot springs. These may represent a "fault relay structure" at the surface, where a fault strand ends and the fault displacement is transferred to a second fault trace. These may be a part of the same fault structure at depth. Furthermore, the ground and spring temperatures vary greatly along the fault trace. This would suggest that the fault zone contains preferential flow paths with comparatively high permeability that allow groundwater to rise more quickly. Increased surface temperatures would suggest a higher flow rate because cooling will happen along these flow paths where the surrounding rock is colder than the upwelling groundwater. Some springs have temperatures as high as 94°C, which is almost boiling, but oxygen isotope data suggest that meteoric recharge occurs in these springs. Circulation depths of up to two to three kilometers are suggested by Sr-isotope data (Anderson and Fairley, 2008). In order to make sense of these data, the main mechanism for moving groundwater from and towards the surface appears to be thermal convection, supported by a component of more shallow topography-driven flow, as seen, for instance, in the groundwater recharge from Borax Lake flowing into the fault zone.

Fairley and Hinds (2004) performed a preliminary quantitative analysis of the data by estimating the potential variability in vertical groundwater fluxes that are reflected in the observed surface temperature distribution using a basic one-dimensional heat flow model. This preliminary study provides information about the distribution of fault permeability, which serves as the foundation for more complex numerical models that explicitly describe flow caused by thermal convection (Fairley, 2009). A boundary of constant temperature (250°C) at 400 m depth is simulated in this model (Figure 3.11), from which groundwater rises upward at a rate determined by a stochastic permeability structure with statistical characteristics similar to those noted in the field. The latter model generates surface temperatures and groundwater discharge rate variability, the statistics of which can be compared with field observations.

One of the best examples of a system powered by thermal convection is the groundwater circulation along faults close to Borax Lake. This location's research demonstrates how conceptualized numerical models combined with comparatively basic field observations, like temperatures of near-surface groundwater, can yield a fairly detailed understanding of the permeability structure of intricate hydrogeological systems, like fault zones.

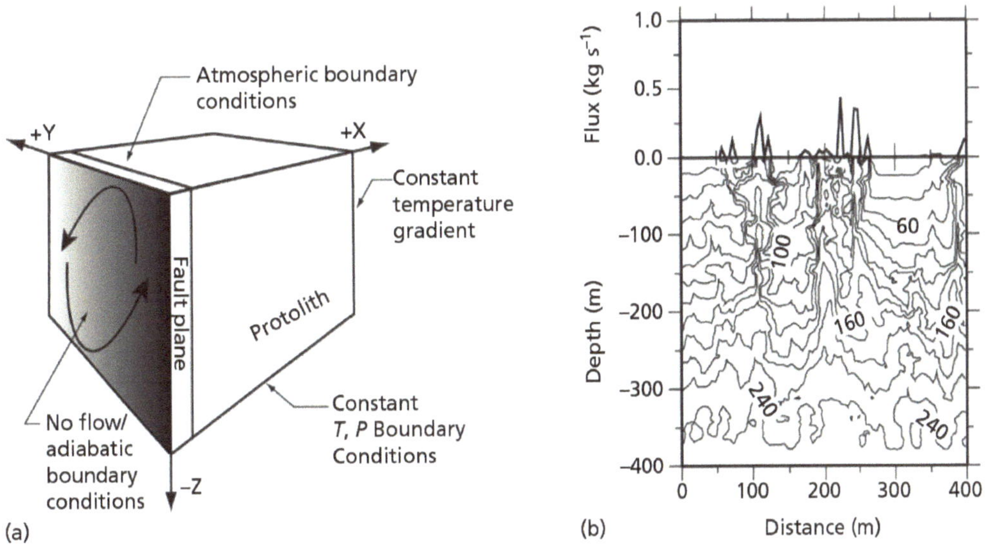

Figure 3.11. *(a) Set-up and boundary conditions of a numerical model describing thermal convection along a fault zone. (b) Temperature distribution and surface groundwater fluxes as calculated in the fault plane for one model of a set of model realizations.*

Source: Kevin M. Hiscock and Victor F. Bense. Hydrogeology Principles and Practice. Wiley-Blackwell. ISBN: 978-1-119-56951-0.

3.5. REGIONAL GROUNDWATER FLOW SYSTEMS DRIVEN PREDOMINANTLY BY VARIABLE-DENSITY FLOW

Two regional-scale systems are examined below, both of which rely heavily on variable-density flow to regulate the system's hydrogeological evolution. First, long-term changes in the distribution of fresh and salt water in coastal regions brought on by millennia of sea level fluctuations are examined. Second, salinity contrasts are significant when determining the locations of brines, which are extremely saline fluids found in deep aquifers in mature sedimentary basins that are frequently far from coastal areas. The effects of this kind of groundwater scenario on fluid flow must be taken into consideration in certain situations. For instance, when designing nuclear waste repositories or evaluating the long-term safety of CO_2 sequestration in deep aquifers, one must take into account the effects of glaciers on groundwater flow on a continental scale.

3.5.1. Fluctuating Sea-Level and Its Impact on the Distribution of Groundwater Salinity in Coastal Areas

Seawater can enter freshwater aquifers through tongues or coastal flooding, which allows the seawater to sink by free convection into underlying aquifers. Conditions like those shown in Figure 3.12(a) will usually hold true when the coastline is in a stable position. Aquifers extending landward from the coastline that support the freshwater system fed by meteoric recharge will contain tons of saline water. The extent of the landward ingress of seawater will depend on the density contrast between fresh and saline water as well as the magnitude of the topography-driven seaward flow of fresh water, which is controlled by the rates of meteoric recharge in the coastal zone, for a stable average position of the coastline (not taking tidal fluctuations into account). In Figure 3.12(a) under steady-state conditions and considering a stable position of the coastline, seawater can be assumed to be present below the seafloor. The extent of saline water sag as a tongue at the base of the coastal aquifer is restricted by the coastward flow of fresh groundwater derived from meteoric recharge. In Figure 3.12(b) during sea level low stand after marine regression, the saline groundwater formerly below the sea floor is displaced by fresh groundwater which leads to a new steady-state situation.

In Figure 3.12(c) during and after transgression, unstable transient conditions prevail for a period. Salinization processes in the coastal zone exposed to meteoric recharge during sea-level low stand are complex. Both free convection of seawater sinking into less dense freshwater in the sub-seafloor aquifer as well as the sideways migration of the saline-freshwater interface inland will lead to an eventual, complete resalinization of offshore fresh groundwater as in Figure 3.12(a). However, model simulations illustrate these processes can take many millennia before such a renewed steady-state is reached. Note the difference in length scale that should be assumed between the onshore and offshore domain shown in these sketches.

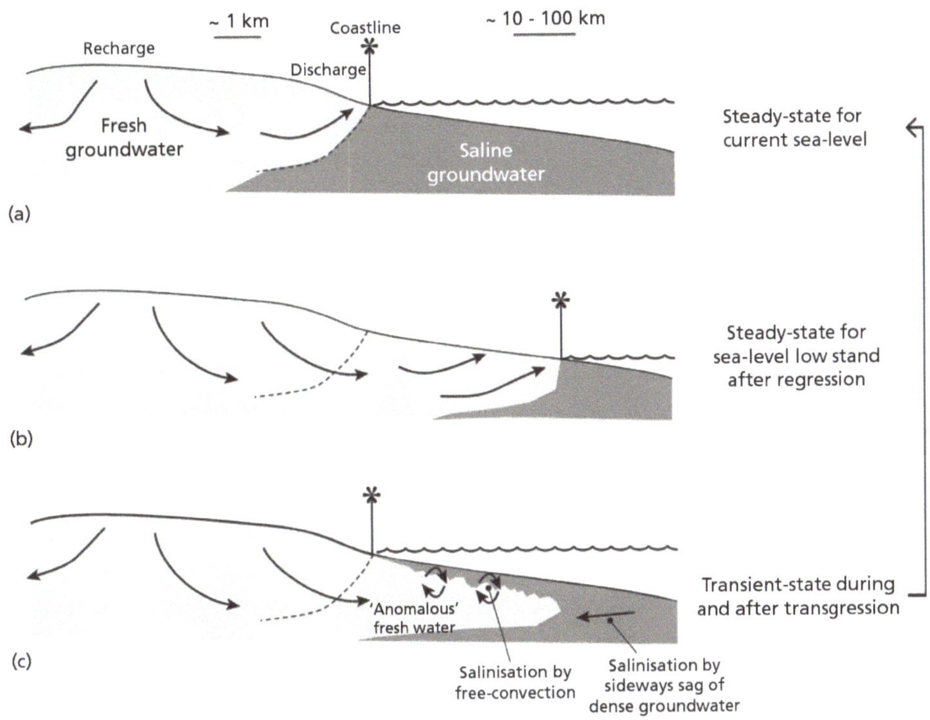

Figure 3.12. *Diagram showing the stages of development of saline groundwater in coastal aquifers.*

Source: Kevin M. Hiscock and Victor F. Bense. Hydrogeology Principles and Practice. Wiley-Blackwell. ISBN: 978-1-119-56951-0.

The salt-fresh interface has been migrating inward and upconing in many coastal areas due to localized groundwater over-abstraction from the fresh water portions of coastal aquifers over several decades. The global process of sea-level fluctuation, which is caused by variations in climate (mainly glaciation) and crustal isostatic forces, has historically resulted in significant changes to the location of the interface between fresh and saltwater on a global scale. Over the past few decades, we have made significant progress in understanding the dynamics of these changes and, consequently, the effects of future sea-level rise on the availability of fresh water resources in coastal areas (e.g., Oude Essink et al., 2010).

Pore waters have been found to have salinities significantly lower than those of seawater in numerous submarine aquifer systems located beneath the continental shelves that flank the world's oceans (Figure 3.13). For instance, drilling for oil and gas on the continental shelf off the coast of New England during the 1960s and 1970s revealed the presence of water that was only 30–70% salinity, which is significantly lower than seawater hundreds of meters below the seafloor (Person et al., 2003).

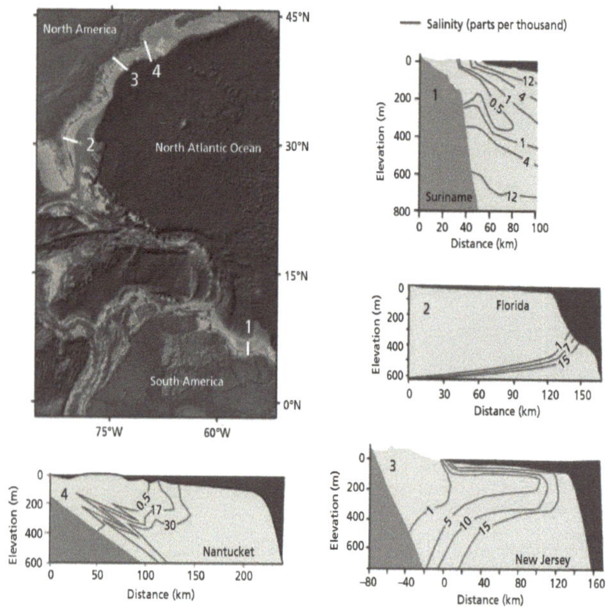

Figure 3.13. *In this diagram, salinity distributions in aquifers on the continental shelf of the North and South Atlantic Ocean illustrate the presence of off-shore fresh to brackish groundwater. By invoking the conceptual model shown in Figure 3.12, it can be hypothesized that these bodies of 'anomalous' fresh water were emplaced during past sea-level low stand and reflect transient hydrogeological conditions on the continental shelf.*

Source: Kevin M. Hiscock and Victor F. Bense. Hydrogeology Principles and Practice. Wiley-Blackwell. ISBN: 978-1-119-56951-0.

It is difficult to understand why there is relatively fresh water beneath the seafloor if equilibrium conditions are the only thing taken into account.

It was suggested (e.g., Kohout et al., 1977) that these comparatively fresh water bodies beneath the current seafloor offshore of New England represent hydrogeological conditions associated with low sea level during Late Pleistocene glacial maxima. Environmental isotopes offer more proof in support of this theory. The unusually fresh waters on the continental shelf are typically between 7 and 20 ka old, according to groundwater age determinations, which supports the theory that these aquifers were replenished during the most recent global sea-level low-stand. Additionally, the abundances of ^{18}O and ^{2}H in these groundwaters frequently suggest that recharge occurred during comparatively cool climate conditions, such as those found during ice-house conditions (e.g., Morrissey et al., 2010). Since then, similar "anomalous" freshwater bodies have been discovered in coastal aquifers throughout Europe (e.g., summarized in Edmunds and Milne, 2001).

The following is the hydrogeological mechanism that explains fresh water hundreds of kilometers offshore beneath the sea floor. Approximately 18,000 years ago, during the Last Glacial Maximum (LGM), the sea level dropped by approximately 120 meters, leaving the continental shelves mostly exposed. For instance, the coastline moved 400 km or more seaward over this time in most of the Atlantic Ocean. As a result, at that time, direct precipitation was present in the aquifers on the continental shelves (Figure 3.12b). When these factors combined with hydrogeological factors that support groundwater recharge (high effective rainfall and transmissive aquifers) lasted long enough, it's possible that large amounts of fresh water displaced the saline water in these aquifers during this time, moving the salt-fresh interface southward.

Sea levels rose as a result of the melting of continental ice sheets brought on by eventual climate change, and the sea began to breach the surface. Seawater would have begun to infiltrate the freshwater bodies that formed during the sea-level low stand when sea level transgression over the continental shelf occurred during the Holocene. This may occur from above, where fresh water in the aquifers beneath the seafloor begins to be superimposed by seawater. Due to variations in density, seawater may begin to naturally seep into the aquifer by means of free-convection, concurrently causing the salt-fresh interface to move landward at a deeper depth. It seems, nevertheless, that the re-salinization process is still ongoing. Because of this, the current situation is out of balance, and "anomalous" water—water on the continental shelf that has a salinity significantly lower than seawater—occurs there (Figure 3.12c).

There are various methods available to obtain a more quantitative understanding of the above-described process and, for example, to determine what kind of conditions would favor the preservation of fresh water on continental shelves. As previously mentioned, free-convection—the process of seawater intruding over an aquifer holding fresh groundwater—is one of the primary ways that aquifers on continental shelves become salinized. The fresh groundwater beneath the sea floor will eventually turn saline in the absence of free-convection, but it will do so much more slowly through the process of molecular diffusion. An alternative method involves employing numerical models (Kooi and Groen, 2001; Post and Kooi, 2003; Cohen et al., 2010), which facilitate the examination of a broader array of parameters governing the durations during which comparatively fresh water in aquifers situated on the continental shelves may remain undisturbed following transgression. These parameters include things like the rate at which sea levels are rising and the effect of aquitards near the seabed, which regulate the various ways in which seawater intrusion can occur during transgression.

In contrast to desalination using seawater, analysis using the kind of methodology previously described combined with field observations yields estimates of the amount of relatively low-salinity water that may be used for drinking water purposes after relatively minor treatment. These result in, for instance, Person et al., (2007b) and Cohen et al., (2010), that there is a comparatively fresh groundwater resource of about $\sim 1.7 \times 10^3$ km^3 on the continental shelf offshore of New England. To put this into perspective, this resource could provide roughly 1035 years of fresh water after relatively minor treatment, compared to complete desalination of seawater, for a city the size of New York with a current water use of $\sim 4.5 \times 10^6$ m^3 day^{-1}.

3.5.2. Brines in Continental Aquifers

Saline water does not only occur in coastal regions on a regional scale. Brine formation occurs deep within sedimentary basins due to the common phenomenon of increasing solute concentrations in aquifers with depth. Even at relatively shallow depths, groundwater in certain continental sedimentary basins has a density significantly higher than that of seawater. Typically, this kind of groundwater is referred to as brine. The Williston, Illinois, and Michigan basins in North America are well-known examples of sedimentary basins with brines at relatively nearsurface depths. These basins are inter-cratonic. The pore waters in these basins have salinities that are up to five or six times higher than seawater's ($35 gL^{-1}$ versus $200 gL^{-1}$; Figure 3.14). The dissolution of evaporite deposits, such as gypsum and halite, in sedimentary sequences may account for the extremely high salinity of the pore waters. These deposits are not known to exist, for instance, in the Illinois Basin. As a result, it has been suggested that these brines are actually seawater that had partially evaporated prior to being encased as pore fluids (e.g., McIntosh and Hanor, 2007). The brines are thought to be roughly the same age because they are hosted by sedimentary rocks from the Devonian and Silurian ages in the majority of North America's continental basins. This suggests that the fluids have been mostly immobile since those ancient geological periods.

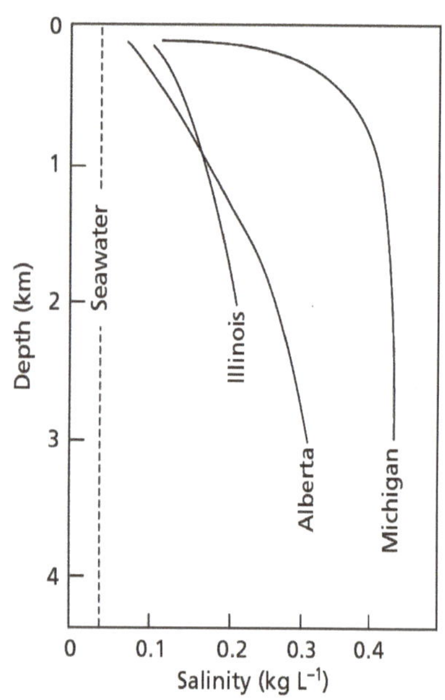

Figure 3.14. *Graph showing depth versus salinity relationships for three sedimentary basins: the Illinois, Alberta and Michigan Basins in North America. For reference, sea water salinity is indicated by the dashed line.*

Source: Kevin M. Hiscock and Victor F. Bense. Hydrogeology Principles and Practice. Wiley-Blackwell. ISBN: 978-1-119-56951-0.

Free convection is unlikely because salinity (and density) increase with depth (Figure 3.14). In order to dilute and displace the brines, topographic gradients may force new meteoric groundwater into these saline aquifers. Nevertheless, in numerous basins, the water-table gradients' hydrogeological forcing is insufficient to eliminate the brines. As a result, it would seem reasonable to assume that even over geological time scales, these deep brines are essentially stagnant due to their density. Because of this, deep aquifers holding these brines have been regarded as reasonably safe locations for the long-term storage of nuclear waste or even the sequestration of CO_2.

3.6. Regional Groundwater Flow Systems Driven Predominantly by Shifting Topography and Stress Changes

Water tables that fluctuate are frequently a major factor in regional fluid flow. When there is enough recharge, the water table will resemble the topography of the surface. But these topographic gradients can shift over geological time. It is currently thought that these changes, which take place over geological timescales, are largely responsible for explaining the features of fluid flow systems that operate at regional to continental scales.

Topographic gradients will shift significantly over time due to two main geological processes, which are covered here. These are the melting and replenishing of continental ice sheets, as well as orogeny and the ensuing erosion. To accurately interpret and potentially predict the location of hydrocarbon accumulations, fresh water bodies in otherwise saline aquifers, and the presence of ore bodies, one must have a solid understanding of the long-term development of fluid flow systems on a continental scale as a result of plate tectonic processes and glaciation.

3.6.1. Mountain Building and Erosion

Mantle convection-induced continental drift forces continental plates to collide, creating orogens. Crust thickening occurs vertically during orogeny, whereas crust shortening happens laterally. As a result, the rock that forms mountain chains was initially formed in sedimentary basins where thermal convection and compaction-driven flow are dominant (Figure 3.15a). Weathering processes then cause the ensuing mountain belt to erode. Tens of Ma are the geological timescales on which this cycle occurs.

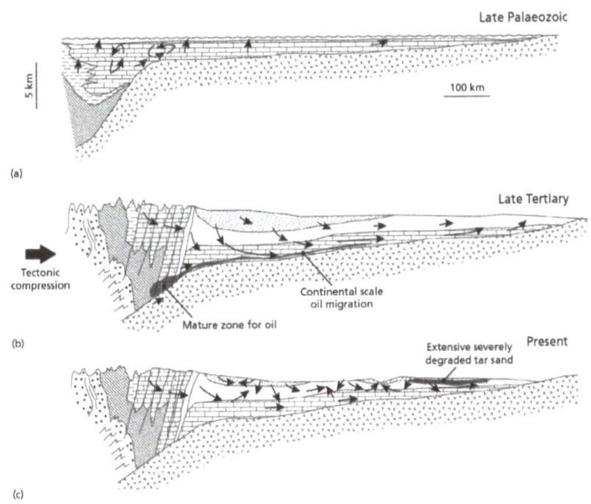

Figure 3.15. *Diagram showing the schematic evolution of the Western Canada Sedimentary Basin as an example of how tectonic compression and mountain building lead to episodic continental-scale fluid migration. (a) During the late Palaeozoic, flow is driven by compaction and thermal convection in a thick limestone aquifer. (b) During the Late Tertiary, uplift and mountain building result in topography-driven flow becoming the dominant driving fluid flow. The extra 'push' provided by tectonic compression means that continental-scale fluid migration pathways are established. These processes result in the long-range migration of oil produced at depth in the mountain belt where the temperature and pressure conditions favor the production of oil. (c) At present, only relatively local-scale topography-driven flow systems exist as a result of erosion and the absence of ongoing tectonic compression. However, the tar sand deposits found near the surface are interpreted as being emplaced during Late Tertiary times when continental-scale transport of hydrocarbons occurred. The occurrence of hydrocarbons near the surface leaves them prone to oxidation and biodegradation.*

Source: Kevin M. Hiscock and Victor F. Bense. Hydrogeology Principles and Practice. Wiley-Blackwell. ISBN: 978-1-119-56951-0.

A mountain chain's development of significant topographic gradients is predicted to cause an episodic mobilization of deep crustal fluids, particularly when combined with higher pore fluid pressures brought on by tectonic compression during the mountain-building process (Figure 3.15b). The explanation for the occurrence of lead-zinc ore deposits, also known as Mississippi-Valley-type (MVT) ore deposits, and hydrocarbon accumulations at basin margins is that they resulted from the expulsion of fluids that were originally deep-seated and had high temperatures and salinities during orogeny and for some time after. With topographic erosion, these continental-scale migration routes vanished, giving rise to groundwater flow systems with a more localized character and a decline in tectonic compression (Figure 3.15c). The metals contained in these brines precipitated out as sulfides such as galena (PbS) and sphalerite (ZnS) when these reactive, high-salinity fluids were placed into a shallower environment that contained sulfur from organic sources over the course of hundreds of kilometers (literally, at a continental scale). These kinds of deposits are now a major contributor to the world's lead and zinc reserves and can be found in the United States and Ireland's Carboniferous basins. Bio-degradation and oxidation have significantly increased the viscosity of the original buoyant oil where it has migrated from a depth to an environment close to the surface (Figure 3.15b), leaving behind a substance that resembles tar. The currently economically viable tar sand deposits located at the eastern end of the Western Canada Sedimentary Basin serve as an example of the latter kind of deposit (e.g., Head et al., 2003).

3.6.2. Impact of Glaciations on Regional Hydrogeology

For the majority of the Pleistocene (~2.5Ma until ~12ka), ice-house conditions were the norm throughout the Northern Hemisphere. This was followed by the current warm interglacial period, known as the Holocene. On a timescale of approximately 20–50 ka, ice-house conditions predominated for about half of the Late Pleistocene, interspersed with warm periods. The Last Glacial Maximum (LGM), which occurred around 18 ka, is the term used to describe the maximum amount of ice that was built up during the last glacial period, which lasted from approximately 90 to 12 ka. Ice sheets covered most of Scandinavia in Europe and extended as far north as 40° N across the continent of North America during the LGM (Eurasian ice-sheet). There is a wealth of evidence indicating that, in comparison to interglacial periods, past glaciations completely reorganized regional hydrology and hydrogeology. It is currently believed that substantial amounts of fresh water recharge from beneath ice sheets occurred during glacial periods in many significant aquifer systems located in northern latitudes. Evidence from geochemistry and environmental isotopes suggests that groundwater recharge rates were significantly elevated during glaciations in many glaciated areas, similar to the situation on the continental shelves discussed above. This resulted in the emplacement of pockets of fresh water into, for example, the otherwise very saline groundwater aquifers of the intercratonic basins in the Canadian Shield, such as the Michigan, Illinois, and Williston Basins. Permafrost conditions are likely to have prevailed in the pro-glacial area ahead of the ice sheet snout, preventing groundwater recharge across those areas, even though significant groundwater recharge may have occurred from underneath ice sheets. Again, supported by geochemical and isotope data is the latter effect (see Jiráková et al., 2011).

When there is a glacial epoch, the primary factors governing groundwater recharge and discharge conditions at the regional level are the thickness and topography of the ice sheet surface along with the availability of sub-glacial water. This is starkly different from the hydrogeological conditions in temperate climates, where water availability for recharge is primarily limited by meteorological conditions and surface topography-controlled water table gradients determine groundwater recharge rates. It will become clear from understanding these long-term changes in recharge conditions that fresh water abstracted from aquifers that were recharged under ice-house conditions may not be quickly replenished under current climate conditions.

Evaluating the factors governing the amounts and rates of groundwater recharge from beneath an ice sheet is crucial to comprehending hydrogeological conditions during glaciation. Ice sheets can be classified as mostly frozen to the ground (dry-based or polar ice sheets) or as having a noticeable amount of liquid water at their base (wet-based or temperate ice sheets). The temperature at the ice bed is controlled by several factors, which determine whether an ice sheet is wet or dry. At the base of an ice sheet, liquid water may originate directly from the in situ melting of the ice or it may be present indirectly through surface melt water or melt water from within the ice sheet (englacially). For the latter mechanism to work, the ice must have crevasses and/or other fracture systems that effectively connect the surface to the ice sheet base (Figure 3.16).

Because the ice sheet will act in part as a thermal blanket, the heat needed for melting at the base of the sheet can originate from both mechanical friction (drag) between the ice and the substratum and a supply of geothermal heat from the rocks underneath the ice sheet. Thus, the likelihood of liquid water forming at the base of the ice sheet will be primarily determined by surface temperature conditions, ice thickness, and ice flow dynamics. Not all of the water beneath an ice sheet with a wet base will seep through. Water can either seep below the surface into bedrock aquifers or flow along the interface between the ice sheet and the substratum (Figure 3.16).

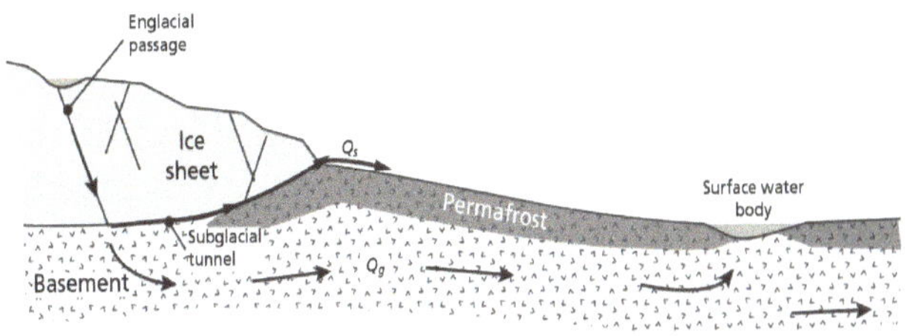

Figure 3.16. *Diagram showing the hydrogeology of ice-sheet margins. Englacial passages in ice sheets can deliver liquid water to the ice-sheet base where it can potentially recharge subglacial aquifers. In such a case, groundwater flow (Qg) can occur from underneath the ice sheet to areas of groundwater discharge in the proglacial area where discontinuities exist in the otherwise relatively impermeable permafrost. Examples of discontinuities include sufficiently insulating surface-water bodies such as lakes or large rivers. Otherwise, subglacial water will be routed along the ice-sheet bed via subglacial tunnels and exit at the toe of the ice sheet (Qs).*

Source: Kevin M. Hiscock and Victor F. Bense. Hydrogeology Principles and Practice. Wiley-Blackwell. ISBN: 978-1-119-56951-0.

Sub-glacial drainage via subsurface groundwater recharge will be largely regulated by the permeability of the rocks beneath the ice sheet. Subglacial water infiltration would not be possible if the underlying strata contained permafrost, which would virtually completely eliminate permeability. The pro-glacial region that the ice sheet spreads over typically contains permafrost. The permafrost beneath most of wet-based ice sheets will eventually thaw, with the exception of the area close to the ice sheet's tip, because temperatures beneath them are almost at the melting point. The foregoing should make it evident that a variety of environmental factors and processes must be taken into account when assessing the likelihood and rates of groundwater recharge from beneath ice sheets. For the LGM and idealized model representations of hydrogeological conditions at the regional to continental scale, many of these are unknown (see Bense and Person, 2008; Lemieux et al., 2008) consider these processes to provide an understanding of the spatial and temporal variability of highly elevated rates of groundwater recharge that will have existed beneath continental ice sheets during previous glaciations. The current distributions of "anomalous" fresh water occurrences recharged during glacial times can be better understood, and the water resource they represent can be managed more effectively, with the help of such idealized model results.

Although it is obvious how important it is to comprehend glaciation as a geological process when it comes to regional groundwater that is relatively shallow (a few hundred meters or less), glaciation also has an impact on the hydrogeology of much deeper aquifers. Similar to sediments in subsiding sedimentary basins, underlying strata are mechanically loaded during the start of a glacial period due to the locally increasing weight of the ice sheet during its accretion. Accordingly, compaction and the resulting overpressure can be brought on by the thickening of an ice sheet's overall stress. The ice sheet's imposition of high hydraulic heads at the surface can nevertheless have no direct effect on groundwater movement in very deep aquifers, but it can do so by creating hydraulic head gradients. Similar to how erosion affects fluid pressures, the opposite effect happens when the glacial period ends, creating sub-hydrostatic pressure regimes (under-pressure). The resulting over- and under-pressures from loading and unloading the ice sheet will partially cancel each other out. However, pressures will begin to decrease in between the stages of ice sheet buildup and retreat, leaving an anomalous pressure field in the deeper aquifers as a result of the glaciations. Perturbating pore pressures will take millennia to dissipate, especially in relatively thick low-permeability rock sequences. According to numerical simulations (Bense and Person, 2008), it is possible that the hydrogeological conditions that existed during the ice-house period that ended about ~12ka are still reflected in the current pore pressure distributions in deep aquifers beneath areas that were once covered by ice sheets.

Consideration of the safety of CO_2 sequestration or nuclear waste burial in such locations depends on the long-term stability of fluids in aquifers at several kilometers below the surface.

When evaluating the safety of nuclear waste repositories at high latitudes, one of the primary concerns is the possible hydrogeological effects of future glaciations on deep aquifer systems (e.g., Canada and Sweden), where it is ideal for groundwater movements to stay at a minimum for the next few millennia. It will be much easier to assess whether such conditions can be anticipated with future climate changes and possibly renewed glaciation if we have a better understanding of the effects that past glaciations have had on regional hydrogeological conditions.

3.6.2.1. Reversal of Regional Groundwater Flow Patterns Forced by Ice-Sheet Waxing and Waning in the Williston Basin, Western Canada

In south western Canada and the neighboring states of the United States, the Williston Basin is a thick (up to 5 km) sedimentary basin of Phanerozoic origin (Figure 3.17).

It is reasonable to presume that compaction-driven flow has no bearing on basin data analysis because basin subsidence ceased at least 500 Ma. Therefore, it can be said that the current groundwater flow in this transboundary aquifer system is mostly driven by topography, with recharge taking place over the topographic highs near the basin's southernmost point, such as South Dakota's Black Hills and comparable highs in Montana and Wyoming. The primary source of groundwater discharge in Manitoba, Canada, is the outcropping Devonian limestone aquifers.

As a result, a regional groundwater flow pattern from south-west to north-east is suggested by the basin hydrogeological architecture (Figure 3.18a). Brine deposits can be found in the basin's deepest aquifers in the center. These brine deposits presently surface as springs in the basin's primary discharge area in Canada.

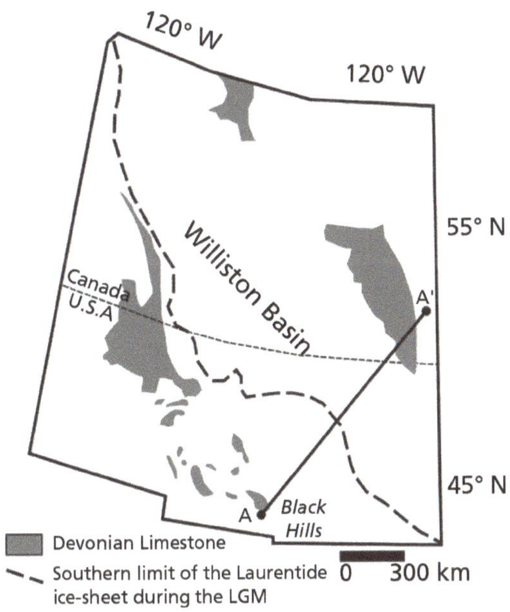

Figure 3.17. *Location map of the Williston Basin showing the Devonian limestone outcrops in the groundwater recharge and discharge areas. Also shown are the southern limit of the Laurentide ice-sheet during the Last Glacial Maximum (LGM) and the orientation of the hydrogeological cross-section shown in Figure 3.18.*

Source: Kevin M. Hiscock and Victor F. Bense. Hydrogeology Principles and Practice. Wiley-Blackwell. ISBN: 978-1-119-56951-0.

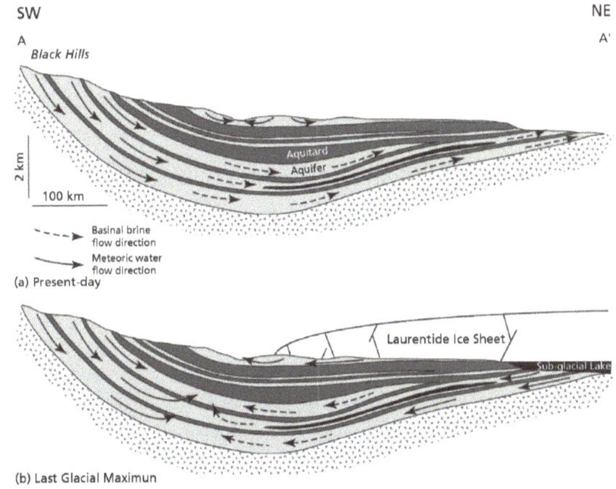

Figure 3.18. *Conceptual hydrogeological cross-section depicting: (a) the present-day flow field in the Williston Basin; and (b) the reversal of the regional groundwater flow system as inferred by Grasby et al., (2000) mainly from geochemical data (Figure 3.19).*

Source: Kevin M. Hiscock and Victor F. Bense. Hydrogeology Principles and Practice. Wiley-Blackwell. ISBN: 978-1-119-56951-0.

A vast dataset on the chemistry of groundwater and stable isotopes for the Williston Basin can be found thanks to samples collected from test wells, water supply wells, oil wells, and springs. These data were compiled by Grasby et al., (2000) indicates that the system contains approximately three different types of groundwaters, each with unique geochemical signatures (Figure 3.19). These are:

1. recent groundwater derived from present-day recharge (low-salinity, $\delta^{18}O \sim -12‰$);
2. brine (high-salinity; $\delta^{18}O \sim +3‰$); and
3. low-salinity groundwater ($\delta^{18}O \sim -24‰$).

The low salinity and isotope signature of the third type of groundwater indicate that it is derived from Pleistocene melt water, with recharge temperatures significantly lower than those found in current climate conditions. The main hydrodynamic mechanisms in the Williston Basin can be identified thanks to the classification of three distinct types of groundwater that are present there. The Pleistocene meteoric water and the deep basinal brines are the specific locations where mixing between the three main water types occurs, as shown in Figure 3.19.

The zone of mixing between Pleistocene water and brines has been identified spatially, and it is found to be within a specific zone on the basin's eastern flank. Grasby et al., (2000) interpret that the distribution of Late Pleistocene water in the Williston Basin today and its mixing relationship to the deeper brines, when combined with additional geomorphological evidence and patterns of the degree of biodegradation of crude oil in the basement, which indicate the influx of oxygenated water, strongly suggest that during the Late Pleistocene a significant westward influx of groundwater occurred on the east flank of the basin. However, since current topographic gradients and the presence of springs identify the eastern part of the Williston Basin as an area of groundwater discharge rather than one of recharge, it would be necessary to completely reverse current hydraulic gradients in order for this groundwater recharge regime to have existed in the past. Thus, Grasby et al., (2000) hypothesized that the Laurentide ice sheet's southern border, which covers the northeastern half of the Williston Basin, must have had the hydraulic capacity to achieve this kind of regional-scale groundwater flow direction reversal, as shown in Figure 3.19(b). In this scenario, there is a significant recharge of the aquifers in the current discharge area, and the "push" of inflowing fresh water from underneath the ice sheet—possibly from a sub-glacial lake—mobilizes the deep basinal brines and creates a zone of intense hydrodynamic mixing.

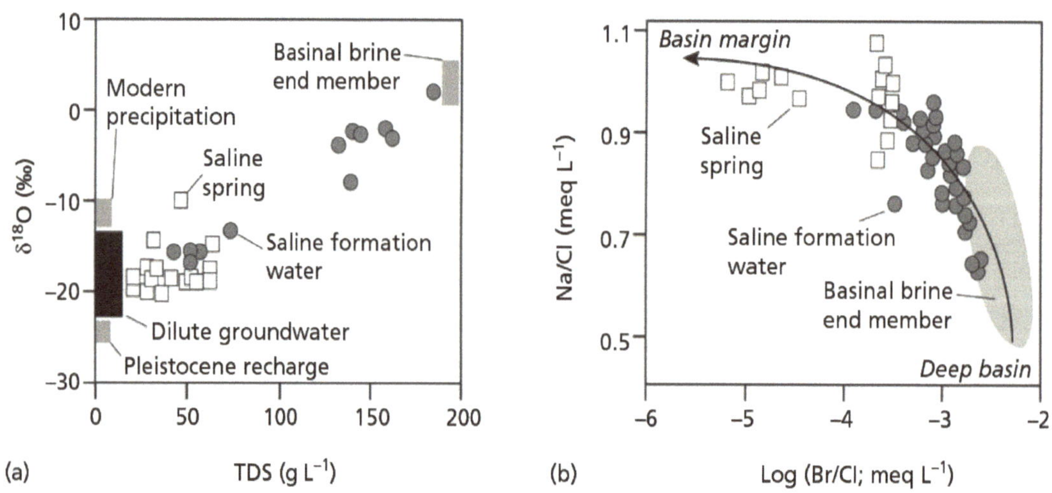

Figure 3.19. *(a) Plot of groundwater oxygen-isotope ratios and salinity (total dissolved solids, TDS) interpreted to show that regional-scale mixing of glacial meltwater and basinal brines has occurred in the current groundwater discharge area of the Williston Basin. (b) The mixing of groundwater in the current discharge area is further illustrated by Na:Cl and Br:Cl ratios.*

Source: Kevin M. Hiscock and Victor F. Bense. Hydrogeology Principles and Practice. Wiley-Blackwell. ISBN: 978-1-119-56951-0.

3.7. COUPLING AND RELATIVE IMPORTANCE OF PROCESSES DRIVING FLUID FLOW

This chapter shows how the rates and directions of groundwater movement can be significantly impacted by geological processes that occur over periods of thousands to millions of years, including mountain building, sea level fluctuation, subsidence in sedimentary basins, and glaciation. Specifically, in the context of regional-scale systems, it can be challenging to interpret groundwater flow patterns and the distribution of salinized and freshwater if these processes are not taken into consideration. It is necessary to take into account more than just the groundwater systems' present geological forcing (e.g., current climate) as well as historical circumstances that continue to leave an impression on the system. This is due to the fact that groundwater systems react slowly to changes in boundary conditions and that geological processes frequently occur over millennia or longer timescales. A lot of the hydrogeological patterns that exist today are a result of a temporary state where groundwater fluxes are reflecting the adaptation to the current environment following a change in boundary conditions in the past.

The interpretation of groundwater flow patterns is complicated by more than just the common transient state of deep groundwater flow systems. The fact that these can occur concurrently makes it challenging to determine the relative importance of each and to determine which simplifications in the analysis of hydrogeological data are warranted. This is another challenge in the analysis of groundwater flow systems for geological processes. An instance of process interference can be observed in the accumulation and subsequent retreat of continental ice sheets, which can affect regional fluid flow in a manner similar to that of mountain formation and erosion. This has sparked debate in the literature (e.g., Neuzil, 2012) regarding the relative importance of deglaciation, erosional unloading, and/or a combination of the two processes in explaining the underpressures found in thick shale units in the Williston Basin.

This chapter's field examples and mechanisms demonstrate the significant influence of geological processes on the long-term and deep groundwater flow. These processes have consequences for how humanity uses and exploits deep aquifer systems, whether as places to store materials that are undesirable at the surface (like excess CO_2 or nuclear waste) or as sources of drinking water (like the Nubian aquifer in northeast Africa and offshore groundwater).

CASE STUDY

GROUNDWATER CASE STUDY: RESOURCE ASSESSMENT

Overview

A client of G3SoilWorks was considering purchasing a multi-million dollar piece of property in San Antonio, Texas that allegedly contained two large springs with massive amounts of fresh water. The current landowner and drilling company claimed that the two groundwater springs they found were interconnected to the Trinity and Edwards aquifers. The client planned on purchasing the water resource and selling the product to the water district. Multiple local firms and individual hydrogeologists in Texas confirmed to the client that the land was good and should be purchased. The client called G3SoilWorks to confirm whether the property was connected to the aquifers and whether it would successfully provide water over an extended period of time without impacting other aquifers.

Previous to the arrival of G3SoilWorks, the drilling company unloaded 10,000 gallons of hydrochloric acid down the well to dissolve the calcium carbonate in the limestone in an attempt to increase the size of the well opening and improve water yield. While that increased the speed at which the water could be pumped, it did not improve the amount of water available from the well. In order to confirm whether water was coming from sustainable ground water, G3SoilWorks applied monitoring devices on the two different wells. The first well quickly dewatered and the ground water never returned. The pump test for the second well showed that the water curve dropped and then rose instead of leveling out.

Problem

- Water curve shows a decline and then increase in groundwater pump test.
- Standard water pump tests show the curve stabilizing or decreasing based on supply and sustainability.
- G3SoilWorks needed to confirm why the curve increased instead of stabilizing from available water.

Solution

After confirming with the landowner and drilling company's hydrogeologist that the ground and surface water did not communicate, Larry Fanning investigated the land surrounding the springs. Within an hour he found a stream that dried out near a sinkhole. They found that groundwater was funneling into a sinkhole, which supplied the spring with water. G3SoilWorks was able to trace the ground water near the sinkhole back to the spring, and further discovered that the water pumped from the spring was supplying this stream, essentially creating a chain.

Conclusion

G3SoilWorks was able to provide evidence to the client against purchasing this land, for a reasonable amount. The client was spared from spending multi-millions on purchasing land that did not provide promised quantities of fresh ground water.

CLASS ACTIVITY

Discuss the four essential elements or pass around four containers with water, soil, air, and light. Ask students what are in the containers. When a student answers discuss that element.

SUMMARY

- Groundwater plays an important role in many geologic processes. For example, the fluid pressures that build up on faults are now recognized to have a controlling influence on fault movement and the generation of earthquakes.
- Subsurface flow systems are responsible for the transfer of heat and chemical constituents through geologic systems, and as a result, groundwater is important in such processes as the development of geothermal systems, the thermodynamics of pluton emplacement, and the genesis of economic mineral deposits. At depth, groundwater flow systems control the migration and accumulation of petroleum.
- Topography-driven groundwater flow is in many areas of the world the dominant mode of groundwater circulation, consideration of the additional impact of geological processes on groundwater flow is often important to fully appreciate the dynamics of groundwater systems on relatively long timescales (millennia).
- Groundwater recharge is a hydrogeological process leading to topography-driven groundwater flow. The rates of groundwater recharge will vary with fluctuating climatic conditions but there are also several geological processes that modify water table gradients on geological timescales such as the build-up and retreat of ice-sheets, and mountain building and erosion.
- Variable density flow is of increasing concern in groundwater hydrology, where fluid density varies in space and time as a function of pressure, temperature and concentrations of multiple species.

- Density gradients in groundwater will cause buoyancy effects and the associated hydraulic gradients will drive groundwater flow. The classic setting in which to consider variable-density flow is that of coastal areas where saline water from the sea interfaces with fresh water originating from meteoric recharge over the land surface.
- The regional-scale occurrence of saline water is not constrained to coastal areas. It is a common phenomenon that solute concentrations in aquifers increase with depth as a result of which brines occur deep in sedimentary basins.
- Undulating water tables are often an important driving mechanism for regional fluid flow. When sufficient recharge occurs, the water table will mimic surface topography. However, over geological time, these topographic gradients can change. Such changes occurring on geological timescales are now believed to play a fundamental role in explaining the characteristics of regional- to continental-scale fluid flow systems.

REVIEW QUESTIONS

1. What are groundwater systems?
2. What is the geological process of groundwater?
3. What is the groundwater recharge?
4. What do sedimentary basins contain?
5. What is variable density flow?

REFERENCES

1. Anderson, T. R., & Fairley, J. P. (2008). Relating permeability to the structural setting of a fault-controlled hydrothermal system in southeast Oregon, USA. *Journal of Geophysical Research, 113*, B05402. https://doi.org/10.1029/2007/JB004962.
2. Bense, V. F., & Person, M. A. (2006). Fault zones as conduit-barrier systems to fluid flow in siliciclastic sedimentary aquifers. *Water Resources Research, 42*, W05421. https://doi.org/10.1029/2005WR004480.
3. Bense, V. F., & Person, M. A. (2008). Transient hydrodynamics in inter-cratonic sedimentary basins during glacial cycles. *Journal of Geophysical Research, 113*, F04005.
4. Bethke, C. M., & Corbet, T. (1988). Linear and non-linear solutions for one-dimensional compaction flow in sedimentary basins. *Water Resources Research, 24*(3), 461–467.
5. Cohen, D., Person, M., Wang, P., et al. (2010). Origin and extent of fresh paleowater's on the Atlantic continental shelf, USA. *Ground Water, 48*(1), 143–158.
6. Domenico, P. A., & Schwartz, F. W. (1998). *Physical and Chemical Hydrogeology* (2nd ed.). John Wiley & Sons.

7. Edmunds, W. M., & Milne, C. J. (Eds.). (2001). *Paleowater's of Coastal Europe: Evolution of Groundwater Since the Late Pleistocene*. Geological Society, London, Special Publications 189.

8. Fairley, J. (2009). Modeling fluid flow in a heterogeneous, fault-controlled hydrothermal system. *Geofluids, 9*(2), 154–166.

9. Fairley, J., & Hinds, J. (2004). Field observation of fluid circulation patterns in a normal fault system. *Geophysical Research Letters, 31*, L19502. https://doi.org/10.1029/2004GL020812.

10. Garven, G. (1995). Continental-scale groundwater flow and geologic processes. *Annual Review of Earth and Planetary Sciences, 23*, 89–117.

11. Ge, S., & Garven, G. (1994). A theoretical model for thrust-induced deep groundwater expulsion with application to the Canadian Rocky Mountains. *Journal of Geophysical Research, 99*(B7), 13,851–13,868.

12. Grasby, S., Osadetz, K., Betcher, R., & Render, F. (2000). Reversal of the regional-scale flow system of the Williston Basin in response to Pleistocene glaciation. *Geology, 28*(7), 635–638.

13. Hanor, J. S., & McIntosh, J. C. (2007). Diverse origins and timing of formation of basial brines in the Gulf of Mexico sedimentary basin. *Geofluids, 7*(2), 227–237.

14. Harrison, W. J., & Summa, L. L. (1991). Paleohydrology of the Gulf of Mexico Basin. *American Journal of Science, 291*(2), 109–176.

15. Head, I. M., Jones, D. M., & Larter, S. R. (2003). Biological activity in the deep subsurface and the origin of heavy oil. *Nature, 426*(6964), 344–352.

16. Ingebritsen, S. E., Sanford, W. E., & Neuzil, C. E. (2006). *Groundwater in Geologic Processes* (2nd ed.). Cambridge University Press.

17. Jiráková, H., Huneau, F., Celle-Jeanton, H., et al. (2011). Insights into palaeorecharge conditions for European deep aquifers. *Hydrogeology Journal, 19*(8), 1545–1562.

18. Kohout, F. A., Hathaway, J. C., Folger, D. W., et al. (1977). Fresh groundwater stored in aquifers under the continental shelf: Implications from a deep test, Nantucket Sound, Massachusetts. *Water Resources Bulletin, 13*(2), 373–386.

19. Kooi, H. (1999). Competition between topography- and compaction-driven flow in a confined aquifer: Some analytical results. *Hydrogeology Journal, 7*(3), 245–250.

20. Kooi, H., & Groen, J. (2001). Offshore continuation of coastal groundwater systems; Predictions using sharp-interface approximations and variable-density flow modeling. *Journal of Hydrology, 246*(1–2), 19–35.

21. Lemieux, J. M., Sudicky, E., Peltier, W., & Tarasov, L. (2008). Dynamics of groundwater recharge and seepage over the Canadian landscape during the Wisconsinian glaciation. *Journal of Geophysical Research, 113*, F01011. https://doi.org/10.1029/2007JF000838.

22. Morrissey, S. K., Clark, J. F., Bennett, M., et al. (2010). Groundwater reorganization in the Floridan aquifer following Holocene sea-level rise. *Nature Geoscience, 3*, 683–687.

23. Neuzil, C. E. (2012). Hydromechanical effects of continental glaciation on groundwater systems. *Geofluids, 12*(1), 22–37.
24. Oude Essink, G. H. P., van Baaren, E. S., & de Louw, P. G. B. (2010). Effects of climate change on coastal groundwater systems: A modeling study in the Netherlands. *Water Resources Research, 46*. https://doi.org/10.1029/2009WR008719.
25. Person, M. A., McIntosh, J., Bense, V. F., & Remenda, V. (2007b). Pleistocene hydrology of North America: The role of icesheets in reorganizing groundwater flow systems. *Reviews of Geophysics, 45*, RG3007. https://doi.org/10.1029/2006RG000206.
26. Person, M., Dugan, B., Swenson, J. B., et al. (2003). Pleistocene hydrogeology of the Atlantic continental shelf, New England. *Geological Society of America Bulletin, 115*(11), 1324–1343.
27. Person, M., Mulch, A., Teyssier, C., & Gao, Y. (2007a). Isotope transport and exchange within metamorphic core complexes. *American Journal of Science, 307*(3), 555–589.
28. Person, M., Raffensperger, J., Ge, S., & Garven, G. (1996). Basin-scale hydrogeological modeling. *Reviews of Geophysics, 34*(1), 61–87.
29. Post, V. E. A., Groen, J., Kooi, H., Person, M., Ge, S., & Edmunds, W. M. (2013). Offshore fresh groundwater reserves as a global phenomenon. *Nature, 504*(7478), 71–78.
30. Post, V., & Kooi, H. (2003). On rates of salinization by free convection in high-permeability sediments: Insights from numerical modeling and application to the Dutch coastal area. *Hydrogeology Journal, 11*(5), 549–559.
31. Schincariol, R. A., & Schwartz, F. W. (1990). An experimental investigation of variable density flow and mixing in homogeneous and heterogeneous media. *Water Resources Research, 26*(10), 2317–2329. https://doi.org/10.1029/90WR01161.

CHAPTER 4

Chemical Hydrogeology

LEARNING OBJECTIVES

After studying this chapter, you will be able to:

- Explain the properties of water
- Discuss on chemical composition of groundwater
- Focus on sequence of hydrochemical evolution of groundwater
- Define the groundwater sampling and graphical presentation of hydrochemical data
- Identify the concept of chemical equilibrium.

KEY TERMS FROM THIS CHAPTER

Bond angle	Chemical hydrogeology
Groundwater	Human inhalation
Hydration	Hydronium ion
Ion exchange	Mineral dissolution
Mineral surfaces	Oxygen atom
Proton transfer	Surface tension
Total dissolved solids (TDS)	Vapour pressure
Viscosity	Water density

INTRODUCTION

Chemical hydrogeology, an interdisciplinary field merging chemistry and hydrogeology, focuses on the chemical characteristics of groundwater and the processes that govern its composition. Groundwater, a critical resource for drinking water, agriculture, and industry, interacts with various geological materials, leading to complex chemical reactions. These reactions influence the water's quality and suitability for different uses. The primary components of chemical hydrogeology include the study of mineral dissolution and precipitation, ion exchange, redox reactions, and the transport of contaminants. Understanding these processes requires a thorough knowledge of both the geologic framework and the chemical environment of aquifers.

Mineral dissolution and precipitation are fundamental processes in chemical hydrogeology. Groundwater often becomes mineralized by dissolving minerals from the surrounding rock formations. For example, calcite dissolution can increase calcium and bicarbonate concentrations in water, influencing its hardness. Conversely, precipitation can remove ions from the water, forming solid minerals and altering the water's chemistry. Ion exchange, another crucial process, involves the replacement of ions in the water with those adsorbed onto mineral surfaces. This process can significantly alter the water's ionic composition, as seen in softening hard water by exchanging calcium and magnesium ions with sodium ions.

Redox reactions, involving the transfer of electrons between species, play a vital role in the chemical dynamics of groundwater. These reactions can change the oxidation state of elements, impacting their solubility and mobility. For instance, the reduction of nitrate to nitrogen gas can decrease nitrate concentrations in groundwater, mitigating pollution. Similarly, the oxidation of iron and manganese can lead to the formation of insoluble oxides, which precipitate out of the water. The transport of contaminants is another critical aspect of chemical hydrogeology, encompassing the movement of pollutants through groundwater systems. Contaminants can originate from various sources, including agricultural runoff, industrial discharges, and leachate from waste disposal sites. Their transport is influenced by groundwater flow patterns, chemical reactions, and the physical properties of the aquifer materials.

Advanced analytical techniques, such as mass spectrometry and chromatography, are employed to analyze groundwater samples, providing detailed information on its chemical composition. Modeling tools are also essential, enabling scientists to simulate groundwater flow and chemical reactions, predict contaminant behavior, and develop remediation strategies. Effective management of groundwater resources requires an integrated approach that considers both the hydrogeological and chemical perspectives. By understanding the chemical processes that govern groundwater systems, scientists can better protect and manage this vital resource, ensuring its sustainability for future generations.

4.1. PROPERTIES OF WATER

One oxygen atom is bonded asymmetrically to two hydrogen ions with a bond angle of 105° in Figure 4.1, which depicts the chemical structure of water. The shape is a function of the bonding-related geometry of the electron orbits. The bonding electrons are drawn towards oxygen and away from the hydrogen atom by oxygen due to its significantly higher electronegativity than hydrogen. Electronegativity is a measure of an atom's propensity to attract an extra electron. Hence, each hydrogen atom has a partial positive charge (δ^+) and the oxygen carries a partial negative charge (δ^-), resulting in a dipole—a small separation between electrical charges with the same sign and equal magnitude. Because of this, water molecules with opposing charges are drawn to one another and form clusters of molecules through a process called hydrogen bonding. With lowering temperature, the clusters get bigger until they reach a maximum at 4°C. The size of the clusters results in a more open structure and less density in the water when it cools from 4°C to 0°C. The water will continue to expand upon freezing. As a result, ice is less dense than liquid water. Values of surface tension, water density, viscosity, and vapour pressure over a 100°C temperature range. Water is not just H_2O, as Table 4.1 illustrates; rather, it is a mixture of six molecules depending on which hydrogen and oxygen isotopes combine to form the water molecule. There are eighteen possible combinations, the most popular being $^1H_2\ ^{16}O$. Both the combined molecular form of hydrogen and oxygen and their ionic form are present in pure water. The ions are formed when water dissociates as follows:

$$H_2O \rightleftharpoons H^+ + OH^- \qquad (1)$$

Proton transfer, or the movement of protons (H^+ ions) from the liquid to solid phases in rock-water interactions, is a crucial aspect of carbonate chemistry and describes how the H+ ion is typically represented as H_3O^+, or the hydronium ion.

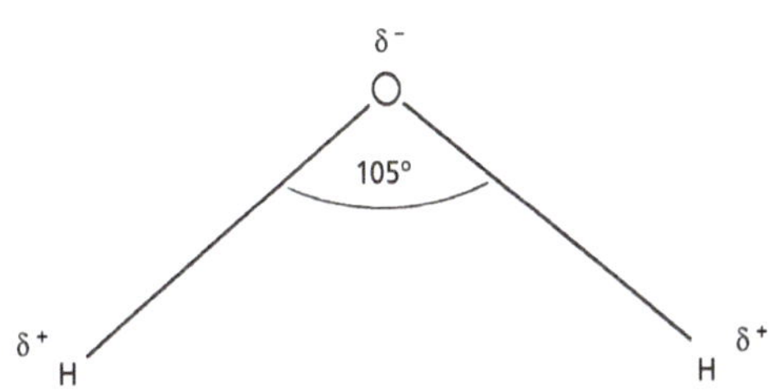

Figure 4.1. *The structure of the water molecule showing the dipole created by the partial negative charge on the oxygen atom and partial positive charge on each hydrogen atom.*

Source: https://www.geokniga.org/bookfiles/geokniga-hydrogeologyprinciplesandpractice.pdf.

Table 4.1. The Relative Abundance of Hydrogen and Oxygen Isotopes in the Water Molecule

Isotope		Relative abundance	Type
^{1}H	Proteum	99.984	Stable
^{2}H	Deuterium	0.016	Stable
^{3}H	Tritium	$0–10^{-15}$	Radioactive*
^{16}O	Oxygen	99.76	Stable
^{17}O	Oxygen	0.04	Stable
^{18}O	Oxygen	0.20	Stable

Source: https://www.geokniga.org/bookfiles/geokniga-hydrogeologyprinciplesandpractice.pdf.

Water is a good solvent for ions because of its polarity; as Figure 4.2 illustrates, the water molecules are drawn to the ions by electrostatic forces and form a cluster with either oxygen or hydrogen oriented toward the ions. Hydration is the term for this phenomenon, which stabilizes the solution. Polar solvents such as water easily dissolve crystalline solids like sodium chloride (NaCl) and break down the ionic crystal into a solution of separately charged ions:

$$Na^{+}Cl^{-} \xrightarrow{H_2O} Na^{+}_{(aq)} + Cl^{-}_{(aq)} \qquad (2)$$

Anions are negatively charged ions like chloride, whereas cations are positively charged atoms like sodium. The total charges on one side of an equation must equal the total charges on the other side when writing a chemical equation. NaCl is an electrically neutral compound on the left side of Equation 2, and the aqueous sodium and chloride ions on the right side each carry a single, opposing charge such that the charges cancel, or balance, each other out. The degree of hydration rises as the dissociated ion's electrical charge increases and as the ionic radius decreases. Cations are typically more strongly hydrated than anions due to their smaller size. Hydration has the effect of making an ion larger, which lowers its mobility and modifies the rates of chemical reactions. Furthermore, complexation of cations results in the formation of complex ions with a fixed composition and excellent chemical stability. This is especially true for transition metals like iron, copper, zinc, cobalt, and chromium, where the ions form chemical bonds with water molecules.

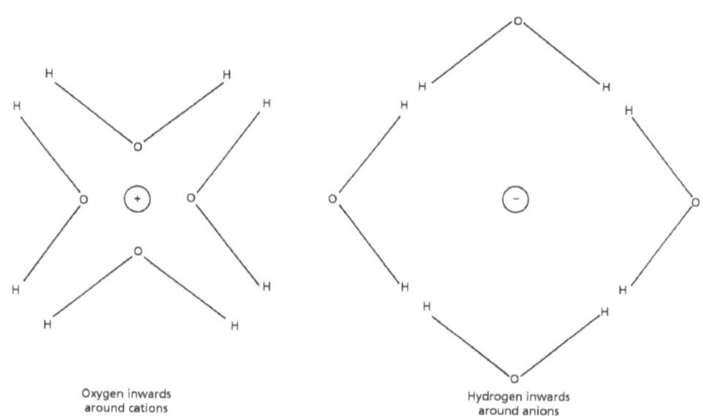

Figure 4.2. *Water molecules surrounding a positively charged cation and a negatively charged anion to form hydrated ions.*

Source: https://www.geokniga.org/bookfiles/geokniga-hydrogeologyprinciplesandpractice.pdf.

4.2. CHEMICAL COMPOSITION OF GROUNDWATER

Groundwater and the geological materials of rocks and soils interact chemically and biologically to produce a diverse range of dissolved organic and inorganic constituents. The different makeup of rainfall and atmospheric dry deposition over groundwater recharge areas, the alteration of atmospheric inputs by evapotranspiration, the differential uptake by biological processes in the soil zone, and the mixing with seawater in coastal areas are other significant factors to take into account.

The six major ions sodium (Na^+), calcium (Ca^{2+}), magnesium (Mg^{2+}), chloride (Cl^-), bicarbonate (HCO_3^-), and sulphate (SO_4^{2-}) are the principal dissolved components of groundwater, as Table 4.2 illustrates and as is the case with freshwaters in the terrestrial aquatic environment. Regardless of whether the water is diluted rainwater or has a salinity higher than seawater, these cations and anions typically make up more than 90% of the total dissolved solids content. Potassium (K^+), dissolved iron (Fe2+), strontium (Sr^{2+}), fluoride (F^-), and potassium (K^+) are examples of minor ions. Aqueous solutions also frequently include trace elements and metal species. Some normally minor ions may become equivalent to major ions in concentrations due to human activity-induced contamination of groundwater. One such is nitrate, which can reach concentrations in soil water and groundwater above 50 mg L^{-1} when nitrogenous fertilizers are applied excessively.

Table 4.2. *Chemical Composition of Groundwater Divided into Major and Minor Ions, Trace Constituents and Dissolved Gases*

Groundwater composition	
Major ions (>5 mg L^{-1})	
Bicarbonate	Sodium
Chloride	Calcium
Sulphate	Magnesium
Minor ions (0.01–10.0 mg L^{-1})	
Nitrate	Potassium
Carbonate	Strontium
Fluoride	Iron
Phosphate	Boron
Trace constituents (<0.1 mg L^{-1})	
Aluminium	Manganese
Arsenic	Nickel
Barium	Phosphate
Bromide	Radium
Cadmium	Selenium
Caesium	Silica
Chromium	Silver
Cobalt	Thorium
Gold	Tin
Iodide	Titanium
Lead	Uranium
Lithium	Vanadium
	Zinc
Dissolved gases (trace to 10 mg L^{-1})	
Nitrogen	Methane
Oxygen	Hydrogen sulphide
Carbon dioxide	Nitrous oxide

Source: https://www.geokniga.org/bookfiles/geokniga-hydrogeologyprinciplesandpractice.pdf.

After Freeze and Cherry (1979).

Because organic matter is converted to carbon dioxide during infiltration through the soil zone, organic compounds are typically found in groundwater at very low concentrations of less than 0.1 mg L^{-1}. Biogeochemical processes can produce anaerobic groundwater conditions and the production of dissolved gases like nitrogen (N_2), hydrogen sulfide (H_2S), and methane (CH_4) in environments rich in organic carbon, such as river floodplains and wetlands. Other dissolved gases come from nitrous oxide (N_2O) produced by biogeochemical processes in soils and groundwater, as well as oxygen (O_2) and carbon dioxide (CO_2), primarily from atmospheric sources. As uranium (U) and thorium (Th) decay, radon ($^{222}R_n$) gas is frequently found in groundwater and can build up to unfavorable concentrations in caverns, mines, and unventilated dwellings. In addition to being common in granitic rocks, uranium can also be found in silicates, phosphates, and crustal rocks. Examples of these materials include uranite (UO_2), zircon ($ZrSiO_4$), and apatite ($Ca_5(PO_4)_3(OH, F, Cl)$). Human inhalation of radon and its decay products, such as polonium (218Po and ^{216}Po) and ultimately lead (Pb) isotopes, can be hazardous.

Groundwater contains trace amounts of the inert gases argon (Ar), helium (He), krypton (Kr), neon (Ne), and xenon (Xe), which can be dissolved and provide important information about the temperature and age of groundwater recharge.

This information can then be used to interpret the hydrochemical and hydrogeological conditions in aquifers. One commonly used technique for classifying groundwaters is the total dissolved solids (TDS) content, which expresses the degree of salinization of the groundwater (Table 4.3). A straightforward determination of TDS by weighing the solid inorganic and organic residue left after evaporating a measured volume of filtered sample to dryness can assist in determining the hydrochemical properties of a local aquifer in the absence of any specialized analytical equipment for measuring individual dissolved components. Because groundwater is an electrolytic solution with dissolved components present in ionic form, measuring a solution's electrical conductivity (EC) can also provide a relative indication of the amount of dissolved salts.

For any investigation, it is possible to relate the TDS value to electrical conductivity (usually expressed in units of micro-siemens centimetre^{-1}) as follows:

$$\mathrm{TDS}\,(\mathrm{mg\ L^{-1}}) = k_e \cdot \mathrm{EC}\,(\mu\mathrm{S\ cm^{-1}}) \tag{3}$$

Table 4.3. *Simple Classification of Groundwater Based on Total Dissolved Solids (TDS) Content*

Category	TDS (mg L^{-1})
Freshwater	0-1000
Brackish water	1000-10,000
Saline water (seawater)	10,000-100,000 (35,000)
Brine water	>100,000

Source: https://www.geokniga.org/bookfiles/geokniga-hydrogeologyprinciplesandpractice.pdf.

where the correlation factor, k_e , is typically between 0.5 and 0.8 and can be determined for each field investigation. The electrical conductivity for fresh groundwater is of the order of 100s μS cm^{-1} while rainwater is of the order of 10s μS cm^{-1} and brines 100,000s μS cm^{-1}. Given that ionic activity, and therefore electrical conductivity, increases with temperature at a rate of about a 2% per °C, measurements are usually normalized to a specific temperature of 25°C and recorded as SEC_{25}.

4.3. SEQUENCE OF HYDROCHEMICAL EVOLUTION OF GROUNDWATER

The idea that groundwater salinity distribution follows a specific hydrological and geochemical law—which can be defined as the cycle of metamorphism of natural waters in the crust of weathering—was first proposed by Chebotarev (1955) in a trio of seminal papers, drawing on almost ten thousand chemical analyses of natural waters. According to Chebotarev (1955), groundwaters range in hydrochemical facies from bicarbonate at outcrop to saline at depths in the Earth's crust.

This distribution of groundwaters is dependent on the interaction between rock and water within the hydrogeological environment. Because of the close relationship between mineralogy and flow regime—which determine the occurrence, sequence, rates, and progression of reactions—Hanshaw and Back (1979) described the chemistry of groundwater. Regarding carbonate aquifers, Hanshaw and Back (1979) introduced the conceptual model illustrated in Figure 4.3, which illustrates the alterations in groundwater chemistry that occur from the moment a carbonate aquifer forms until the aquifer system develops.

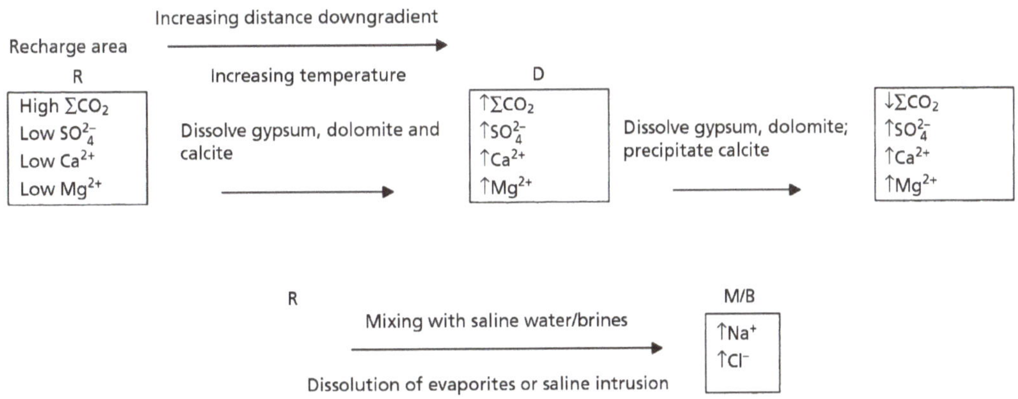

Figure 4.3. *Schematic model showing the evolution of chemical character of groundwater in carbonate aquifers. In areas of recharge (R), the high concentrations of CO_2 and low dissolved solids content cause solution of calcite, dolomite and gypsum. As the concentrations of ions increase and their ratios change downgradient (D), groundwater becomes saturated with respect to calcite which begins to precipitate. Dedolomitization (dissolution of dolomite to form calcite with a crystalline structure similar to dolomite) occurs in response to gypsum solution with calcite precipitation. Where extensive accumulations of evaporite minerals occur, their dissolution results in highly saline brines (B). Another common pathway is caused by mixing with seawater (M) that has intruded the deeper parts of coastal aquifers.*

Source: https://www.geokniga.org/bookfiles/geokniga-hydrogeologyprinciplesandpractice.pdf.

When carbonate sediments first emerge from the marine environment, freshwater flushes the salt from the sediments, causing the salinity to drop and Ca-HCO_3 to take over the hydrochemical facies. At this point, the rock aquifer is formed by the carbonate sediments being selectively dissolved, recrystallized, cemented, and possibly dolomitized.

The dissolution of dolomite and high-magnesium calcite causes Mg^{2+} to gradually increase as recharge moves downgradient (R → D in Figure 4.3), while Ca^{2+} stays largely constant. In this chemical evolution, HCO_3^- stays largely constant while SO_4^{2-} rises as gypsum dissolves. Extremely saline waters or brines are the result of coastal conditions or large-scale evaporite mineral accumulations (R → M/B in Figure 4.3).

The idea of hydrochemical facies, or water type, created by Chebotarev (1955), is helpful in understanding groundwater chemistry and identifying hydrochemical processes. When cation and anion concentrations fall within specific bounds, a unique groundwater zone is referred to as a hydrochemical facies. Figure 4.4 illustrates the usual variations in hydrochemical facies along a groundwater flowpath. Rainfall that has been diluted and treated with Na-Cl water that contains CO_2 seeps into the soil zone, where it dissolves more CO_2 that has been created by organic matter decomposition. Chemicals for agriculture, such as fertilizers, are applied when necessary and add additional Na^+, K^+, Cl^-, NO_3^-, and PO_4^{3-}. CaMg-HCO_3 water type is produced when the dissolved CO_2 in the soil and unsaturated zone produces a weakly acidic solution of carbonic acid, H_2CO_3. This solution then dissociates and aids in the dissolution of calcium and magnesium carbonates. Groundwater below the water table gradually loses oxygen, nitrate, and sulfate due to bacterial respiration and the mineralization of organic matter, as it moves away from the oxygen reservoir found in the soil and unsaturated zone. Metal sulphides are formed when F_e and M_n become mobilized under conditions of increasing reduction. The water changes to a Na-HCO_3 water type when there is distributed clay material in the aquifer due to ion exchange, which replaces Ca^{2+} in solution with Na^+. A region of static water and aquifer diagenesis are reached before mixing with saline water in the deeper, more confined portion of the aquifer creates a Na-Cl water type.

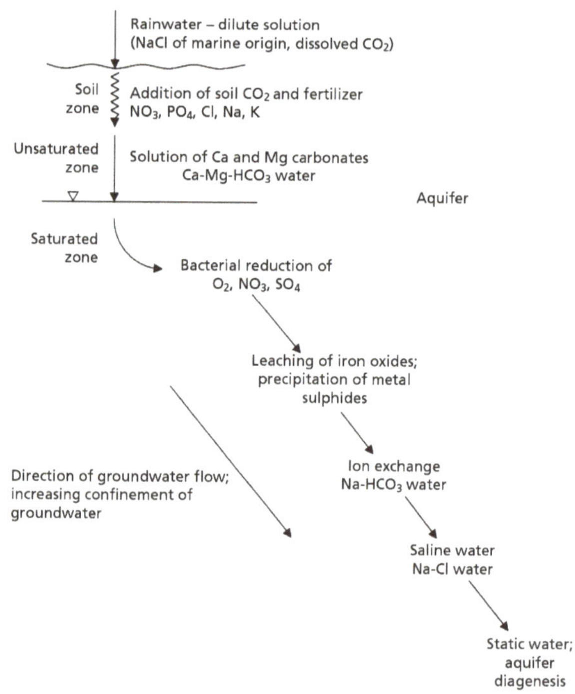

Figure 4.4. *Schematic diagram of the evolution of groundwater along a flowpath from recharge area to confined section showing important hydrochemical processes that affect the chemical composition of groundwater. The processes shown are not wholly sequential in that ion exchange can occur without substantial redox reactions.*

Source: https://www.geokniga.org/bookfiles/geokniga-hydrogeologyprinciplesandpractice.pdf.

4.4. GROUNDWATER SAMPLING AND GRAPHICAL PRESENTATION OF HYDROCHEMICAL DATA

Gathering a raw water sample that is typical of the aquifer's hydrochemical conditions is the goal of field sampling. A sampling device of some kind is needed to collect a groundwater sample for analysis. Examples of these include bailers, depth samplers, gas lift samplers, and pumps (inertial, suction, or submersible) in boreholes or open holes, as well as specially designed piezometers or nested piezometers. Before sampling, care must be taken to flush out any standing water in the well or borehole by removing up to three or four times the volume of the well or borehole. Regularly pumped abstraction boreholes yield a sample that is representative of a large volume of aquifer, while discrete samples that are representative of a small volume of aquifer are obtained through depth-specific sampling from a multilevel system or from a section of the borehole column that is isolated by inflatable packers. The reader is referred to Price and Williams (1993) and Lerner and Teutsch (1995) for a comparison of the performances of different depth sampling techniques.

High-density polyethylene bottles with screw caps that are new or cleaned make it easy to gather groundwater samples. Prior to filling sampling bottles, the bottle needs to be rinsed with the water sample two or three times. If the sample contains suspended particles or sediment, it should also be filtered through a 0.45 μm membrane filter. When analyzing trace metals, a 0.1 μm filter is advised. A few filled sample bottles are usually gathered at one location, one of which is kept to stabilize the dissolved metals in solution. After the sample has been collected and filtered to lower the pH to 2, a few drops of concentrated nitric or acetic acid are added as part of the conventional preservation procedure before analysis. To determine whether the pH after acidification is sufficient, a test with pH paper can be used. After being collected, samples need to be kept cold in a cool box and sent to the lab as soon as possible—ideally the same day—to be stored at 4°C to prevent bacterial growth and nutrient species degradation (NO_3^-, SO_4^{2-}, and PO_4^{3-}). Samples shouldn't be frozen since this can cause certain elements to precipitate.

In order to gather information representative of the hydrochemical conditions in the aquifer, the well-head chemistry should be measured while at the sampling location as part of the sampling methodology. The groundwater sample will undergo changes in temperature, pH, redox potential (Eh), and dissolved oxygen (DO) content as it is exposed to ambient conditions at the ground surface, as well as during storage and laboratory analysis. It is also necessary to measure the alkalinity and electrical conductivity (EC) on-site. Measurements on a flowing sample inside an isolation cell or flow cell are recommended to limit the exsolution of gases from the groundwater, such as the loss of CO_2, which will raise the pH value of the sample, and to prevent the sample from mixing with atmospheric oxygen and influencing the Eh and DO values. A plastic tube from the sampling tap enters the flow cell's base, and an overflow tube is located at the top of the cell. Access holes on the flow cell lid allow different measurement probes to be inserted. Before recording the final values, it is usually necessary to give the Eh and DO electrodes 20 to 30 minutes to stabilize in the flow cell. Lloyd and Heathcote (1985) and Appelo and Postma (1994) are recommended reading for additional details on measuring hydrochemical parameters and gathering samples, including sampling springs and pore waters.

Before attempting any hydrochemical interpretation, it is first necessary to check the quality of laboratory chemical analyses and that the condition of electroneutrality has been met whereby the sum of the equivalent weight of cations equates to the sum of the equivalent weight of anions. This check is commonly carried out by calculating the ionic balance error of the major ions where:

$$\text{Ion balance error (\%)} = \{\Sigma\text{cations} - \Sigma\text{anions})/(\Sigma\text{cations} + \Sigma\text{anions})\} \times 100 \quad (4)$$

With current analytical tools, an ionic balance error of less than 5% should be possible, and definitely less than 10%. Greater errors indicate that one or more analyses may be flawed, which is unacceptable. To find observable patterns and trends in hydrochemical data, a number of techniques have been developed for visual inspection. It is feasible to identify hydrochemical facies and get a basic understanding of the hydrogeological processes influencing the chemistry of groundwater by classifying chemical analyses. Plotting distribution diagrams, bar charts, pie charts, radial diagrams, and pattern diagrams are among the most basic techniques (as presented by Stiff, 1951). Although they are simple to build, other methods, such as Schoeller (named after Schoeller, 1962), trilinear (Piper, 1944), and Durov (Durov, 1948) diagrams, are used instead because they are more practical for the graphical presentation of large numbers of analyses.

Groundwater concentrations of Na^+ and Cl^- in the Milk River aquifer system, situated in the southern region of the Western Canadian Sedimentary Basin, are depicted in the distribution diagram presented in Figure 4.5.

The Tertiary–Cretaceous period of the Earth's history, which is primarily composed of mudstones and shale, gave rise to multiple sandstone units, including the artesian Milk River aquifer. The aquifer is covered by up to 400 meters of younger rocks in the northern section, but it breaks out in the southern part where recharge takes place. With a considerable amount of upward leakage through confining beds, groundwater flow is typically northward. Na^+ and Cl^- concentrations exhibit significant regional variations, with the south experiencing the lowest concentrations due to freshwater recharge, as depicted in Figure 4.5. A significant characteristic is the distinct northward trend of fresh groundwater, which is defined by the 100 mg L^{-1} Cl^- contour and aligns with a firmly established zone of aquifer permeability. This hydrochemical pattern is understood to be a wide area of mixing that arises from the pre-existing, more saline formation water being flushed out by meteoric recharge water (Schwartz and Muehlenbachs, 1979).

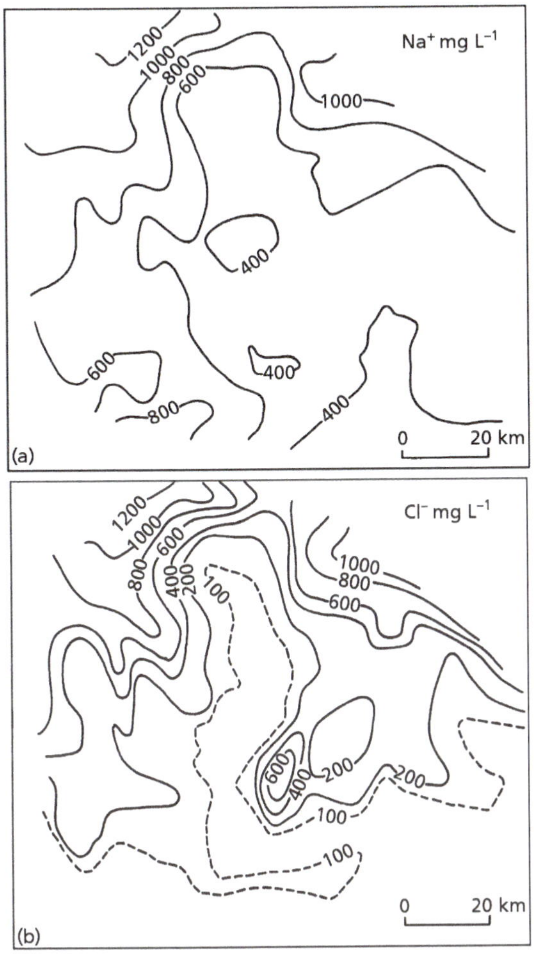

Figure 4.5. *Distribution diagrams of (a) Na+ and (b) Cl– concentrations in groundwaters from the Milk River aquifer system, Alberta.*

Source: https://www.geokniga.org/bookfiles/geokniga-hydrogeologyprinciplesandpractice.pdf.

Concentrations are shown on the Schoeller diagram as meq L^{-1}. The chemistry of groundwater in crystalline rocks, where the near-surface groundwater is normally replenished by precipitation and snowmelt, is illustrated in Figure 4.6. Silicates are the minerals that form crystalline rocks; the main cations of these minerals are Ca^{2+}, Mg^{2+}, Na^{+}, and K^{+}. The main weathering agent in North America is carbonic acid, which is mostly derived from CO_2 dissolved in the soil zone. The water usually reaches a Ca⁻ or Na-HCO_3 composition under these circumstances. Significant variations in the aquifer rock's composition are the cause of variations in this general type of water (Trainer, 1988).

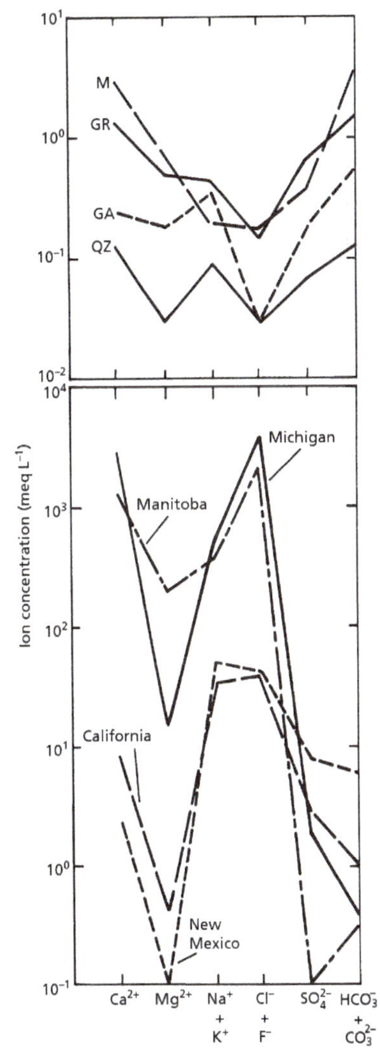

Figure 4.6. *Schoeller (semi-logarithmic) diagram illustrating near-surface groundwater chemistry in crystalline rocks.*

Source: https://www.geokniga.org/bookfiles/geokniga-hydrogeologyprinciplesandpractice.pdf.

Bar charts, pie charts, radial diagrams, and pattern diagrams for the primary ion analyses listed in Table 4.4 are displayed in Figure 4.7 to demonstrate basic techniques for presenting hydrochemical data. The eastern part of the Northern Rocky Mountain Province, in south-west Montana's Milligan Canyon area, is the subject of these analyses. The region is a wide synclinal basin with older, deformed rocks underlayering volcanic breccia and andesitic lava on the northern rim and folded carbonate rocks of Mesozoic and Paleozoic age on the southern rim. The Tertiary sediments of siltstones, limestones, and sandstones that contain deposits of gypsum ($CaSO_4 \cdot 2H_2O$) and anhydrite ($CaSO_4$), as well as unconsolidated alluvial and aeolian deposits, fill the basin. Several igneous intrusives from the Upper Cretaceous and early Tertiary are also found there. Although upward groundwater flow from the Madison limestone aquifer beneath the basin also contributes, groundwater flow is primarily from west to east. The Tertiary basin yields high levels of groundwater.

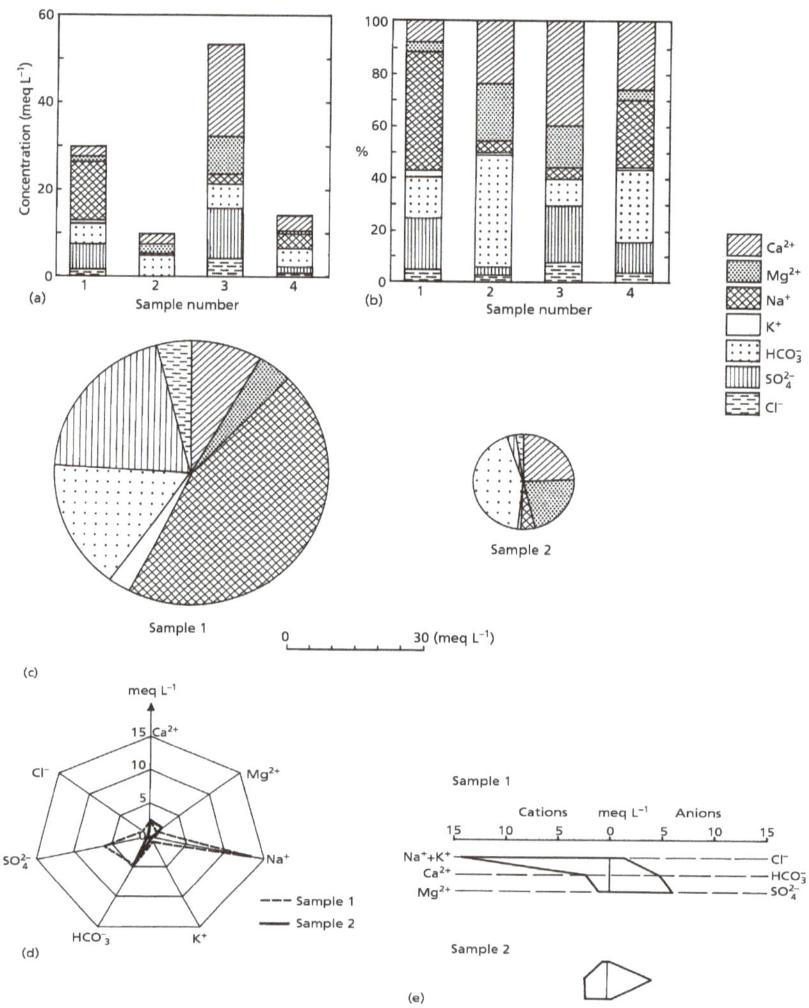

Figure 4.7. *Simple graphical representation of major ion analyses of groundwaters.*

Source: https://www.geokniga.org/bookfiles/geokniga-hydrogeologyprinciplesandpractice.pdf.

Table 4.4. *Major Ion Analyses of Four Groundwater Samples*

Water type	Sample number	Ca^{2+}	Mg^{2+}	Na^{+}	K^{+}	HCO_3^-	SO_4^{2-}	Cl^-
Ion concentration (mg L^{-1})								
Ca-HCO_3	2	47	25.2	10	3.1	251	29	7
Ca-SO_4	3	426	103	56	5.4	328	1114	143
Na-HCO_3	4	73	6.4	83	4.5	236	158	17
Na-SO_4	1	49	13.7	312	29.2	286	566	43
Ion concentration (meq L^{-1})								
Ca-HCO_3	2	2.35	2.10	0.43	0.08	4.11	0.30	0.20
Ca-SO_4	3	21.30	8.58	2.43	0.14	5.38	11.60	4.09
Na-HCO_3	4	3.65	0.53	3.61	0.12	3.87	1.65	0.49
Na-SO_4	1	2.45	1.14	13.57	0.75	4.69	5.90	1.23

Source: https://www.geokniga.org/bookfiles/geokniga-hydrogeologyprinciplesandpractice.pdf.

Figures 4.8 and 4.9 display the same primary ion analyses for the Milligan Canyon region as trilinear and Durov diagrams. These graphical presentation techniques allow samples with very different total ionic concentrations to occupy the same position in the diagrams because the concentrations of individual samples are plotted as percentages of the total cation and/or anion concentrations. Additionally, samples that plot on a straight line within the central diamond field of the trilinear diagram (Figure 4.8) indicate the mixing of groundwaters between two endmember solutions, such as freshwater and saline water. The Durov diagram (Figure 4.9) provides more hydrochemical interpretations. Two more parameters can be used to represent the central square field by extending lines from it to the neighboring scaled rectangles.

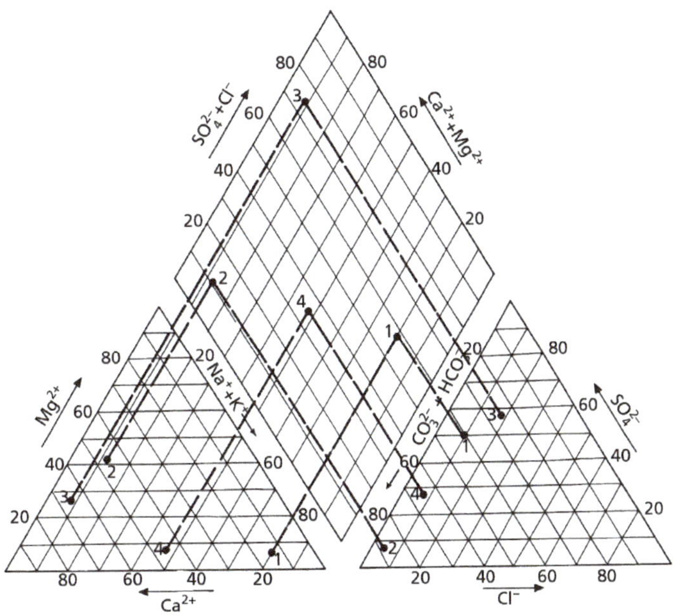

Figure 4.8. *Trilinear diagram of major ion analyses of groundwaters.*

Source: https://www.geokniga.org/bookfiles/geokniga-hydrogeologyprinciplesandpractice.pdf.

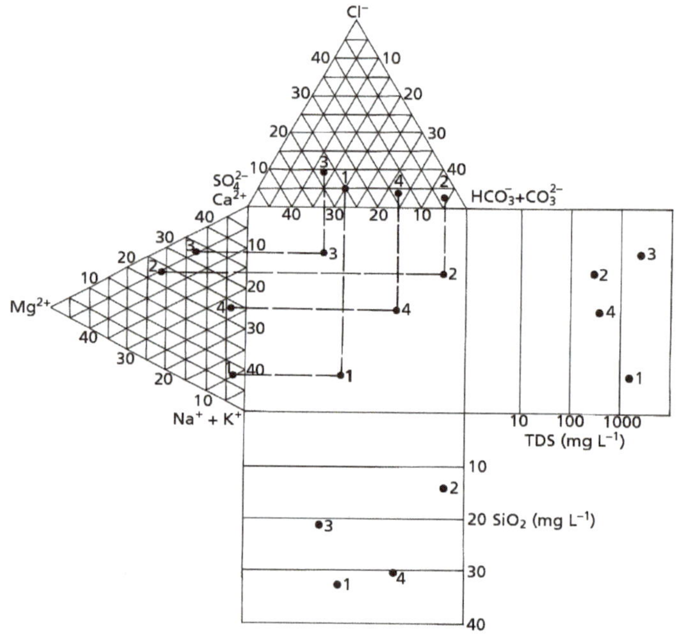

Figure 4.9. *Durov diagram of major ion analyses of groundwaters.*

Source: https://www.geokniga.org/bookfiles/geokniga-hydrogeologyprinciplesandpractice.pdf.

After plotting the chemical data in a variety of ways, the next step in the hydrochemical interpretation is to determine the hydrochemical facies that are present and create maps and cross-sections that illustrate the distribution of water types across the region. Figure 4.10 displays the overall hydrochemical and hydrogeological interpretation for the Milligan Canyon groundwater example. The samples are grouped based on the hydrochemical facies present ($Ca\text{-}HCO_3$, $Ca\text{-}SO_4$, $Na\text{-}HCO_3$, and $Na\text{-}SO_4$) in Figure 4.10a, which displays the trilinear diagram's central diamond field. It is evident that an elongate groundwater body with a distinct $Ca\text{-}SO_4$ water type is located in the center of the basin when the spatial distribution of the hydrochemical facies is plotted in Figure 4.10b and compared with the groundwater flow direction and regional geology obtained from a potentiometric map. According to the hydrochemical interpretation provided by Krothe and Bergeron (1981), groundwater with a $Ca\text{-}HCO_3$ character is produced by recharge taking place in the structurally high areas surrounding the basin's rim, which are formed by older carbonate rocks. The recharge water changes chemically as it passes through the Tertiary deposits due to the increased residence time and potential for anhydrite and/or gypsum to dissolve. This results in the $Ca\text{-}SO_4$ water type and high SO_4^{2-} and TDS concentrations. At the point where groundwater discharges out of the basin and the HCO3- ion becomes dominant once more, groundwaters with high concentrations of Na^+ are concentrated in the northeast and east central regions of the basin. Sulfate reduction, mixing between $Ca\text{-}HCO_3$ and $Ca\text{-}SO_4$ waters at the basin's exit, and potential ion exchange between Ca^{2+} and Na^+ linked to montmorillonite clay in the Tertiary sediments and the weathering of the Elkhorn Mountain Volcanics, which are rich in Na-plagioclases, are the three factors that collectively account for these chemical changes.

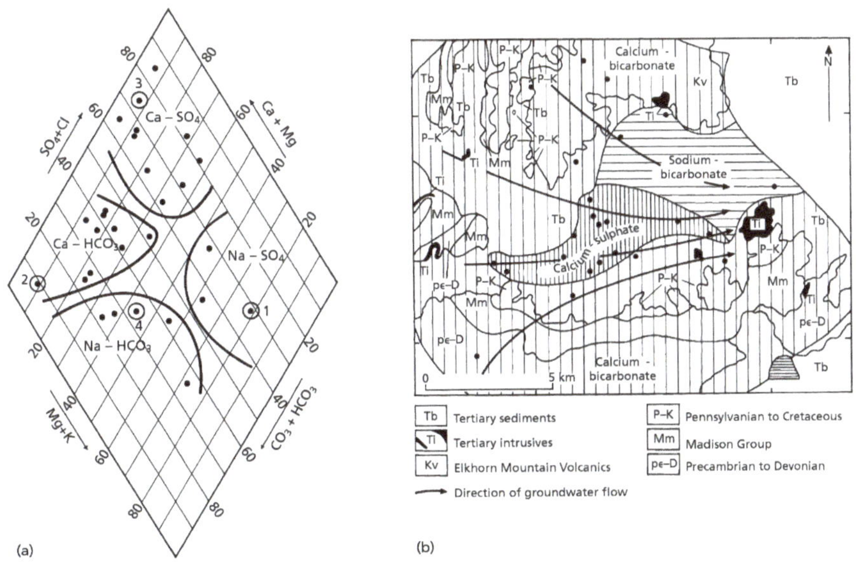

Figure 4.10. *(a) Identification using a trilinear diagram and (b) mapping of the distribution of hydrochemical facies.*

Source: https://www.geokniga.org/bookfiles/geokniga-hydrogeologyprinciplesandpractice.pdf.

The majority of the above-discussed graphical techniques have been used in regional hydrochemical studies. With the exception of more straightforward diagrams like bar charts, these techniques are not always suitable for contaminated groundwater investigations because of the large spatial variation in contaminant species concentrations between background and contaminated groundwaters. In this instance, visualizing the "hot spots" of contamination on a site map by charting the contaminant concentrations as pie charts can be helpful. As an illustration, see Figure 4.11.

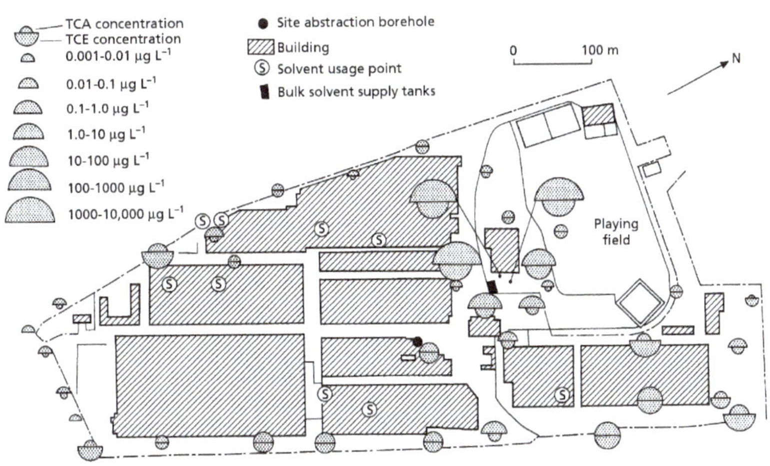

Figure 4.11. *Site plan showing points of solvent use and soil gas concentrations of TCE (trichloroethene) and TCA (1,1,1-trichloroethane) at an industrial site in the English Midlands.*

Source: https://www.geokniga.org/bookfiles/geokniga-hydrogeologyprinciplesandpractice.pdf.

4.5. CONCEPT OF CHEMICAL EQUILIBRIUM

A common idea in aqueous chemistry, hydrochemical processes in groundwater can be understood as slowly moving towards chemical equilibrium. A system at equilibrium will respond to perturbations by minimizing the imposed change, according to the Le Chatelier Principle, which can be used to qualitatively characterize shifts in the system's equilibrium. For example, consider groundwater flowing through a limestone aquifer composed of calcite:

$$CaCO_3 + H_2CO_3 \rightleftharpoons Ca^{2+} + 2HCO_3^{2-} \quad (5)$$

Until equilibrium is reached, the chemical reaction represented by Equation 5 will move either to the left (mineral precipitation) or to the right (mineral dissolution). To put it another way, a tendency for the reaction to shift to the left and precipitate calcite would reduce any increase in Ca^{2+} or HCO_3^- in solution. Chemical equilibrium, which can be viewed from both a kinetic and an energetic perspective, defines the boundary conditions that chemical processes will move toward. First, the kinetic approach is explained.

4.5.1. Kinetic Approach to Chemical Equilibrium

The equilibrium relationship, which characterizes the equilibrium chemical mass activities of a reversible reaction, is frequently referred to as the law of mass action. The effective concentration of the reacting substances determines the rate of reaction.

For the reaction:

$$aA + bB \rightleftharpoons cC + dD \quad (6)$$

the law of mass action expresses the relation between the reactants and the products when the chemical reaction is at equilibrium such that:

$$K = \frac{[C]^c[D]^d}{[A]^a[B]^b} \quad (7)$$

where the square brackets indicate the thermodynamically effective concentration, or activity.

Equation 7 is a statement of chemical equilibrium where K is the thermodynamic equilibrium constant (or stability constant). Values of K depend on temperature with solute concentrations expressed in terms of activities. An equilibrium constant greater than unity suggests that equilibrium lies to the righthand side of the equation describing the chemical reaction and that the forward reaction is favoured.

4.5.2. Energetic Approach to Chemical Equilibrium

According to this theory of chemical equilibrium, the reactant mixture with the lowest energy has the most stable composition. In contrast to the kinetic approach, this more rigorous thermodynamic treatment takes into account entropy, or S, a measure of a system's disorder, and enthalpy, or H, the heat content at constant pressure. An accurate indicator of the amount of energy released or absorbed in a reaction is the change in enthalpy (ΔH, expressed in J mol^{-1}). Most reactions have a change in entropy (ΔS, expressed in J mol^{-1} K^{-1}) that increases disorder through processes like compound splitting into its constituent ions. The change in entropy for a reversible process is calculated by dividing the heat absorbed by a reaction by the absolute temperature (T), at which the heat is absorbed.

The total energy released, or the energy change in going from reactants to products, is termed the Gibbs free energy, G (measured in kJ mol^{-1}). If energy is released, in which case the products have lower free energy than the reactants, G is considered negative. The change in Gibbs free energy is defined as:

$$\Delta G = \Delta H - T\Delta S \qquad (8)$$

Enthalpy and free energy values of zero are conventionally assigned to elements in their standard state, which is 25°C and 1 atmosphere pressure. The majority of geochemistry and aqueous chemistry textbooks, such as Krauskopf and Bird (1995) and Stumm and Morgan (1981), provide standard state thermodynamic data, denoted by the superscript o and tabulated as values of standard free energies, enthalpies, and entropies. The tabulated values can be combined using basic arithmetic to determine the values of ΔGo for various reactions. In principle, a reaction with a negative ΔGo value will release energy and proceed spontaneously, much like water flowing down a hydraulic gradient. Energy must be supplied for the opposite reaction to occur. For example, consider the reaction of aqueous carbon dioxide (H_2CO_3) with calcite (Equation 5). Relevant data for this reaction at standard state are ΔGo (H_2CO_3) = −623.1 kJ mol^{-1}, ΔGo ($CaCO_3$) = −1128.8 kJ mol^{-1}, ΔGo (Ca2+) = −553.6 kJ mol^{-1} and ΔGo (HCO_3^-) = −586.8 kJ mol^{-1}. Therefore, the change in standard free energy for this reaction, $\Delta G^\circ = \Sigma\Delta G^\circ_{products} - \Sigma\Delta G^\circ_{reactants}$ = −1727.2 − (−1751.9) = 24.7 kJ mol^{-1}. Since an energetically favoured reaction proceeds from reactants to products, the relationship between ΔG and the equilibrium constant, K, for a reaction is given by:

$$\Delta G = -RT \log_e K \qquad (9)$$

where R is the universal gas constant relating pressure, volume and temperature for an ideal gas (8.314 J mol^{-1} K^{-1}). Hence, one useful application of the energetic approach to chemical equilibrium is the use of thermodynamic data to derive equilibrium constants, K, using Equation 9. Now, for the reaction given in Equation 5, and using Equation 9 at standard conditions (T = 298 K), then $\log_e K = -\Delta G^\circ /RT = -(24.7 \times 10^3)/(8.314 \times 298) = -9.97$. Hence, K, the thermodynamic equilibrium constant for the reaction of dissolved carbon dioxide with calcite at 25°C is equal to 4.68×10^{-5} or, expressed as the negative logarithm to base 10 of K (pK) = 4.33 and $K = 10^{-4.33}$.

4.6. CARBONATE CHEMISTRY OF GROUNDWATER

Water's chemical composition is significantly influenced by acids and bases. Regarding the hydrochemistry of the majority of natural waters, the carbonate system is the most significant acid-base system. The fate of various contaminants, such as metal species, can be influenced by groundwater and carbonate mineral interactions in rock-water interactions. Understanding the chemistry of carbonate materials and the ways in which water has interacted with these minerals in an aquifer are necessary for interpreting groundwater ages determined by the carbon-14 dating method. In a carbonate system, the effective concentration of dissolved CO_2 in water serves as the fundamental control on the reaction rates. The amount of CO_2 in the atmosphere is approximately 0.03%, but in the soil zone, this rises to several percent of the soil atmosphere due to CO_2 produced during the decomposition of organic matter. Groundwater can react with carbonate minerals, usually dolomite and calcite, when it percolates through the soil zone and replenishes the aquifer.

CO_2 dissolves in water forming small quantities of weak carbonic acid, as follows:

$$CO_{2(g)} + H_2O \rightleftharpoons H_2CO_3 \tag{10}$$

From the law of mass action, the equilibrium constant for this reaction is:

$$K_{CO_2} = \frac{[H_2CO_3]}{[H_2O][CO_{2(g)}]} \tag{11}$$

Henry's law states that a dissolved gas's partial pressure in diluted solutions, measured in atmospheres, is equal to its molality, or activity in the case of diluted solutions.

Also, given that the activity of water is unity except for very saline solutions, then Equation 11 becomes:

$$K_{CO_2} = \frac{[H_2CO_3]}{P_{CO_2}} \tag{12}$$

Carbonic acid is polyprotic (i.e., it has more than one H^+ ion) and dissociates in two steps:

$$H_2CO_3 \rightleftharpoons H^+ + HCO_3^- \quad (13)$$

$$HCO_3^- \rightleftharpoons H^+ + CO_3^{2-} \quad (14)$$

From the law of mass action, dissociation constants can be expressed as follows:

$$K_{H_2CO_3} = \frac{[H^+][HCO_3^-]}{[H_2CO_3]} \quad (15)$$

$$K_{HCO_3^-} = \frac{[H^+][CO_3^{2-}]}{HCO_3^-} \quad (16)$$

Using a mass balance expression for the dissolved inorganic carbon (DIC) in the acid and its dissociated anionic species, expressed in terms of molality, then:

$$DIC = (H_2CO_3) + (HCO_3^-) + (CO_3^{2-}) \quad (17)$$

Equations for the relative concentrations of H_2CO_3, HCO_3^-, and CO_3^{2-} as a function of pH are obtained by rearranging Equations 13–17 and using an arbitrary value of unity for DIC, as illustrated graphically in Figure 4.12. Figure 4.12 illustrates that HCO_3^- is the predominant carbonate species throughout the majority of the groundwater's typical pH range of 6–9, which helps to explain why HCO_3^- is one of the main dissolved inorganic species in groundwater.

To calculate actual concentrations of inorganic carbon species in groundwater, first consider the dissolution of calcite by carbonic acid (Equation 5).

With reference to Equation 10, if the partial pressure of carbon dioxide (Pco_2) in the infiltrating groundwater increases, then reaction proceeds further to the right to achieve equilibrium. Now, at 25°C, substitution of Equations 12, 15 and 16 into the equation expressing the equilibrium constant for the dissociation of calcite:

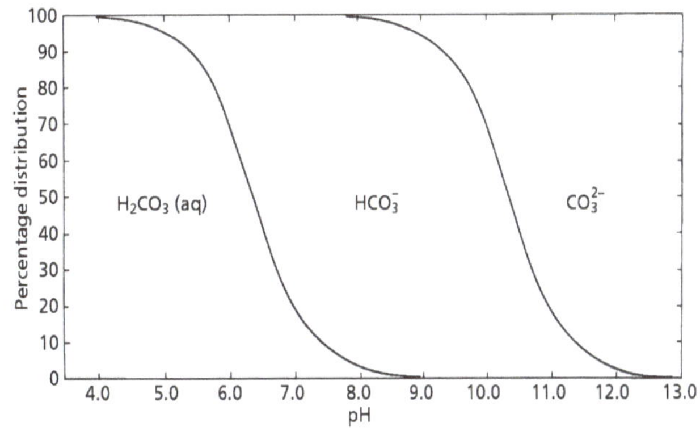

Figure 4.12. *Percentage distribution of inorganic carbon species in water as a function of pH at 25°C and 1 atmosphere pressure.*

Source: https://www.geokniga.org/bookfiles/geokniga-hydrogeologyprinciplesandpractice.pdf.

$$K_{calcite} = [Ca^{2+}][CO_3^{2-}] \quad (18)$$

yields:

$$[H^+] = 10^{-4.9}\{[Ca^{2+}]Pco_2\}^{1/2} \quad (19)$$

To obtain the solubility of calcite for a specified Pco_2, an equation of electroneutrality is required for the condition $\Sigma zm_c = \Sigma zm_a$ for calcite dissolution in pure water where z is the ionic valence and mc and ma are, respectively, the molalities of the cation and anion species involved:

$$2(Ca^{2+}) + (H^+) = (HCO_3^-) + 2(CO_3^{2-}) + (OH^-) \quad (20)$$

When combining Equations 19 and 20 with Equations 12, 15, and 16 yields a polynomial expression in terms of two variables and the activities, the concentrations of H^+ and OH^- are insignificant when compared to the other terms in Equation 20 with regard to the groundwater Pco_2 values. Iterative computer solutions are possible for a given Pco_2 value.

The calculations for equilibrium calcite dissolution in water under conditions where the supply of carbon dioxide is unrestricted are graphically displayed in Figure 4.13.

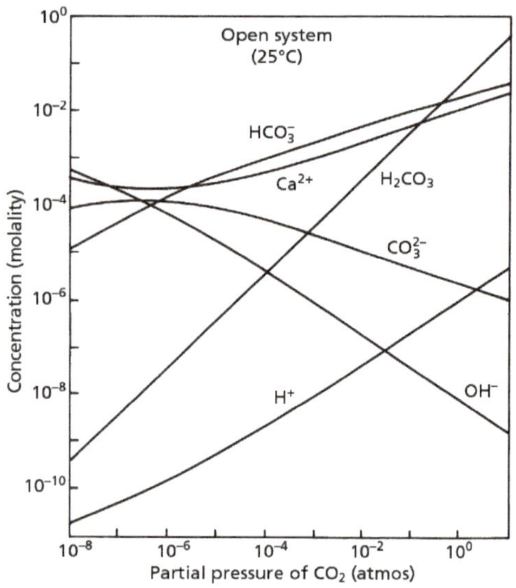

Figure 4.13. *Dissolved species in water in equilibrium with calcite as a function of Pco2 at 25°C. Note the parallel lines at the top right side for Ca2+ and HCO3– (just 0.301 unit apart) that demonstrate that at high Pco2 values, such as found in groundwater, these are the major species formed by dissolving calcite.*

Source: https://www.geokniga.org/bookfiles/geokniga-hydrogeologyprinciplesandpractice.pdf.

Figure 4.13 shows that the equilibrium value of H^+, or pH, varies strongly with Pco_2, and that the solubility of calcite is strongly dependent on Pco_2. So, precise measurement of pH in the field without any carbon dioxide loss from sample exsolution is necessary for the accurate computation of the inorganic carbon species in groundwater. Pco_2 is always greater than 10^{-4} atmospheres in the groundwater environment, and it is once more observed that HCO^{3-}, not CO_3^{2-}, is the predominant ionic species of DIC in groundwater. Changes in the distribution of carbonate species along a groundwater flow path are illustrated through a description of the carbonate chemistry of a limestone aquifer in the west of England.

There are two ways to think about the chemical evolution of the carbonate system: "open" and "closed". Figure 4.14 illustrates that carbonate dissolution occurs under "closed" conditions, when carbonate dissolution takes place without replenishing CO_2, while open conditions result in a constant Pco_2.

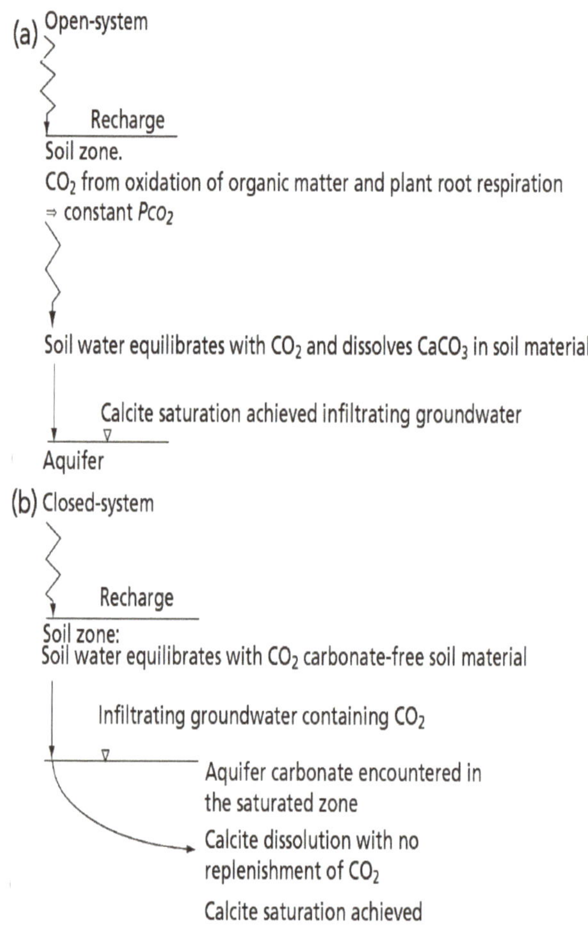

Figure 4.14. *Schematic representation of the development of open and closed systems of calcite dissolution in soil–aquifer systems.*

Source: https://www.geokniga.org/bookfiles/geokniga-hydrogeologyprinciplesandpractice.pdf.

It is feasible to model the chemical evolution of groundwater dissolving carbonate using the above-mentioned theory. Steps along the paths are computed by hypothetically dissolving small amounts of carbonate material (calcite or dolomite) for a given temperature and starting condition for P_{CO_2} until the water becomes saturated (lines 2 and 3 at 15°C in Figure. 4.15). Lower temperatures will shift the saturation line to higher solubilities, while higher temperatures will result in saturation at lower solubilities. Figure 4.15 shows that the pH values at saturation for closed-system dissolution are higher and the HCO_3^- concentrations are lower. Actually, very small amounts of dolomite and calcite have a significant impact on the carbonate chemistry of groundwater passing through the soil zone, favoring open-system conditions and causing a pH that is always between 7 and 8.

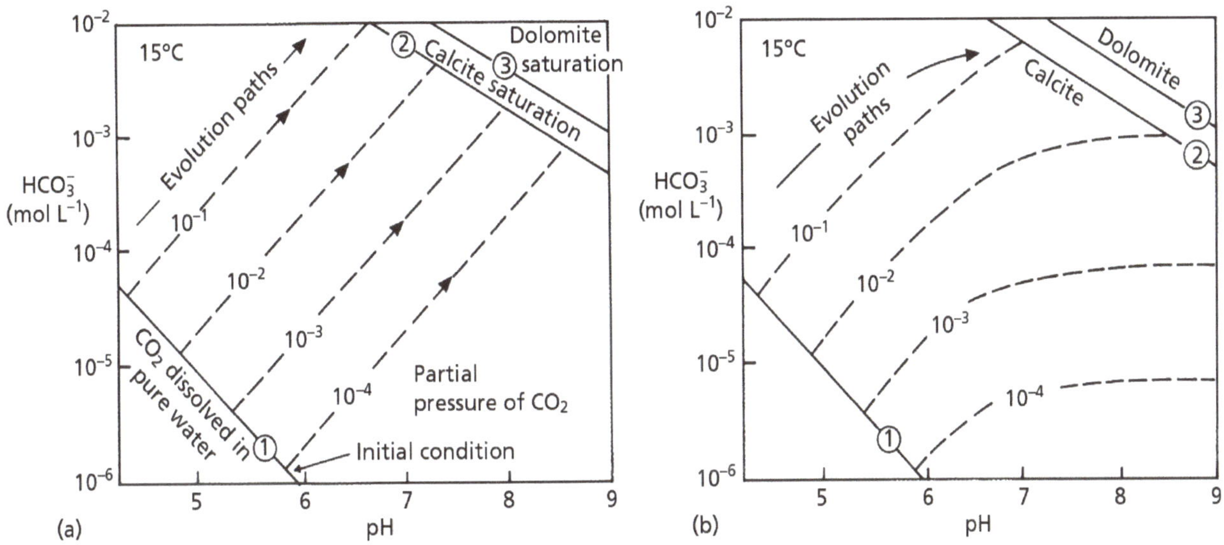

Figure 4.15. *Chemical evolution paths for water dissolving calcite at 15°C for (a) open-system dissolution and (b) closed-system dissolution. Line 1 represents the initial condition for the water containing dissolved CO2; line 2 represents calcite saturation; and line 3 represents dolomite saturation if dolomite is dissolved under similar conditions.*

Source: https://www.geokniga.org/bookfiles/geokniga-hydrogeologyprinciplesandpractice.pdf.

Figure 4.16 presents the results of chemical modeling of recharge to the Chalk aquifer in north Norfolk, eastern England, to illustrate carbonate dissolution pathways in a limestone aquifer.

The Chalk aquifer in this region is covered in glacial deposits, which include two different types of till deposits and outwash sands and gravel that are identified by their combination of contained clay, sand, and carbonate fractions. Hiscock (1993) showed that it is possible to model the evolution of the carbonate chemistry for both open- and closed-systems in this aquifer system. The initial conditions for the two models are two shallow well waters in the glacial deposits that are both undersaturated with regard to calcite.

By means of the computer application mix_2 (Plummer et al., 1975) until saturation is attained, carbonate is added to the groundwaters in tiny increments. Table

4.5 provides the initial and final chemical compositions of the well waters, and Figure 4.16 illustrates the evolution paths taken in each case.

The fact that the chemistry of the model and the real Chalk groundwaters, which were taken from beneath the glacial deposits, agrees shows that calcite saturation can occur in an open system.

According to Langmuir (1971), the distribution of points around the phase boundary in Figure 4.16 is caused by either changes in groundwater chemistry that affect Pco_2 or variations in soil Pco_2 at the time of groundwater recharge.

The primary cause for the Chalk aquifer in north Norfolk is fluctuations in soil Pco_2. Recharge entering carbonate-depleted soils in sandy till cover areas results in calcite saturation for lower soil Pco_2 values, ranging from $10^{-2.5}$ to $10^{-2.0}$ atmospheres. In contrast, calcite saturation occurs for higher soil Pco_2 values, in the range of $10^{-2.1}$ to $10^{-2.0}$, in carbonate-rich soils developed in chalky, clay-rich till (Hiscock, 1993).

While this framework is helpful in understanding the chemical evolution of the carbonate system, it ignores a number of important factors, such as seasonal variations in Pco_2 and soil temperature; processes that affect the concentrations of Ca^{2+} and Pco_2 such as adsorption, cation exchange, and gas diffusion and dispersion; and the process of incongruent dissolution, which results in a non-stoichiometric dissolution where one of the dissolution products is a mineral phase that shares an ionic component with the dissolving phase. The calcite dissolution discussed above was predicated on the separate dissolution of dolomite and calcite.

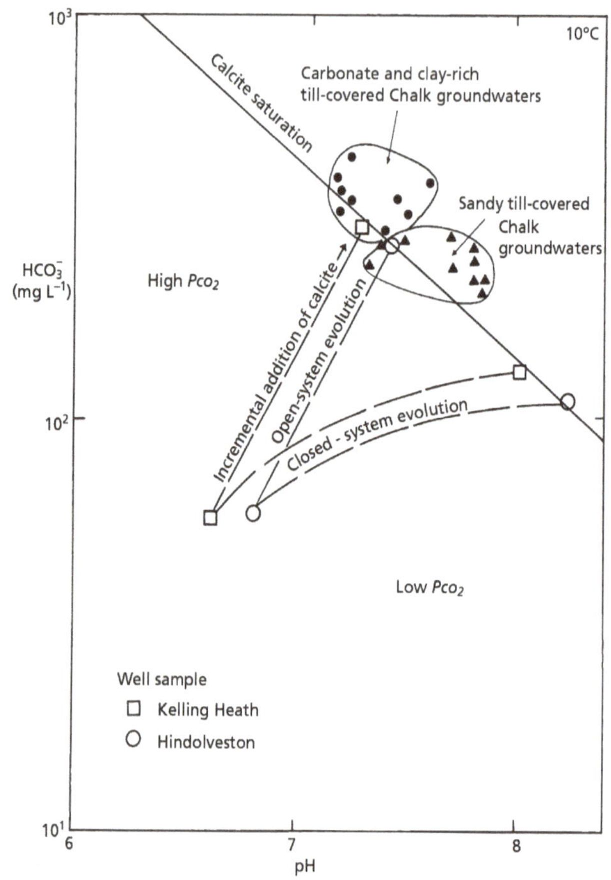

Figure 4.16. *Graphical representation of the modelled open- and closed-system carbonate dissolution paths. Starting conditions are represented by well samples contained in glacial till deposits. The two fields showing Chalk groundwaters represent saturated conditions with respect to calcite with the differences explained by the greater dissolution of carbonate material at higher Pco2 values in soils developed on the carbonate and clay-rich till deposits.*

Source: https://www.geokniga.org/bookfiles/geokniga-hydrogeologyprinciplesandpractice.pdf.

Table 4.5. *Chemical Models for the Formation of Chalk Groundwaters Beneath Glacial Deposits in North Norfolk, Eastern England, Under the Processes of Open- and Closed-System Dissolution of Calcite*

Well sample name	$CaCO_3$ added ($\times 10^{-3}$ mol L^{-1})	Ca^{2+} (mg L^{-1})	HCO_3^- (mg L^{-1})	pH	Pco_2 (atmos)	$\Omega_{calcite}$
Open-system conditions						
Kelling Heath	0	10.0	56.7	6.62	1.94	−2.47
	2.00	90.2	299.5	7.30	1.94	−0.03
Hindolveston	0	9.4	59.2	6.81	2.12	−2.09
	1.75	79.5	271.5	7.44	2.12	0.02
Closed-system conditions						
Kelling Heath	0	10.0	56.7	6.62	1.94	−2.47
	0.70	36.0	130.5	8.02	3.00	−0.05
Hindolveston	0	9.4	59.2	6.81	2.12	−2.09
	0.44	28.4	111.6	8.24	3.29	0.02

Source: https://www.geokniga.org/bookfiles/geokniga-hydrogeologyprinciplesandpractice.pdf.

However, when both minerals are present in a hydrogeological system, they may dissolve either simultaneously or sequentially. This can result in different equilibrium relationships than those depicted in Figure 4.15. In such cases, comparing the equilibrium constants for calcite and dolomite at the specific groundwater temperature is essential to determine which mineral is dissolving in an incongruous manner.

For instance, assuming the thermodynamic information presented in Table 4.6, Kcalcite = K1/2 dolomite at roughly 20°C. Under these circumstances, no calcite dissolution will occur because the water is already saturated with respect to calcite because the solubility product of dolomite is equal to $[Ca^{2+}][Mg^{2+}][CO_3^{2-}]$ $[CO_3^{2-}]$ and that of calcite is equal to $[Ca^{2+}][CO_3^{2-}]$ $[CO_3^{2-}]$. Kcalcite crystallizes at less than 20°C. < K1/2dolomite and if groundwater dissolves dolomite to equilibrium the water becomes supersaturated with respect to calcite which can then precipitate. In a system where the rate of dolomite dissolution is equal to the rate of calcite precipitation, this is the condition of incongruent dissolution of dolomite. At temperatures higher than 20°C, Kcalcite > K1/2 dolomite, and calcite dissolution will happen if dolomite saturation is reached with groundwater entering a calcite-containing area, increasing Ca^{2+} and CO_3^{2-}concentrations. In order to achieve an incongruent dissolution of calcite, the water will now be supersaturated with respect to dolomite, and dolomite precipitation will occur, albeit slowly. Dolomite dissolves at any temperature when groundwater first dissolves calcite to equilibrium before coming into contact with it because dolomite equilibrium cannot be reached until the water has absorbed a significant amount of Mg^{2+}. However, calcite precipitates and the dolomite dissolution becomes incongruent as the water becomes supersaturated with respect to calcite due to the influx of Ca^{2+} and CO_3^{2-} ions from the dissolution of the dolomite (Freeze and Cherry, 1979).

Long-term incongruent calcite and dolomite dissolution can have a significant impact on the diagenesis—the mineralogical evolution of the aquifer rock—and the chemical evolution of groundwater. One example of this is the dolomitization of calcareous sediments. There have been reports of dedolomitization—the transformation of a dolomite-bearing rock into a calcite-bearing rock—in the context of the Floridan aquifer system, for instance. In the Upper Floridan aquifer, Mg^{2+} concentrations typically range from

1 to 1000 mg L^{-1}, with the highest concentrations found in areas where seawater is present in the aquifer. Mg^{2+} concentrations typically rise in downgradient directions in freshwater aquifers due to dedolomitization; however, there is insufficient data to demonstrate that Mg^{2+} concentrations in the Upper Floridan aquifer are constrained by dolomite formation (Sprinkle, 1989).

Table 4.6. *Thermodynamic Equilibrium Constants for Calcite, Dolomite and Major Aqueous Carbonate Species in Pure Water for a Temperature Range of 0–30°C and 1 Atmosphere Total Pressure*

Temperature (°C)	pK_{CO_2}	$pK_{H_2CO_3}$	$pK_{HCO_3^-}$	$pK_{calcite}$	$pK_{dolomite}$	$K_{calcite}/(K_{dolomite})^{1/2}$
0	1.11	6.58	10.63	8.38	16.18	0.51
5	1.19	6.52	10.56	8.39	16.39	0.63
10	1.27	6.47	10.49	8.41	16.57	0.75
15	1.34	6.42	10.43	8.43	16.74	0.87
20	1.41	6.38	10.38	8.45	16.88	0.97
25	1.47	6.35	10.33	8.48	17.00	1.05
30	1.52	6.33	10.29	8.51	17.11	1.11

Source: https://www.geokniga.org/bookfiles/geokniga-hydrogeologyprinciplesandpractice.pdf.

Note that $pK = -\log 10K$.

4.7. ADSORPTION AND ION EXCHANGE

The adsorption and ion exchange processes in aquifers play a crucial role in altering the natural chemistry of groundwater and are key factors to consider when forecasting the movement of pollutants like heavy metals and polar organic chemicals. These reactions not only impact the ions being exchanged, but also other species through dissolution and precipitation reactions. Ion exchange is a significant method of reducing pollutants like NH_4^+ in aquifers (Carlyle et al., 2004).

Natural materials' hydraulic conductivity may alter as a result of ion exchange reactions (Zhang and Norton, 2002). Groundwater contains ionic species that can react with solid surfaces. Adsorption happens when a positively charged ion in solution is drawn to and held by a solid with a negatively charged surface, as Figure 4.17a illustrates. Different minerals will attract anions or cations to their surfaces depending on the pH of the solution, depending on the point of zero charge (PCZ) of the rock-forming mineral, which is the pH at which the mineral has zero charge (at pH values less than the PCZ the mineral has a net positive charge, at values greater than the PCZ a net negative charge). With the exception of the most acidic solutions, clay minerals have negative surface charges, which attract cations to their surfaces in order to balance the negative charge.

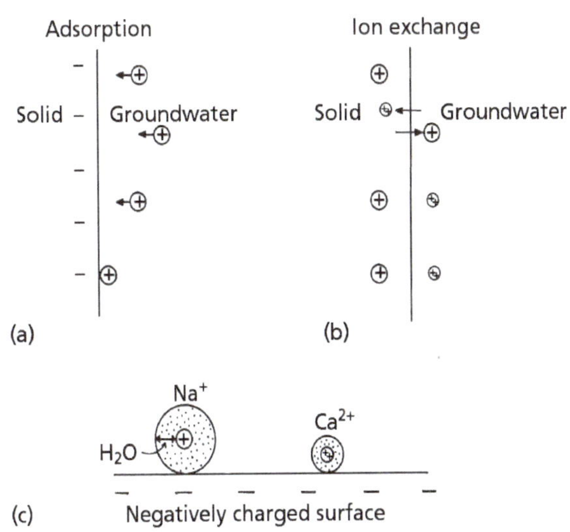

Figure 4.17. *Pictorial representation of (a) adsorption and (b) cation exchange reactions in groundwater. In (c) the divalent Ca^{2+} ions with a smaller hydrated radius are more strongly adsorbed by the negatively charged surface (e.g., clay) than are the monovalent Na+ ions with a larger hydrated radius.*

Source: https://www.geokniga.org/bookfiles/geokniga-hydrogeologyprinciplesandpractice.pdf.

In most groundwaters, the surface charge of iron oxides and hydroxides can be either positive or negative (the PCZ values for haematite and goethite are between 5–9 and 7.3–7.8, respectively; Krauskopf & Bird, 1995). This means that at low pH, anion adsorption (like that of arsenic oxyanions) can occur, and at high pH, release can occur.

Generally speaking, as surface area and grain size decrease, the degree of adsorption rises. Because clays have small grains and a large surface area for sorption reactions, they are usually the most reactive materials. Furthermore, because of crystal lattice defects that cause an excess of negative charge at the surface that cations can adsorb onto, clays have a tendency to be strong adsorbers. Van der Waals force can cause weak adsorption, which is essentially a physical process, or strong adsorption, which is caused by chemical bonding. Due to their higher charge density, which results from their valence and smaller hydrated radius, divalent cations are typically more strongly adsorbed than monovalent ions (Figure 4.17c). Usually, batch or column experiments are used to measure the adsorptive capacity of particular soils or sediments.

Ion exchange happens when ions in an aqueous solution replace ions within a solid's mineral lattice (Figure 4.17b). Clays, soil organic matter, metal oxides, and hydroxides are the main sources of ion exchange sites; each of these materials has a measurably high cation exchange capacity (CEC). Divalent ions have a stronger bond to a solid surface during cation exchange, which causes them to replace monovalent ions in most cases. The interaction between the concentration of cations in solution and the energy of adsorption of the cations at the exchange surface determines the amounts and kinds of cations exchanged.

The monovalent ions have a smaller energy of adsorption and are therefore more likely to remain in solution. As a result of a larger energy of adsorption, divalent ions are more abundant as exchangeable cations. Ca^{2+} is typically more abundant as an exchangeable cation than is Mg^{2+}, K^+ or Na^+. The energy absorption sequence is: $Ca^{2+} > Mg^{2+} > K^+ > Na^+$ and this provides a general ordering of cation exchangeability for common ions in groundwater.

Experimental values of CEC are determined, and laboratory findings are expressed in terms of meq $(100\ g)^{-1}$. CEC is typically ascertained by extracting the cations using a solution that contains a known cation, typically NH_4^+, from soils or aquifer materials. Usually used for this purpose, ammonium acetate (CH_3COONH_4), the salt of a weak acid and weak base, has its pH adjusted to the value best suited to the investigation. Talibudeen gives an overview of techniques for identifying exchangeable cations (1981). Table 4.7 provides the surface area and CEC of different clays as well as Fe and Al oxy-hydroxides.

Table 4.7. *Surface Area and Cation Exchange Capacity (CEC) Values for Clays and Fe and Al Oxyhydroxides*

	Surface area ($m^2\ g^{-1}$)	CEC (meq $100\ g^{-1}$)
Fe and Al oxy-hydroxides (pH ~8.0)	25–42	0.5–1
Smectite	750–800	60–150
Vermiculite	750–800	120–200
Bentonite	750	100
Illite	90–130	10–40
Kaolinite	10–20	1–10
Chlorite	-	<10

Source: https://www.geokniga.org/bookfiles/geokniga-hydrogeologyprinciplesandpractice.pdf.

The eastern English Jurassic Lincolnshire Limestone aquifer provides an example of cation exchange in groundwater. The oolitic limestone that makes up the Lincolnshire Limestone aquifer is between 10 and 30 meters deep, and it contains varying amounts of finely divided iron minerals, as well as scattered clay and organic matter that serve as reactive exchange sites for Na^+ and Ca^{2+}.

Thick marine clays limit the aquifer's downward movement. According to a hydrochemical survey (Figure 4.18), cation exchange happens about 12 km downslope from the aquifer outcrop, and after another 12 km, the exchange of Ca^{2+} with Na^+ reduces the Ca^{2+} to a minimum of <4 mg L^{-1} (Edmunds and Walton, 1983). The limited cation exchange capacity of the limestone is exhausted, which explains the lack of cation exchange closer to the aquifer outcrop. Due to incongruent dissolution, the concentrations of Sr^{2+} and, to a lesser extent, Mg^{2+}, keep rising for about 22 km from the outcrop.

In order to restore carbonate equilibrium, calcite dissolves when Ca^{2+} is removed through cation exchange. However, cation exchange reactions become dominant and both Sr^{2+} and Mg^{2+} concentrations start to decrease once the Ca^{2+}/Sr^{2+} and Ca^{2+}/Mg^{2+} equivalents ratios drop below a certain critical level (~20: 1 and 1: 1, respectively) (Edmunds & Walton, 1983).

One characteristic of saline water intrusion in coastal regions is cation exchange reactions. The cation exchangers in the aquifer have mostly Ca^{2+} adsorbed on their surfaces, as freshwater in coastal areas is usually dominated by Ca^{2+} and HCO_3^- ions from the dissolution of calcite. The predominant ions in seawater are Na^+ and Cl^-, so materials from aquifers that come into contact with it will have Na^+ bonded to their exchange surfaces. When seawater intrudes a coastal freshwater aquifer, the following cation exchange reaction can occur:

$$Na^+ + {}^1/_2Ca\text{-}X \rightarrow Na\text{-}X + {}^1/_2Ca^{2+} \qquad (21)$$

where X indicates the exchange material. As the exchanger takes up Na+, Ca2+ is released, and the hydrochemical water type evolves from Na-Cl to CaCl. The reverse reaction can occur when freshwater flushes a saline aquifer:

$$ {}^1/_2Ca^{2+} + Na\text{-}X \rightarrow {}^1/_2Ca\text{-}X + Na^+ \qquad (22)$$

where Ca^{2+} is taken up from water in return for Na^+ resulting in a Na-HCO_3 water type. The chemical reactions that occur during freshwater and saline water displacements in aquifers can be identified from a consideration of conservative mixing of fresh and saline water end-member solutions and comparing with individual water analyses. For conservative mixing:

$$c_{i,mix} = f_{saline} \cdot c_{i,saline} + (1 - f_{saline})c_{i,fresh} \qquad (23)$$

where c_i is the concentration of ion i; $_{mix,\ fresh}$ and saline indicate the conservative mixture and end-member fresh and saline waters; and f saline is the fraction of saline water. Any change in the sample composition as a result of reactions, for example cation exchange, other than by simple mixing ($c_{i,react}$) is then simply found from:

$$c_{i,react} = c_{i,sample} - c_{i,mix} \qquad (24)$$

As an example calculation, the data shown in Table 4.8 are for samples 0, 14 and 19 of Figure 4.18 representing fresh, mixed and saline groundwaters present in the Lincolnshire Limestone aquifer. To calculate how much Na+ has been added to the mixed groundwater sample by cation exchange, Equation 23 can be re-written as:

$$c_{Na,mix} = f_{saline} \cdot c_{Na,saline} + (1 - f_{saline})c_{Na,fresh} \tag{25}$$

Now, using the mixed and saline Cl– concentration values (samples 14 and 19) to indicate the fraction of the saline water (f_{saline} = 114/1100), and assuming a freshwater end-member Cl– concentration value equal to zero, then substituting the values from Table 4.8 into Equation 25:

$$c_{Na,mix} = (114/1100)920 + (1 - 114/1100)14$$
$$= 107.9 \text{ mg L}^{-1}$$

Similarly, Equation 24 can be re-written to calculate the amount of Na+ released in the cation exchange reaction:

$$c_{Na,react} = c_{Na,sample} - c_{Na,mix} \tag{26}$$

Now, using the result for $c_{Na,mix}$ found above:

$$c_{Na,react} = 280 - 107.9 = 172.1 \text{ mg L}^{-1}$$
$$(\text{or } 7.5 \text{ meq L}^{-1})$$

A similar calculation for Ca^{2+} removed from the mixed groundwater sample results in:

$$c_{Ca,mix} = (114/1100)9.5 + (1 - 114/1100)135$$
$$= 122.0 \text{ mg L}^{-1}$$

and

$$c_{Ca,react} = 17 - 122.0 = -105.0 \text{ mg L}^{-1}$$
$$(\text{or } 5.3 \text{ meq L}^{-1})$$

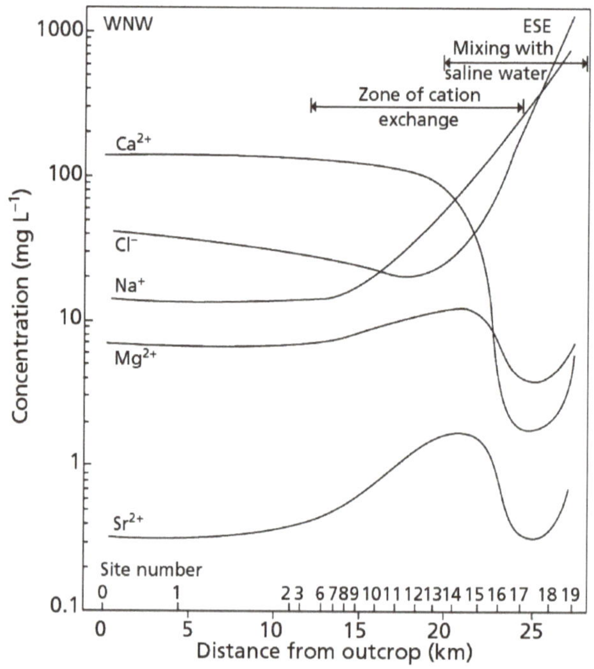

Figure 4.18. *Hydrogeochemical trends in the Lincolnshire Limestone for Ca2+, Mg2+, Na+, Sr2+ and Cl–. The trend lines are for 1979 and illustrate the effect of cation exchange between Ca2+ and Na+ and the onset of mixing with saline water in the deeper aquifer.*

Source: https://www.geokniga.org/bookfiles/geokniga-hydrogeologyprinciplesandpractice.pdf.

Table 4.8. *Selected Hydrochemical Data for the Lincolnshire Limestone Aquifer to Illustrate Cation Exchange*

	Na^+ (mg L^{-1})	Ca^{2+} (mg L^{-1})	Cl^- (mg L^{-1})	Na^+/Cl^-
Freshwater (Sample 0, Ropsley)	14	135	42	0.51
Mixed water (Sample 14, Pepper Hill)	280	17	114	3.74
Saline water (Sample 19, Deeping St Nicholas)	920	9.5	1100	1.27

Source: https://www.geokniga.org/bookfiles/geokniga-hydrogeologyprinciplesandpractice.pdf.

CASE STUDY

1,4-DIOXANE TREATMENT USING GAC-AOP-SORPTIVE RESIN

The site is located at a former chemical blending and distribution facility situated in a commercial and industrial area of Santa Fe Springs, California. The former facility covered approximately 10 acres and operated from the early 1960s until 2002. The site lies within the Montebello Forebay of the Central Groundwater Basin, a subdivision of the Coastal Plain of Los Angeles County, where the uppermost groundwater is present under nonpressure conditions. The on-site hydrostratigraphy includes the four investigated units (from shallowest to deepest): Bellflower aquiclude, Artesia/Gage aquifer, "upper aquitard," and Hollydale aquifer. Groundwater beneath the site is first encountered at about 75 feet below ground surface (bgs) within the Artesia/Gage aquifer, which is impacted by site contaminants.

The primary chemicals of concern for on-site groundwater treatment activities have included 1,1-dichloroethane (DCA), 1,2-DCA, cis-1,2-dichloroethene (DCE), trans-1,2-DCE, 1,1-DCE, tetrachloroethene, vinyl chloride, 1,4-dichlorobenzene, 1,2-dichloropropane, 1,4-dioxane, benzene, ethylbenzene, and toluene. A groundwater extraction system designed to remove and hydraulically contain groundwater impacted in the Artesia/Gage aquifer has been operating at the site since 1988 under the regulatory authority of the Los Angeles Regional Water Quality Control Board. From 1988 to 2014, treated water was discharged to the storm drain under a National Pollutant Discharge Elimination System (NPDES) permit.

In 2012, a Waste Discharge Requirements (WDR) permit was issued for reinjecting treated water back into the Artesia/Gage aquifer to help stabilize the saturated thickness and flush the residual contaminant mass in the subsurface to the extraction wells. The 1,4-dioxane limit specified in the WDR was 3 micrograms per liter (μg/L). The remediation well field currently comprises four extraction and two reinjection wells screened in the Artesia/Gage aquifer.

The groundwater treatment system initially consisted of a bioreactor, air stripper, sand filter, and two liquid-phase granular activated carbon (LGAC) vessels used to treat the dissolved petroleum hydrocarbons, halogenated and nonhalogenated compounds, alcohols, ketones, and glycols. In 2006, 1,4-dioxane was analyzed for and detected in the monitoring well field at approximately μg/L. Treating 1,4-dioxane with the initial system was costly, and additional actions were taken in 2008 to meet the NPDES discharge limits (3 μg/L for 1,4-dioxane). In 2008, an advanced oxidation process (AOP) system was added to the treatment system, which effectively removed the 1,4-dioxane. After additional system modifications, the bioreactor and air stripper units were taken offline, and the treatment system then consisted of only the AOP and LGAC vessels to meet the NPDES discharge limits. By 2015, drought conditions caused the groundwater pumping yield to drop below 7 gallons per minute (gpm), and the AOP system (designed to optimally treat groundwater at 50 gpm) became too costly to operate. Therefore, the AOP system was shut down, and a synthetic media system was investigated and installed,

Regulatory Issues

Regulatory orders required that the groundwater extraction and treatment system continue to operate. As concentrations of chemicals decreased since implementation of the AOP system in 2008, and operation of the AOP was costly, in 2015, a cost-benefit analysis was performed to assess options for either replacing the AOP system or implementing a new technology. Based on the cost-benefit analysis performed, it was determined that the best alternative to replacing the AOP unit was a system that uses synthetic media filtration, with two media vessels operating in series (lead-lag operation). The synthetic media system was necessary to meet NPDES discharge limits of 3 µg/L for 1,4-dioxane, in addition to the overall objective of vertically and laterally containing chlorinated volatile organic compounds (CVOCs) and 1,4-dioxane migration.

Technology/System Description

In July 2016, a full-scale skidded synthetic media system (AMBERSORB™, designed and constructed by Emerging Compounds Treatment Technologies [ECT2] of Portland, Maine) was mobilized to the site to begin treating the impacted groundwater. The M-25 Ground Water Treatment System (GWTS) contained 22 cubic feet of synthetic media in two identical stainless-steel vessels. The GWTS had a nominal treatment capacity of 24 gpm, sufficient to treat the influent, which had decreased to 7 gpm as a result of drought conditions. The two vessels were aligned in a lead-lag flow configuration during normal operations. The influent groundwater passed through bag filter pretreatment before entering the media vessels. Once the lead vessel reached treatment capacity based on the volume processed, the lag vessel was switched to the lead position and the contaminant of concern loaded vessel was regenerated. The GWTS included integrated regeneration equipment, comprised of an electric steam boiler, electric superheater, condenser, subcooler, and condensate tank. The condenser was cooled by a dry cooler located adjacent to the shipping container. The subcooler was cooled by a chiller located inside the shipping container. The chiller condenser and steam condenser shared the dry cooler cooling water circuit. These components support the in situ regeneration of the resin, a process by which superheated steam is passed through the resin bed to desorb 1,4-dioxane and volatile organic compounds (VOCs). The contaminant-laden steam is then condensed and cooled for subsequent processing and/or removal. The condensate tank held the total volume of condensate liquid from a vessel regeneration. The condensate was decanted first through an internal tank weir, allowing dense nonaqueous phase liquid (DNAPL) product to settle in the bottom of the internal weir section. The remaining condensate was pumped from the tank to the owner's GAC treatment system and then recycled back to the influent of the GWTS. Figure 1 shows a process flow diagram of a generic resin system.

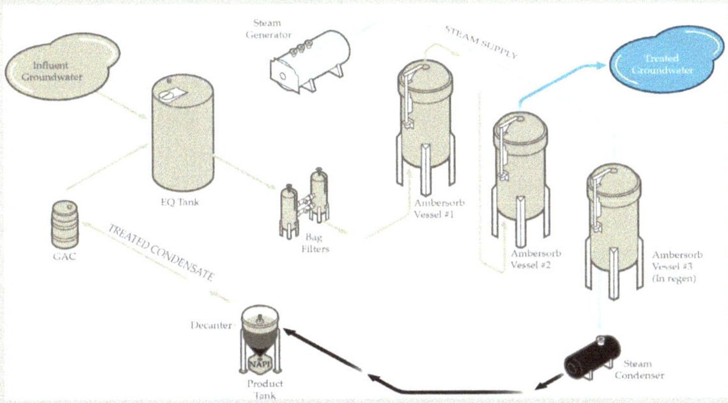

Figure 1. *Process flow diagram of generic regenerable resin system.*

Source: https://14d-1.itrcweb.org/case-studies/.

Assessment

The synthetic resin system was successful from both treatment endpoint and operational efficiency perspectives. Effluent 1,4-dioxane concentrations from both the lead and lag vessels were consistently nondetect at less than 1.0 µg/L, even though influent concentrations were 5 to 10 times higher than anticipated during system design. Influent and effluent 1,4-dioxane concentrations are shown in Figure CS3-2. System runtime between lead vessel regenerations also exceeded the design basis of 45-day loading cycles, stretching as long as 90 days. The switch from the AOP system to the synthetic resin system was favorable, resulting in simpler operation, substantial operation and maintenance (O&M) savings, and consistent compliance with NPDES and WDR permit limits. The system continues to operate with regenerations occurring once every 2 to 3 months, with an overall system uptime of >99%.

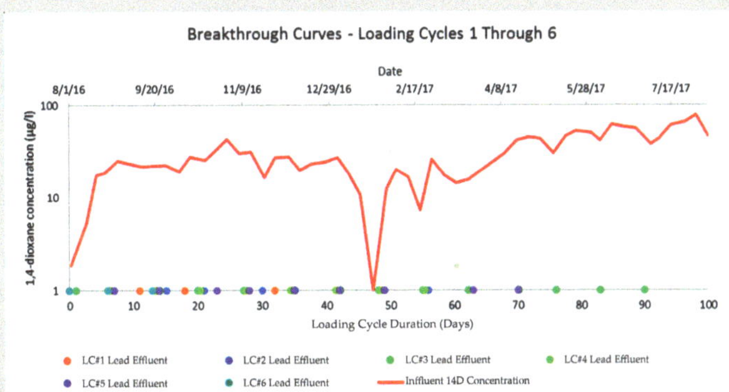

Figure 2. *Correlation between loading cycle duration and influent and lead effluent 1,4-dioxane concentrations.*

Source: https://14d-1.itrcweb.org/case-studies.

Costs

Based on pilot-test results, the O&M "routine" services would consist of system regeneration, compliance monitoring, water softener maintenance, product disposal, and utilities. Potential "nonroutine" services would include three-way valve actuator troubleshooting and superheater control panel repairs and replacement. Since the operational costs are relatively sensitive to extracted flow volumes, separate estimates were considered for operating the synthetic resin system at 7 gpm and 24 gpm. Based on these conditions, the projected "routine" and "nonroutine" costs for the synthetic resin system operations are shown in the graph below.

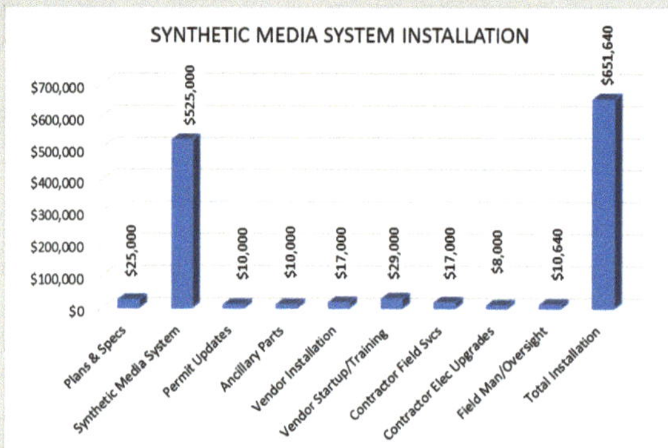

Figure 3. *Summary of installation costs.*

Source: https://14d-1.itrcweb.org/case-studies.

The cost comparison was based on initial capital expenses and projected system operations. The synthetic resin system operating costs can vary significantly with pumping yield, because a higher yield would result in more frequent regeneration events. The cost comparisons were therefore based on pumping yields of 7 gpm and 24 gpm.

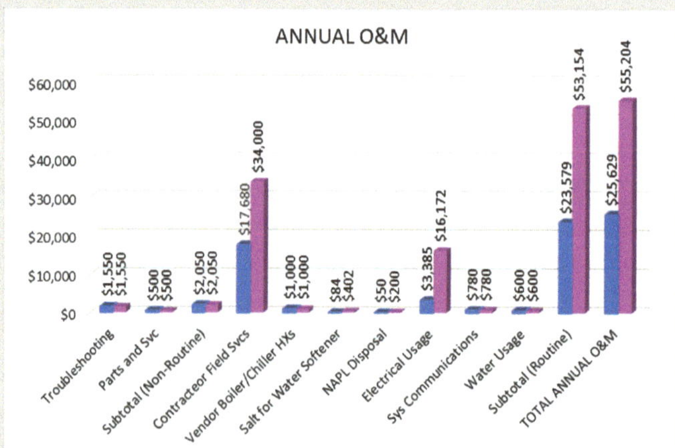

Figure 4. *Summary of operational costs; blue shows costs at 7 gpm, purple shows costs at 24 gpm.*

A lifecycle cost comparison showed that the initial capital cost for the synthetic media system would be higher than the AOP system, but the synthetic media system would have lower operating costs. The expected "break-even" point for costs is on the order of 6 years at 7 gpm and 12 years at 24 gpm. The actual flow rate was 7 gpm. Based on these results, the synthetic media system was selected because it would have fewer mechanical components, less chemical use, less maintenance, and a lower overall cost under the given conditions.

CLASS ACTIVITY:

UNDERSTANDING GROUNDWATER CONTAMINATION THROUGH CHEMICAL HYDROGEOLOGY

Objective: Students will learn about the basic principles of chemical hydrogeology, specifically groundwater contamination, through a hands-on activity. They will simulate groundwater flow and contamination in a simplified model, analyze the results, and discuss the implications of groundwater contamination.

Materials: Sand, Gravel, Water, Small plastic cups, Drinking straws, Measuring spoons, Graph paper, Markers, etc.

SUMMARY

- Chemical hydrology, subdivision of hydrology that deals with the chemical characteristics of the water on and beneath the surface of the Earth. Water in all forms and modes of occurrence is affected chemically by the materials with which it comes into contact.
- Positively charged atoms like sodium are known as cations, while negatively charged ions like chloride are called anions. When writing chemical equations, the sum of charges on one side of the equation must balance the sum of charges on the other side.
- The chemical and biochemical interactions between groundwater and the geological materials of soils and rocks provide a wide variety of dissolved inorganic and organic constituents.
- The equilibrium relationship is often called the law of mass action which describes the equilibrium chemical mass activities of a reversible reaction. The rate of reaction is proportional to the effective concentration of the reacting substances.

- Acids and bases exert significant control over the chemical composition of water. The most important acid–base system with respect to the hydrochemistry of most natural waters is the carbonate system.
- Adsorption and ion exchange reactions in aquifers can significantly influence the natural groundwater chemistry and are an important consideration in predicting the migration of contaminants such as heavy metals and polar organic chemicals.
- Cation exchange reactions are a feature of saline water intrusion in coastal areas. Freshwater in coastal areas is typically dominated by Ca^{2+} and HCO_3^- ions from the dissolution of calcite such that cation exchangers present in the aquifer have mostly Ca^{2+} adsorbed on their surfaces.

REVIEW QUESTIONS

1. How does the polarity of water molecules affect its solvent capabilities in groundwater systems?
2. Describe the common methods used to analyze the chemical composition of groundwater samples.
3. What are the main stages of hydrochemical evolution in groundwater systems?
4. Describe the use of Piper diagrams and Stiff diagrams in the graphical presentation of hydrochemical data.
5. Define chemical equilibrium in the context of groundwater chemistry.
6. What is the difference between the kinetic and equilibrium approaches to chemical reactions in groundwater?
7. How does the concept of Gibbs free energy relate to chemical equilibrium in groundwater?
8. Discuss the role of ion exchange in modifying the chemical composition of groundwater and its applications in water treatment.

REFERENCES

1. Andrews, J. E., Brimblecombe, P., Jickells, T. D., Liss, P. S., & Reid, B. J. (2004). *An Introduction to Environmental Chemistry* (2nd ed.). Blackwell Science.
2. Appelo, C. A. J., & Postma, D. (1994). *Geochemistry, Groundwater and Pollution*. A.A. Balkema.
3. Chapelle, F. H. (1993). *Ground-Water Microbiology and Geochemistry*. John Wiley.
4. Domenico, P. A., & Schwartz, F. W. (1998). *Physical and Chemical Hydrogeology* (2nd ed.). John Wiley.
5. Fetter, C. W. (2001). *Applied Hydrogeology* (4th ed.). Pearson Higher Education.
6. Freeze, R. A., & Cherry, J. A. (1979). *Groundwater*. Prentice-Hall.
7. Krauskopf, K. B., & Bird, D. K. (1995). *Introduction to Geochemistry* (3rd ed.). McGraw-Hill.

8. Lloyd, J. W., & Heathcote, J. A. (1985). *Natural Inorganic Hydrochemistry in Relation to Groundwater: An Introduction*. Clarendon Press.

9. Mazor, E. (1997). *Chemical and Isotopic Groundwater Hydrology: The Applied Approach* (2nd ed.). Marcel Dekker.

10. Stumm, W., & Morgan, J. J. (1996). *Aquatic Chemistry: Chemical Equilibria and Rates in Natural Waters* (3rd ed.). John Wiley.

CHAPTER 5

Environmental Isotope Hydrogeology

LEARNING OBJECTIVES

After studying this chapter, you will be able to:

- Understand stable water isotopes and the physical environment
- Discuss on stable isotopes of water and nitrate for the identification of groundwater flowpaths
- Define age dating of groundwater

KEY TERMS FROM THIS CHAPTER

Anthropogenic nitrogen	Aquifer material
Evapotranspiration	Meteoric water line (MWL)
Geochemical tracers	Groundwater flowpaths
Groundwater recharge	Hydrograph separation
Isotope hydrology	Isotope ratio infrared spectroscopy (IRIS)
Oxygen isotopes	Soil salinization
Subglacial topography	Water isotopes
Water molecul	Isotope ratio mass spectrometry (IRMS)

INTRODUCTION

Based on isotope variations found in natural waters, environmental isotope hydrology is a relatively new field of study. These isotopic properties have been determined over a large temporal and spatial range. Although humans cannot control them, they can be observed and analyzed to provide important regional data on the source, turnover, and transit time of water in the system—information that is frequently not possible to obtain through other means. Comparing the expense of these investigations to that of traditional hydrological studies, they are typically quite inexpensive.

The stable isotopes deuterium (hydrogen-2), carbon-13, oxygen-18, and the radioactive isotopes tritium (hydrogen-3) and carbon-14 are the primary environmental isotopes of hydrological interest. Because the concentrations of hydrogen and oxygen isotopes are typically not affected by interaction with the aquifer material, they make excellent geochemical tracers of water. However, the material of the aquifer may interact with carbon compounds found in groundwater, making it more difficult to interpret carbon-14 data.

5.1. STABLE WATER ISOTOPES AND THE PHYSICAL ENVIRONMENT

Stable water isotopes are a very useful tool for studying the physical world. Analyzing the isotopic composition of water molecules can help address two of the most significant environmental issues that will impact human development in the twenty-first century: water resources and man-induced climate change.

The same chemical species is represented by atoms that have an equal number of protons and electrons. The amount of neutrons separating them, however, can make a difference. Since their masses differ, even though they are isotopes of the same element, they have different physical properties despite having the same chemical properties.

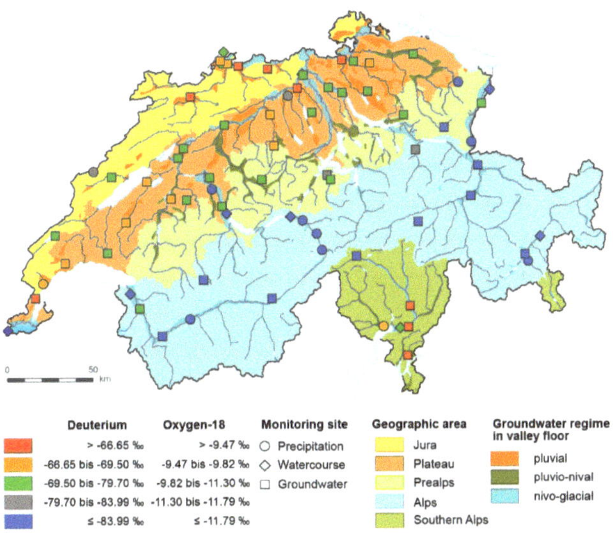

Figure 5.1. *Deuterium and oxygen-18 in precipitation, watercourses and groundwater, geographic area and groundwater regime in valley bottoms.*

Source: https://www.bafu.admin.ch/bafu/en/home/topics/water/info-specialists/state-of-waterbodies/state-of-groundwater/stable-water-isotopes.html.

The two isotopes of hydrogen ($^{1}_{1}\text{H}$ or H, with one proton, and $^{2}_{1}\text{H}$ or D, called deuterium, with one proton and one neutron) and the two stable isotopes of oxygen ($^{16}_{8}\text{O}$ and $^{18}_{8}\text{O}$) are the most abundant isotopes that make up the water molecule. The most common water molecules that result from this are $H_2{}^{16}O$, which is by far the most common, $H_2{}^{18}O$, and HDO. The various terrestrial reservoirs—atmosphere, ocean, ice sheets, lakes, rivers, groundwater, or sea ice—have different ratios of these elements. This is the outcome of isotopic fractionations, which are mostly caused by phase shifts (Figure 5.1).

5.1.1. Isotopic Fractionation during Phase Changes

Water vapor, which is produced when water evaporates, has an equilibrium concentration of fewer $H_2{}^{18}O$ or HDO molecules relative to $H_{216}O$ than the liquid water it originates from. The ratio of the heavy to the light isotope in the liquid phase to the same isotopic ratio in the vapor phase, or equilibrium fractionation coefficient, determines the partition. Since the vapor pressures of the various water molecules are inversely correlated with their mass, these coefficients—whether for deuterium or oxygen 18—have values greater than one.

Water's isotopic composition is typically referenced to Standard Mean Ocean Water, known as SMOW. This composition is measured in delta units (d) per mil.

$$\delta^{18}O = \frac{{}^{18}O/{}^{16}O(\text{sample}) - {}^{18}O/{}^{16}O(\text{SMOW})}{{}^{18}O/{}^{16}O(\text{SMOW})} \times 1000$$

and

$$\delta D = \frac{D/H(\text{sample}) - D/H(\text{SMOW})}{D/H(\text{SMOW})} \times 1000$$

The isotopic composition of standard sea water is thus $\delta^{18}O = \delta D = 0\%$.

Water vapor's d-values are negative because it loses heavy isotopes (D or ^{18}O) during the evaporation of seawater. In contrast to vapor, condensation in the atmosphere leads to an enrichment of heavy isotopes. As a result, following a first precipitation event that removes condensed water, the residual water vapor becomes more negative, which causes subsequent precipitation events to become increasingly impoverished in heavy isotopes. Therefore, as one gets farther away from the vapor source, there is a reservoir effect that causes more precipitation to have negative d-values. In arid tropical regions with high temperatures and low relative humidity, oceanic water evaporation is the primary source of water vapor. Thus, there is a latitude effect: precipitation in the temperate regions ($\delta^{18}O$ values around -5 to -10%) is less depleted in heavy isotopes than rain in the tropics ($\delta^{18}O$ values between 0 and -5%), while snow in the polar regions will be more depleted in heavy isotopes ($\delta^{18}O$ values around -20 to -50%). Thus, the Polar regions have the lowest d-values for precipitation. Likewise, precipitation at higher altitudes indicates a later stage in the evolution of an air mass forced to climb a mountain slope than does precipitation at lower altitudes. This phenomenon is known as the altitude effect. As a result, at higher altitudes, the precipitation d-values are more negative.

For both oxygen 18 and deuterium, the equilibrium fractionation coefficient is inversely correlated with temperature, being larger at lower temperatures. Since the isotopic composition of the ice layers found in polar ice sheets represents past precipitation that has occurred at different temperatures, this effect can be used for palaeoclimatic reconstruction.

A more advanced approach could be to conduct a co-isotopic analysis in both δD and $\delta^{18}O$, rather than focusing on the water's isotopic composition in either oxygen or hydrogen. Rainwater samples are arranged in a δD-$\delta^{18}O$ diagram along a straight line with a slope of approximately 8, known as the Meteoric Water Line (MWL), which is determined by the equation $\delta D = 8\ \delta^{18}O + d$. This equation varies only in the value of the independent term d., which is known as the deuterium excess, from one particular region to another. The deuterium excess has a higher value if the temperature and/or relative air humidity are higher in the vicinity of the vapor source. For example, deuterium excess is approximately 22% in rainwater from the eastern Mediterranean region, 13–15% in the western Mediterranean basin, and 10% in the Atlantic Ocean region west of Portugal. Therefore, under favorable conditions, the deuterium excess value can be used to determine the origin of precipitation.

Let's collect a water sample that falls on the local Meteoric Water Line and allow it to partially evaporate. Due to varying diffusivities in air of the different isotopic water molecules, a kinetic isotope effect will be observed during this phase change. This effect is more pronounced at higher temperatures and lower air relative humidity levels. The remaining water samples will be arranged in a δD - $\delta^{18}O$ diagram on a straight line with a slope less than 8, known as the evaporation line. The distance along this line from a remaining water sample to the intersection with the local MWL depends on the extent of evaporation.

Another process that leads to isotopic fractionation is the phase transition from liquid water to ice. However, the ratio of the enrichment factors 1 for oxygen 18 and deuterium is less than 8. Therefore, it stands to reason that ice samples formed when liquid water freezes will line up on a straight path with a slope that is smaller than the MWL. Research has demonstrated that this so-called freezing slope actually exists. A finite reservoir freezes, causing selective incorporation of heavy isotopes into the ice, which gradually reduces the amount of heavy isotopes in the remaining water. As a result, along the freezing slope, successively formed ice layers exhibit an increasing number of negative d-values. This freezing slope is also influenced by the water's initial isotopic composition; it is lower when the water has more negative d-values at the beginning (Souchez and Jouzel, 1984).

5.1.2. Use of Water Isotopes in the Hydrological Study of a Drainage Basin

Pollutants such as nitrates and phosphates can be found circulating in both ground and surface water within a drainage basin. The relationship between discharge and nitrate concentration can indicate the importance of runoff waters during times of high discharge. Conversely, an inverse relationship between discharge and phosphate concentration may point to the significance of groundwater in transporting these compounds, particularly when river discharge is low. It is valuable to determine the specific contributions of groundwater and runoff water, including surface and subsurface runoff, to river discharge during a specific time period. Through the use of water isotopes, it is often possible to calculate these contributions at a gauging station after accounting for direct rainfall into the river.

Infiltration is the process by which rainwater penetrates from the soil surface into the ground through cracks and pores in the soil. This helps to reduce fluctuations in the seasonal $\delta^{18}O$ signal as the water moves deeper into the ground. There is a critical depth, typically around 10 meters, below which groundwater maintains a constant isotopic composition over time. This lack of isotopic variation at greater depths is a result of the extended residence time of water in both the soil and groundwater layers. The groundwater in the drainage basin of a small river that empties into Lake Erie in North America consistently has a $\delta^{18}O$ value of -11.5%. A summer flood with a δ18O value of -6.8 was observed at the gauging station as a result of heavy rain. Due to the short residence time, this isotopic composition is also that of the runoff waters. The river reached a maximum value of -9.3% during peak discharge and a δ18O value of -10.8% during the flood decrease within the fifteen days that were taken into consideration for the study. Estimating runoff and groundwater contribution at peak discharge and at the point when the river's $\delta^{18}O$ reached -10.8% is possible with a basic isotope mass balance. These contributions to the river discharge from groundwater were 53% and 85%, respectively. A groundwater discharge curve can be created, and the storm or runoff flow can be assessed at each point, as shown in Figure 5.2.

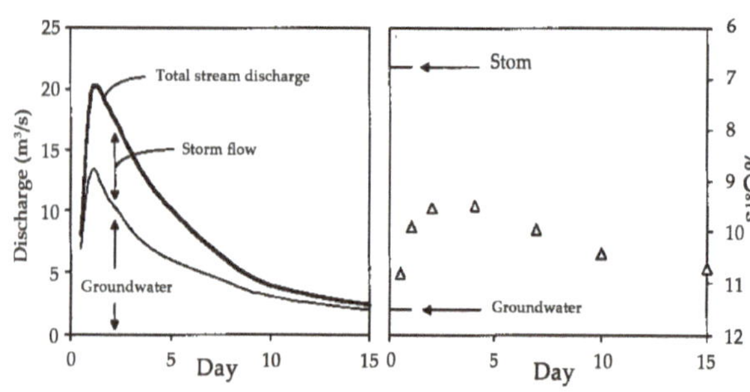

Figure 5.2. *Storm hydrograph separation for a two-component system using δ18O - drainage basin north of Lake Erie.*

Source: https://journals.openedition.org/belgeo/docannexe/image/16199/img-4.jpg.

5.1.3. Evaporation, Transpiration, and Salinization in an Arid Area

The significance of the derived kinetic isotopic effect during evaporation and the physical conditions of the air above the water determine how residual water samples align in a δD - $\delta^{18}O$ diagram on an evaporation line that slopes less than the local MWL. From the results of a pan experiment, where evaporation occurs in a small water reservoir and successive d-values are measured in the residual water, such an evaporation line can be established for a particular environment. Because evaporation promotes an increase in salinity, it is useful to plot the residual waters' δD values against their respective chlorine contents. A section of irrigated fields in the Nile Delta underwent this process. In contrast, there is no apparent isotopic fractionation as a result of transpiration from the vegetation. In fact, it can be demonstrated that the leaf's limited water reservoir is enriched in heavy isotopes, leading to the production of water vapor with an isotopic composition identical to that of soil water when a steady state is reached. Transpiration can thus be understood as a process devoid of obvious isotopic fractionation.

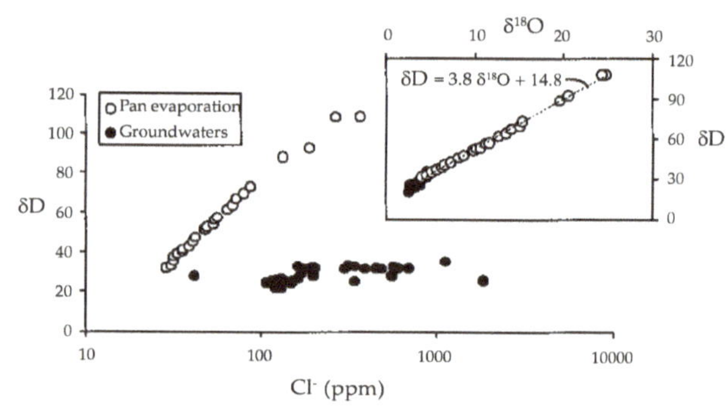

Figure 5.3. *δD (= d^3H) versus chlorine for Nile Delta agricultural drainage and evaporation pan experiment in the same region. Inset plot shows the slope of the evaporation line for the pan data.*

Source: https://journals.openedition.org/belgeo/16199.

As a result, water that has experienced transpiration over time displays an increase in chlorine content along with the same isotopic composition. Water samples from these irrigated fields in Egypt now exhibit this behavior, suggesting that transpiration plays a major role in the rise in dissolved chlorine. Here, plant transpiration—rather than direct evaporation of the water in the irrigation canals—is primarily responsible for the salinization of the soil that occurs after water losses to the vapor phase. Thus, water isotopes have proven to be a useful tool for examining the mechanisms behind soil salinization in this arid region (Figure 5.3).

5.1.4. Recharge of Aquifers

The isotopic composition of aquifers in mountainous areas may be influenced by the altitude effect. For example, the aquifer used by Evian, a company that produces mineral water in the French Alps near the town of the same name, is impacted by this phenomenon. This aquifer is replenished by precipitation from water vapor that ascends the steep slopes bordering the southern edge of Lake Geneva.

The low nitrate content of mineral water is a crucial parameter. It is important to anticipate the impact of human activities like pastoralism on groundwater quality. Avoiding areas affected by unfavorable human economic practices is a top priority for this company. By mapping the distribution of $\delta^{18}O$ in precipitation, recharge zones have been identified and planted with trees to prevent nitrate-producing pastoral activities in these key areas. An additional instance of the efficacious application of stable isotope analyses in aquifer research can be discovered in an entirely distinct setting: the Negev region situated in the Middle East. Similar to other countries in the eastern Mediterranean, this region has a deuterium excess of rain of approximately 22. Nonetheless, the aquifer has a deuterium excess of roughly 10%. Currently, it is very uncommon for rains with such a low deuterium excess to occur; instead, they are only caused by strong south westerly winds that carry 10% of the region's water vapor from the Atlantic. The frequency of these meteorological conditions was significantly higher during the Last Ice Age. Since the aquifer is thought to have formed during that period, the majority of its contents are fossilized water. This isotopic study indicates that there is a significant chance that this aquifer will be overused in this particular instance.

5.1.5. Antarctic ice as Climatic Archive

An incredible record of our surroundings is found in polar ice. Notably, this is the result of the subsequent characteristics.

First, the temperature at which the ice crystals formed determines their oxygen or hydrogen isotopic composition. Polar precipitation exhibits a seasonal effect because the equilibrium fractionation coefficients for deuterium or oxygen 18 are temperature dependent. Compared to summer snow, winter snow has more negative d-values. Coincidentally, by counting the d-values' annual fluctuations, this can be used to date the ice. Therefore, over longer time periods, a colder climate is favorable for more negative d-values, whereas a warmer climate is associated with fewer negative d-values. It is possible to reconstruct the exact temperature of a previous period from the d-values of an ice layer that originates from the snow of that period because there is an excellent correlation between the surface temperature and the d-values of the snow precipitated in Polar Regions. An ice layer's isotopic composition can be thought of as a palaeothermometer of sorts.

Second, air that was originally in the snow cover's pores gradually becomes trapped as gas bubbles inside the ice during the formation process. These bubbles' composition reveals the makeup of the atmosphere at the moment of their formation. The composition of the air present in an ice layer, either as gas bubbles or as gas hydrates (in the high pressure deeper parts of the ice sheet), thus reflects the composition of the atmosphere of the corresponding period because ice sheets are made up of ice from successive glacial and interglacial periods. Reconstructing the atmospheric composition of the past is thus made possible through the analysis of gases extracted from an ice layer, such as greenhouse gases.

The Vostok ice core was extracted through deep drilling near the Russian Vostok station in East Antarctica. Situated on an ice plateau far from the coast, the station sits at an altitude of approximately 3500 meters above sea level. The ice core itself measures about 3700 meters thick, with a climatic record of excellent quality extending to a depth of 3310 meters, where the ice is 420,000 years old. The time scale provided by the core relies on the rate of annual snow accumulation, which is very low at Vostok, resulting in a broad time span. Ice layers become thinner with depth due to deformation, with a single year's accumulation near the surface corresponding to a 25 mm thick layer, compared to a 3 mm layer at 3000 meters depth. As a result, fine resolution of climatic events at depth is not feasible. While the Vostok ice core highlights significant climatic events over the past 420,000 years, it does not reveal variations of low periodicity.

The strong correlation between the temperature inferred from the ice's isotopic composition and the concentration of greenhouse gases (CO_2 and CH_4) is one of the most remarkable aspects of the Vostok ice core's climatic record over the last 420,000 years. The climate system oscillates between an interglacial state, where temperatures are 8 to 10°C lower than they are today, and a glacial state, where temperatures are closer to current values (or 1 to 2°C higher). This oscillation occurs once every 100,000 years. This oscillation is a result of variations in insolation caused by the periodicity of the eccentricity of the earth's orbit around the sun. Significant climatic variations occur on a timescale of 100,000 years due to orbital forcing, but this is not the only factor to take into account. It is true that greenhouse gases have a powerful amplifier. The composition of greenhouse gases in the atmosphere fluctuates between two well-defined values between glacial and interglacial periods: roughly 180 ppmv CO_2 during each glacial period and 280–300 ppmv CO_2 during each interglacial period, while the typical values for CH_4 are 320–350 ppbv and 650–770 ppbv. When it comes to either temperature or greenhouse gas concentrations, a sawtooth pattern is evident: the rise towards interglacial levels is faster than the fall towards glacial minimum values. The Vostok ice core data demonstrate that greenhouse gases unquestionably affect the planet's climate.

In the previous 420,000 years, the atmosphere's current CO_2 concentration of roughly 365 parts per million has never been attained. Therefore, there is little question that man is experimenting in the atmosphere through his actions, something that nature has never done before in this era.

5.1.6. Subglacial Lake Vostok

A subglacial lake called Lake Vostok is located near the ice-bedrock interface near the Russian station mentioned earlier. Through the use of aerial radio echo sounding, it has been found. Certain wavelengths of radar radiation pass through ice. These wavelengths make it possible to identify subglacial water masses like Lake Vostok and to define the subglacial topography. This lake is roughly half the size of Belgium, covering an area of 14,000 km^2, and has a depth of 700 m. Its dimensions are comparable to those of Lake Ontario, one of the Great Lakes in the United States. Its ice ceiling is tilted, with 750 meters below sea level under 4300 meters of ice in the north and only 250 meters below sea level under 3750 meters of ice in the south, close to Vostok Station. It is elongated in a north-south direction. The Vostok ice core no longer exhibits an intact climate and environmental record below 3310 meters below the surface. This deepest region of the ice core has been studied by our research team. Arguments suggest that the isotope climatic signal is modified by intense ice deformation between 3310 and 3539 meters below the surface. This explains why isotope variations (δD in this case) are dampened in this depth interval. The characteristics of the ice abruptly change below 3539 meters. Among other things, the deuterium excess values go from 14 to 7 or 8, the gas content drops to almost zero, and the d-values rise dramatically at the transition. The deuterium excess has sharply decreased at 3539 m depth, and the other features discussed above suggest that Lake Vostok waters have frozen, resulting in the ice below this depth. As previously mentioned, the deuterium excess in an ice sample can be calculated by subtracting eight times the $\delta^{18}O$ value from its δD value. Instead of aligning on the MWL, where ice layers from snow precipitation at the ice sheet surface are aligned, samples of ice from water freezing—as opposed to dry metamorphism of snow—are aligned on a freezing slope in a $\delta D - \delta^{18}O$ diagram. Because of the refreezing of water in a large reservoir such as Lake Vostok, ice samples enriched in heavy isotopes (exhibiting higher d-values) have all had a lower deuterium excess since the freezing slope has a value less than 8.

Additionally, when water freezes, almost no gas remains in the resulting ice. Over 200 meters of lake ice are thought to be present above the lake's liquid surface, according to calculations. Since the lake's ice ceiling is lower at the base of the ice sheet, melting occurs there as part of the water circulation process, where there is increased pressure. The resulting meltwater rises along the ice-water interface's tilt and gets supercooled. Frazil ice crystals form in the water plume and accumulate at the ice sheet's base. This results in lake ice in areas with higher ceilings, i.e., Under Vostok Station, there is a decrease in pressure. Once the presence of lake ice was established, a detailed investigation was conducted to determine whether any microorganisms that could survive in the lake water had found their way into the lake ice. Thus far, disparate teams have only produced contradicting outcomes (Souchez et al., 2000)

5.2. STABLE ISOTOPES OF WATER AND NITRATE FOR THE IDENTIFICATION OF GROUNDWATER FLOWPATHS

Determining the flowpaths of groundwater can yield significant insights into the flow of surface water and the contaminants that interact with it. For instance, pollutants may be released into streams directly, but if they recharge into groundwater, which subsequently flows into streams indirectly, the groundwater may greatly improve the stream's water quality. Specifically, because nitrate is highly mobile and mostly comes from nonpoint source pollution, it can be challenging to identify the source because it is dispersed over a large area through a variety of groundwater flowpaths. Thus, it's critical to comprehend the groundwater flowpath in order to prevent the spread of pollutants and clean up contaminated stream water.

While traditional concentration-based chemical analyses like total nitrogen (TN), total phosphorus (TP), total organic carbon (TOC), biochemical oxygen demand (BOD), and chemical oxygen demand (COD) have been used to track mixed contamination effectively, they are limited in their ability to monitor contaminant movement and physical processes. In contrast, stable isotope analysis is a valuable tool for identifying sources, understanding processes, and determining contributions of different inputs. Specifically, stable water isotopes ($\delta^{18}O$ and δD) are influenced by meteorological processes, providing a unique signature of their origin critical for investigating groundwater sources. The stable isotopes of nitrogen and oxygen in nitrate ($\delta^{15}N$ and $\delta^{18}O$) are also essential for pinpointing sources of nitrate contamination due to their distinct isotopic values across different sources (Xue et al., 2009).

Stable nitrate isotopes have been used to identify nitrate sources since nitrate contaminants became an environmental concern in the 1970s, while stable water isotopes have been used as tracers in hydrograph separation studies since Craig's groundbreaking work. As of right now, nitrate is a prevalent groundwater contaminant that poses a major risk to drinking water supplies and aids in the eutrophication of surface waters. The predominant nitrogen species in groundwater is nitrate, which can be obtained from atmospheric precipitation, livestock manure, synthetic fertilizers, soil organic nitrogen, and sewage effluent. In certain regions, the amount of anthropogenic nitrogen deposited into the atmosphere surpasses the nutrient requirements of the ecosystem. Additionally, the impact of atmospheric deposition on nitrogen export during transient discharge events like precipitation and melting has not been thoroughly studied.

A helpful method for figuring out the water flowpath and the source of nitrates is isotopic hydrograph separation, which uses stable isotopes in water and nitrate. This method has been extensively employed to determine the percentage of various water sources that contribute to stream water, allowing for the deduction of the flowpath and residence time. The flowpath of groundwater into streams or rivers can be identified, in particular, by differentiating between nitrate sources like direct atmospheric deposition and biological assimilation and release in the soil zone. As a result, the dual isotopic technique—the isotopic analysis of nitrogen and oxygen in nitrates—has been applied in numerous studies to determine the origin of nitrate. To illustrate, Böttcher et al., identified the nitrate sources in a downgradient groundwater from an agricultural area, and Durka et al., subsequently identified the nitrate sources in a German state of Bavaria that was left undisturbed. Because nitrate sources like event water (rainfall or snowmelt), soil water, and groundwater have distinct isotopic signatures, it is possible to identify the source of nitrate in stream water using the dual isotope approach.

Understanding how precipitation enters soil water or recharges groundwater before being released into stream water is crucial for researching the hydrograph separation of stable water isotopes. To this end, in order to establish a foundation for comprehending and forecasting the movement of contaminants in the groundwater flowpath, research on stable water isotopes in the atmospheric source must first be carried out. In order to better understand the movement of water components like groundwater, precipitation, snowmelt, and soil water in stream water, numerous studies have been carried out over the past 40 years using the hydrograph separation technique using stable water isotopes or conservative chemical tracers. Especially, Ladouche et al., used hydrograph separation with stable water isotopes, major chemical parameters, and dissolved organic carbon (DOC) to study the components of streamflow.

Dahlke along with others utilized the stable oxygen isotope ($\delta^{18}O$) value of water to show that, in the 30% glaciated sub-arctic catchment of Tarfala, northern Sweden, pre-event water constituted the majority of storm runoff. Later on, Rahman et al., described the daily variation of runoff components in the Alpine watershed using an end-member mixing analysis, and Kim et al., employed isotope and chemical tracers to determine the effect of a granitic watershed with a thin soil layer's pre-event water component.

Measuring the stable isotopic compositions of precipitation, stream water, and groundwater samples is a key component of isotope hydrology. These measurements are then interpreted to help conceptualize and quantify the groundwater flowpath, velocity profile, and hydrogeochemical and biogeochemical reactions. Water movement among reservoirs, such as evapotranspiration, groundwater recharge, discharge, and runoff, has been identified with the use of isotopic approaches in conjunction with more traditional hydrogeological and hydrogeochemical data, such as information on lithology, meteorology, and solute concentrations.

5.2.1. Hydrograph Separation

The process of separating the components of streamflow in a small catchment area or watershed into two or more distinct components is known as hydrograph separation. For instance, Dincer et al., initially proposed the use of isotopic tracers for isotopic hydrograph separation. was created by Sklash and Farvolden, and numerous studies have assessed it. The isotopic hydrograph separation technique is predicated on the idea that, following precipitation, two elements enter the stream: (1) runoff from rainwater, or new water, and (2) groundwater, or old water. A two-component mixing model was used to divide the stream water discharge into components that are groundwater and rainwater. You can apply the mass balance formulas listed below, which were presented by Sklash and Farvolden.

$$Q_t = Q_r + Q_g \tag{1}$$

$$C_tQ_t = C_rQ_r + C_gQ_g \tag{2}$$

$$x = \frac{C_t - C_g}{C_r - C_g} \tag{3}$$

where Q denotes each component's discharge, C is the concentration of an observed tracer or isotopic composition, t, r, and g stand for total discharge, groundwater, and rainwater, respectively, and *xr* is the proportion of stream water that rainwater $(x_r = \frac{Q_r}{Q_t})$. contributes.

The application of these mass balance equations is based on the following four assumptions: (1) the concentration of tracers in rainwater and groundwater differs significantly; (2) the isotopic compositions and concentrations of the tracers for groundwater and rainwater are constant in space and time; (3) for two-component hydrograph separation, the concentrations of each tracer are equivalent in groundwater and vadose water, or else the contribution of vadose water is negligible; and (4) surface storage contributes minimally during the runoff. If these presumptions hold true, the amounts of stream water contributed by groundwater and rainwater can be ascertained using two-component hydrograph separation. If not, three or more component hydrograph separation needs to be done. For instance, hydrograph separation of the three components of runoff, soil water, and groundwater should be used if the amount of vadose water in the saturation zone is significant and needs to be considered. Depending on the geological features, soil water in two-component systems can be interpreted as either groundwater or runoff. Hydrograph separation of the three components (soil water, rain or snowmelt, and groundwater) should be used when evaluating the soil water among the factors contributing to the stream water after rainfall or snowmelt.

Hydrograph separation of the three components is basically expressed in the form of a three-way linear system of equations, which can be interpreted as follows:

$$Q_t = Q_r + Q_g + Q_s \tag{4}$$

$$C_t = C_t\frac{Q_r}{Q_t} + C_g\frac{Q_g}{Q_t} + C_s\frac{Q_s}{Q_t} \tag{5}$$

$$I_t = I_t\frac{Q_r}{Q_t} + I_g\frac{Q_g}{Q_t} + I_s\frac{Q_s}{Q_t} \tag{6}$$

where Q denotes each component's discharge, C represents an observed tracer's concentration, I denotes each component's isotopic composition, and the subscripts t, r, g, and s stand for total discharge, rainwater, groundwater, and soil water, respectively. Since solutions for more than three components are difficult to obtain, matrix operation has been applied to the Equations (4)–(6) in the present work as follows:

$$A = \begin{bmatrix} 1 & 1 & 1 \\ C_r & C_g & C_s \\ I_r & I_g & I_s \end{bmatrix}, X = \begin{bmatrix} \frac{Q_r}{Q_t} \\ \frac{Q_g}{Q_t} \\ \frac{Q_s}{Q_t} \end{bmatrix}, B = \begin{bmatrix} 1 \\ C_t \\ I_t \end{bmatrix}. \quad (7)$$

$$AX = B, X = A^{-1}B \quad (8)$$

A three-component hydrograph separation using both isotopic and chemical compositions is made possible by the introduction of a system of linear equations. The matrix can be solved using MATLAB. These are systems of n equations in n + 1 unknowns that are mathematically underdetermined and lack a unique solution. Nevertheless, recently published studies introduce software (IsoSource model) that computes multiple source proportions using mass balance conservation requirements, even with n isotope systems and >n + 1 sources. The IsoSource model can be used to separate the sources of contaminants in wastewater because it is predicated on the idea of stable isotope mass conservation (Lee et al., 2014).

The second aforementioned assumption states that there shouldn't be any variation in the isotopic compositions of rainwater and groundwater over time or space (i.e., no isotopic fractionation), as this would cause deviation. Consequently, a systematic error in the portion of rainwater contributing to the stream will occur if the isotopic composition of groundwater and rain changes over time. Gaussian error propagation can be utilized to ascertain this systematic error. It is well known that groundwater, or old water, has a fairly stable isotope composition. However, there is much more isotopic fractionation in rain or snowmelt (new water), so hydrograph separation based on the mean isotope value produces errors.

The uncertainty of new water generated from isotopic fractionation can be calculated according to the following equation:

$$\Delta x_r = -\frac{x_r}{C_r - C_g} \Delta c_r \quad (9)$$

where Δx_r is the systematic error when new water (rain or snowmelt) contributes to the stream, and Δc_r is the error in c_r. This is the variation in the tracer concentration or the ratio of stable isotopes in the rain (new water). Therefore, according to Equation (9), the error generated when considering the effect of new water on the stream is inversely proportional to the difference of the tracer concentration between new and old water, and directly proportional to the actual contribution of new water (x_r) to the stream water and the tracer concentration of the new water over time (Δc_r).

5.2.2. Stable Water Isotopes

After leaving the ocean, water moves across the continents, evaporates, cools, condenses, and forms clouds before falling as precipitation (rain or snow).

The precipitated water, which includes runoff, groundwater, and streams, is then re-evaporated and recycled. Based on the isotope data from Hoefs and Coplen et al., Figure 5.4 illustrates how the isotopic composition is altered during the transition from ocean to continent through the hydrologic cycle processes of evaporation and rainout. The heavier isotopes in water, H_2HO and $H_2{}^{18}O$, are preferentially concentrated in the more condensed phase (liquid instead of vapor, solid instead of liquid), while the lighter isotope, $H_2{}^{16}O$, is concentrated in the remaining phase when water undergoes a change in physical phase. As a result, the rainout process continuously fractionates heavy isotopes into the precipitation (Rayleigh-like distillation), gradually decreasing the amount of heavy isotopes in the residual vapor.

As a result, compared to earlier precipitations that originated from the same atmospheric water vapor, later precipitations will have a lower concentration of heavy isotopes. Furthermore, because the isotope composition of water varies among the water cycle's components, isotope measurement enables the identification of water masses' sources and the understanding of their interactions. In particular, isotope composition is a helpful tool for figuring out mixing patterns and relative contribution rates via hydrograph separation because stream water has a complex relationship between rainfall (new water) and groundwater (old water).

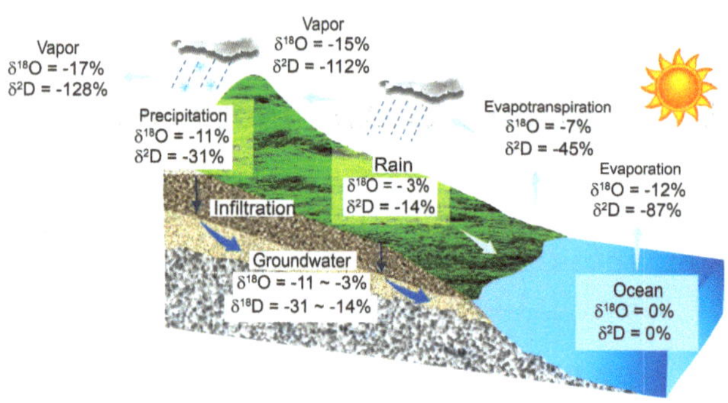

Figure 5.4. *The diagram of isotopic composition change of atmospheric water vapor showing the processes of evaporation and rainout as the air mass proceeds from an ocean to a continent.*

Source: https://www.mdpi.com/2073-4441/12/1/138#B38-water-12-00138.

5.2.2.1. Isotopes in Precipitation

In contrast to other tracers, precipitation events such as rain or snowmelt naturally add stable water isotopes to the watershed. Once inside the watershed, these isotopes are transported by the watershed's natural flow of water. These environmental isotopes, supplied by meteoric processes, can be used to trace and identify the various air and water masses contributing precipitation to the watershed, since the stable isotope compositions of the water only change via the aforementioned mixing and fractionation processes during evaporation and condensation. Furthermore, since precipitation is a primary source of water in the hydrological cycle, it is crucial to comprehend the mechanisms governing the temporal and spatial distributions of precipitation's isotopic composition.

Stable hydrogen and oxygen isotopes generally undergo similar fractionation processes, which accounts for their similar behavior in the hydrological cycle. Friedman was the first to notice that the stable hydrogen and oxygen isotope concentrations present in most meteoric water exhibit covariance as a result of these similarities. This covariance can be explained by the following relationship, which was defined by Craig:

$$\delta^2H = 8\,\delta^{18}O + 10$$

The meteoric water line (MWL), a linear relationship, offers a useful tool for deciphering and tracking the origins of water. Specifically, the global meteoric water line (GMWL) has been defined as an MWL with an intercept of 10 and a slope of 8. The GMWL's slope of 8 can be attributed to the condensation of water vapor in near-equilibrium conditions. The ratio of the fractionation coefficients and variables indicating whether or not evaporation has occurred in the water entering the soil, groundwater, and lakes are related to the slope. The local meteoric water line typically has a slope of less than 8 (usually between 4 and 7) due to the evaporation of soil or lake water.

The y-intercept in a plot of Equation (10), also known as the deuterium excess (or d-excess), is called. Dansgaard defines this as the difference between the isotopic equilibrium during evaporation and the sourced precipitation. It is associated with the relative humidity parameter of the evaporation source. Dansgaard identified four factors—altitude, distance from the coast, latitude, and quantity—that govern this isotope value depletion. The d-excess is helpful in locating moisture source regions because the impacted factors vary depending on the region. Recent findings by Lee et al., examined the findings of earlier research on the significant regional variations in the fractionation of stable water isotopes. For instance, the stable oxygen isotopic value in the New Hampshire region of the United States is 2 to 3%, while the stable hydrogen isotopic value is 10 to 12%. In contrast, the isotopic values in Incheon, Korea, are 20% for oxygen and roughly 60% for hydrogen, and in Jeju Island, Korea, they are 7 to 8% and 50 to 60%, respectively.

The y-intercept in a plot of Equation (10), also known as the deuterium excess (or d-excess), is called. Dansgaard defines this as the difference between the isotopic equilibrium during evaporation and the sourced precipitation. It is associated with the relative humidity parameter of the evaporation source. Dansgaard identified four factors—altitude, distance from the coast, latitude, and quantity—that govern this isotope value depletion. The d-excess is helpful in locating moisture source regions because the impacted factors vary depending on the region. Recent findings by Lee et al., examined the findings of earlier research on the significant regional variations in the fractionation of stable water isotopes. For instance, the stable oxygen isotopic value in the New Hampshire region of the United States is 2 to 3%, while the stable hydrogen isotopic value is 10 to 12%. In contrast, the isotopic values in Incheon, Korea, are 20% for oxygen and roughly 60% for hydrogen, and in Jeju Island, Korea, they are 7 to 8% and 50 to 60%, respectively.

The vapor source of the precipitation, which is usually from adjacent oceanic regions, as well as the processes involved in precipitation formation and air mass trajectory—that is, the impact of the vapor source and rainout processes along the air mass's path—all affect the isotopic compositions of precipitation. The majority of these variables have a connection to isotopic fractionation via diffusion during physical phase transitions like melting, sublimation, condensation, and evaporation.

1. Isotopic Evolution of Snow

In alpine settings, groundwater recharge is primarily facilitated by snowmelt. Water-resource management requires an understanding of the hydrological responses of snowmelt that contributes to the watershed because snow dynamics are highly variable in space and time. Although the isotopic composition of the snowpack profile typically reflects the unique isotopic composition of individual precipitation events, isotopic exchange, snowpack metamorphism, and surface sublimation attenuate the signal in the snow layers provided by these individual events. Two main processes contribute to the isotopic composition of snowmelt that is produced from a snowpack: (1) sublimation and molecular exchange between vapor and the snowpack, and (2) meltwater infiltration and exchange with snow and meltwater within the snowpack.

Regarding the first process, Cooper noted an exception where the snowpack does not contain the well-mixed conditions of a water body, but Moser and Stichler indicated that the isotopic fractionation associated with sublimation of snow surfaces behaves similarly to that of evaporating water. In the second process, as the melting progresses, the meltwater gradually becomes enriched in heavy isotopes after initially being depleted of them in comparison to the remaining snowpack. As liquid water seeps through the snowpack, isotopic exchange between the ice and liquid water occurs, leading to this isotopic evolution. Hydrograph separations based on the isotope composition of the bulk snow will therefore be incorrect because the isotopic compositions of snowmelt and bulk snow are typically different. The isotopic evolution of a snowpack is important for hydrological and climatic research because, in temperate regions, snowmelt contributes significantly to surface runoff and groundwater.

It is reasonable to anticipate complex changes in isotopic compositions to transpire between accumulation and melting, as evidenced by studies conducted on both artificial and natural snowpack. Sublimation and exchange with atmospheric water vapor significantly modify the isotopic composition of the upper snow layers. The predictable result of evaporation is enrichment in $\delta^{18}O$ and $\delta^{2}H$ in the snowpack. Once theoretical fractionation models are adjusted to account for the less-than-well-mixed conditions of the natural snowpack, they can reasonably predict the effects of simple evaporation. These models were developed for evaporation from well-mixed water bodies. But unlike simple surface evaporation, isotopic change in the snowpack is more complex and depends on a number of variables, including air temperature, relative humidity, vegetation cover, soil temperature, and the amount of time the snowpack is on the ground. Mast and associates demonstrated that while the majority of the water in Andrews Creek was fresh water from snowmelt (based on hydrograph separation using $\delta^{18}O$), a significant amount of that water had been transported via subsurface flowpaths before coming into contact with the stream, and significant interaction had taken place with soil or materials that resembled soil (based on hydrograph separation using dissolved silica). It has been discovered that a combination of the microbial cycle, nitrate flushing, and nitrates directly from rain or snowmelt causes the highest nitrate concentrations in springs and streams.

5.2.2.2. Stable Water Isotope Measurements

Isotope ratio mass spectrometry (IRMS) is the primary method used to determine the stable isotope composition of water. This method uses mass differences to calculate the relative isotope ratios of molecular compounds. A beam of charged molecules is created (often by thermally ionizing gaseous samples) and bent in a magnetic field to produce a mass spectrum. In general, chemical pretreatment is needed for stable isotope analysis of water using IRMS. For instance, hydrogen isotopes require reduction with metals like uranium, zinc, platinum, and chromium, while oxygen isotopes require ion-exchange between H_2O and CO_2 via bicarbonate reactions. As a result, CO_2 is used to analyze the oxygen isotope composition and H_2 is used to analyze the hydrogen isotope composition. Alfred Nier created the first dual-inlet mass spectrometer in the late 1940s. However, the traditional offline methods for sample preparation take a lot of time, and the investigator's skill level determines the analytical precision. Due to these factors, the traditional dual inlet technique was modified to create the continuous-flow isotope ratio mass spectrometer. This device delivers a trace amount of the gas for analysis in the form of a stream of helium carrier gas.

Recently, laser-based methods for the analysis of stable water isotopes have been developed: isotope ratio infrared spectroscopy (IRIS). This method looks at the properties of vibration-rotation transition-related water absorption in the near-infrared wavelength range. These transitions are dependent on the substitution of ^{18}O and ^{2}H in H_2O gas molecules. The isotope ratio can be determined because of the direct relationship between these molecular motions and the proportion of isotopes. Off-axis integrated cavity output spectroscopy (OA-ICOS) and wavelength-scanned cavity ring-down spectroscopy (WS-CRDS) are two subtypes of the IRIS technique. The IRIS technique has several advantages over conventional IRMS, including ease of preparation and operation, relative portability for field application, and applicability with relatively small amounts of water samples (ppb, ppt). On the other hand, the presence of O-H bonding dissolved organic molecules has the drawback of reducing analytical performance because of spectral interferences with water molecules.

5.2.3. Stable Nitrate Isotopes

5.2.3.1. Pretreatment Method for Nitrate Isotope Analysis

Because the various sources of nitrates have isotopically distinct compositions of $\delta^{15}N$ and $\delta^{18}O$, dual isotope analysis of nitrates ($\delta^{15}N$ and $\delta^{18}O$) can be an effective tool for identifying nitrate sources and nitrate cycling mechanisms in stream water. A number of pretreatment techniques have been developed in the last few decades to concentrate dissolved nitrates in preparation for dual isotope analysis. Before recently, nearly all nitrates for $\delta^{15}N$ and $\delta^{18}O$ analyses were made by modifying the silver nitrate technique, which involves concentrating samples on anion exchange resins, eluting them, and purifying them to yield $AgNO_3$. In order to produce N_2 gas for $\delta^{15}N$ analysis by IRMS, the $AgNO_3$ obtained from freeze drying is combined with a catalyst made of CuO/Cu wire/CaO and heated to 850 °C in a sealed reactor. In the meantime, $\delta_{18}O$ is examined by combining graphite (graded for spectroscopic analysis) with $AgNO_3$ to produce CO gas through pyrolysis or CO_2 gas through total combustion. Studies from alpine, agricultural, and urban settings have successfully employed and published the combined techniques. The ion exchange technique mentioned above has the following benefits: minimal isotope fractionation of nitrate during ion exchange; ease of transport and storage; and direct field applicability. The lengthy and expensive sample preparation process, as well as interference from the presence of other anions (Cl^-, SO_4^{2-}) in the sample, are drawbacks of the ion exchange method. Furthermore, a sizable sample size is necessary for accurate analysis.

Denitrification, which involves inoculating a pure culture of denitrifying bacteria devoid of the enzyme necessary to reduce nitrate beyond N_2O, is an additional nitrate pretreatment technique. After that, IRMS analyzes the gas. Compared to the silver nitrate method, microbial denitrification requires less sample and saves time and money on sample preparation. However, this method takes a long time to cultivate the microorganisms, and the presence of toxic substances (pesticides, heavy metals, and antibiotics) in the sample affects the microorganisms' activity. Furthermore, the N_2O gas's composition may be distorted by the presence of NO_2^-. McIlvin and Altabet described an advanced method of chemically reducing nitrate to N_2O gas in order to solve these problems. This method uses a 1:1 azide and acetic acid solution to convert nitrate (NO_3^-) to nitrite (NO_2^-) via cadmium reduction, and finally to nitrous oxide (N_2O). Analyses of the N_2O gas follow the same protocol as the microbial denitrification method. With just a small amount of sample needed for analysis, this chemical reduction method can drastically cut down on the time and expense associated with sample preparation. Furthermore, in contrast to microbial denitrification, it is unaffected by any harmful materials present in the sample. However, as with microbial denitrification, there is a chance of exposure to hazardous chemicals (sodium azide, cadmium) during the sample pretreatment, and incorrect data may be obtained because of the NO_2^-in the sample.

More recently, besides IRMS, measurements of $\delta^{15}N$ and $\delta^{18}O$ from the headspace N_2O gas are analyzed in a N_2O triple isotope analyzer using laser absorption spectroscopy after N_2O produced by conversion of NO_3^- by earlier mentioned pretreatment. The instrument measures N_2O concentrations (0.3–20 ppm), and $\delta^{15}N_\alpha$, $\delta^{15}N_\beta$, $\delta^{15}N_{bulk}$, $\delta^{17}O$, $\delta^{18}O$, and H_2O values in air to precisions of 0.03 ppb for N_2O, less than ±1% (SEM) for N and less than ±2% (SEM) for O isotopes over 300 s of measurement integration. However, the laser spectrometry technique is lower precision and accuracy than IRMS technique (less than 0.2% for $\delta^{15}N$-NO_3^-, 0.5% for $\delta^{18}O$-NO_3^-).

5.2.3.2. Identification of Contaminant Source Using Nitrate Isotopes

The nitrate $\delta^{15}N$ and $\delta^{18}O$ isotope ratios analyzed reveal unique values for each source of contamination. The $\delta^{15}N$ value in atmospheric NO_3^- typically falls within the range of -15% to +15%, reflecting a wide variation due to various factors such as chemical reactions, seasons, weather conditions, anthropogenic activities, pollution proximity, and distance from the ocean. Synthetic nitrogen fertilizers exhibit δ15N values between -4% and +4%, while soil nitrogen ranges from -2% to +5%. Conversely, manure and sewage may show enriched $\delta^{15}N$ levels, reaching +10% to +20% due to specific processes like volatilization of ammonia and oxidation of residual waste.

The $\delta^{15}N$ serves as a crucial indicator for nitrates in the environment, yet its utility alone is limited in distinguishing nitrogen sources and cycles given the overlapping values among precipitation, soil, fertilizer, manure, and sewage. Therefore, the combination of $\delta^{15}N$ and $\delta^{18}O$ analysis enhances the accuracy of identifying and separating nitrate sources, reducing the uncertainty associated with nitrogen isotopes (Figure 5.5).

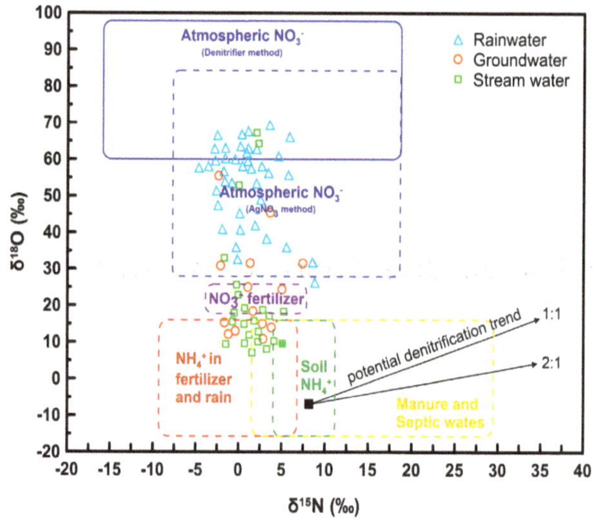

Figure 5.5. *A plot of the δ15N and δ18O values of nitrate from various N sources. The nitrate in stream samples (green squares) was largely derived from groundwater sources. (Kendall, C. 1998).*

Source: https://www.mdpi.com/2073-4441/12/1/138#B79-water-12-00138.

According to the traditional theory, one oxygen atom in newly formed nitrate in soil comes from dissolved atmospheric oxygen (O_2), while the other two oxygen atoms come from the surrounding water sources. If these oxygens are included without any fractionation, and the $\delta^{18}O$ values of water and atmospheric sources are known, the $\delta^{18}O$ value of microbial nitrate can be calculated as follows:

$$\delta^{18}O_{NO_3} = 2/3\ \delta^{18}O_{H_2O} + 1/3\ \delta^{18}O_{O_2} \quad (11)$$

The δ18O values of synthetic nitrate fertilizer are 22 ± 3%; those of soil nitrogen converted from ammonium via nitrification are between −10 and +10%; and those of manure and sewage are below 15%. In contrast, the $\delta^{18}O$ values of atmospheric-derived nitrates are typically high, ranging from +20 to +70%. For each source of contamination, nitrate thus exhibits a unique isotopic composition of nitrogen and oxygen, making it helpful in identifying the source of contamination. Using the previously discussed hydrograph separation, it is possible to identify not only the source of contamination but also the contribution of said source. However, isotopic fractionation caused by mineralization, absorption/desorption, nitrification, denitrification, volatilization, assimilation (uptake), and leaching from the soil zone modifies the $\delta^{15}N$ and δ18O values of nitrate. The lighter isotopes (^{14}N and ^{16}O) are preferred by common microbial organisms over the heavier isotopes (^{15}N and ^{18}O). As a result, the microbial products are isotopically depleted and the remaining nitrates are enriched in ^{15}N and ^{18}O. For instance, the nitrogen gases produced during denitrification—the process by which microorganisms transform nitrate into nitrogen gases—are lighter than the residual nitrates (low $\delta^{15}N$ and $\delta^{18}O$). Consequently, denitrification raises the values of $\delta^{15}N$ and $\delta^{18}O$ in the residual nitrates, and there is a positive correlation between the enrichment ratios of $\delta^{15}N$ and $\delta^{18}O$, ranging from 1.3:1 to 2.1:1. This suggests that the enrichment factor can be used to estimate the initial isotope composition even in the case of isotope fractionation by denitrification.

5.2.3.3. Movements of Nitrate from Surface to Stream

Precipitation travels from the surface to the stream after it hits the ground, progressively changing the composition of water isotopes. There are usually two flow pathways involved in these processes: direct and indirect. The vertical movement of dissolved nitrate through the soil profile into the groundwater, where it can be flushed out and contribute to the stream water, is the indirect pathway. The direct pathway is the runoff of surface water from rainfall or melting snowpack into the stream water. These processes have as possible sources groundwater, nitrification, mineralization of soils beneath snowpacks, atmospheric precipitation and snowmelt, or a combination of these. As was already mentioned, groundwater nitrates that come from soil nitrification have very different $\delta^{18}O$ values than nitrates from atmospheric sources. As a result, the $\delta^{18}O$ value of nitrate is comparable to the atmospheric source if rainfall and surface water run off straight into the stream. On the other hand, the isotope composition of stream water is comparable to that of groundwater or soil water if precipitation seeps into the soil layer, recharges to groundwater, and is then released into the stream.

Numerous earlier studies have demonstrated that the primary source of stream water is groundwater (old water) obtained through indirect pathways. Through the analysis of nitrate's $\delta^{15}N$ and $\delta^{18}O$ values, Kendall et al., concluded that groundwater is the primary source of nitrates in stream water, with atmospherically derived nitrate from the snowpack contributing only a small amount. Ohte, among others. investigated using the $\delta\ ^{18}O$ values of nitrate, with precipitation, soil water, and groundwater serving as the three end members, the nitrate sources of a headwater stream at the Sleepers River Research Watershed in Vermont, USA, during snowmelt. The findings showed that the input to the groundwater progressively increased to become the main source of nitrate as it was replenished by meltwater and precipitation during snowmelt. As shown in Figure 5.6, Piatek et al., analyzed $\delta^{15}N$ and $\delta\ ^{18}O$ values of nitrate in the stream and compared them to those of snow and groundwater in the Arbutus Watershed of New York State, NY, USA, to indicate that stream water, atmospherically-derived solutions, and groundwater had overlapping nitrate $\delta^{15}N$ values.

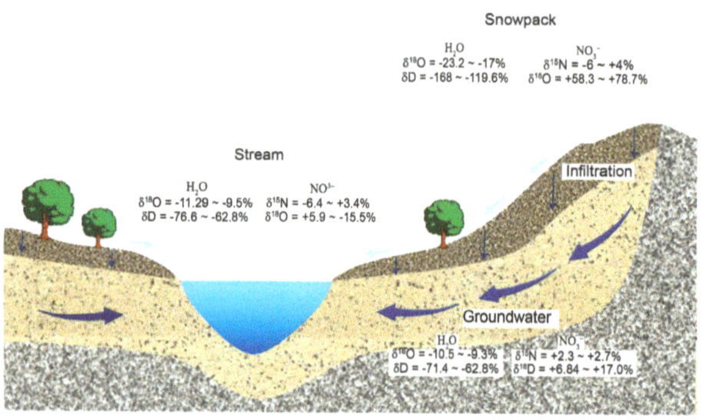

Figure 5.6. *Isotopic compositions of water and nitrate in snowpack, groundwater, stream.*

Source: https://www.mdpi.com/2073-4441/12/1/138#B79-water-12-00138.

Though they showed comparable ranges in groundwater and stream water, the δ18O values of nitrates were much lower than those of atmospheric solutions. Apart from these investigations, Barnes et al., showed a seasonal variation in the rate of atmospheric nitrate contribution and determined that 1–3% of the summer and 10–18% of the winter/spring exported stream NO_3^- comes from direct atmospheric deposits on average. The development of effective and successful abatement strategies, such as ecosystem management, NO_x emission controls, and potential links between nitrogen exports and climate change, depends on this kind of information. Moreover, because precipitation, groundwater, and soil have different isotope values, $\delta^{15}N$ and $\delta^{18}O$ values of nitrate can be used to identify the source of nitrate and the flowpath process through hydrograph separation.

5.2.3.4. Implications of the Flowpath of Water and Nitrates

Since nitrate contaminants typically have distinct isotope compositions, the stable isotopes of nitrate, $\delta^{15}N$ and $\delta^{18}O$, can be used to trace the sources of nitrate in water bodies. A multilateral investigation of precipitation, land-use type and area utilization rates, synthetic fertilizers, animal wastes, the presence of point sources (septic tanks and landfills), and the presence of sewer systems is required to improve the dependability of contaminant tracking. Hydrogeological data can also be used to track the sources of contaminants and evaluate variations in the level of contaminants. Examples of these data include dissolved oxygen (DO), electrical conductivity (EC), redox potential (Eh), matrix characteristics, groundwater flow rate and direction, and nitrate concentrations.

Isotope fractionation by nitrification, denitrification, and the presence of multiple contaminants continue to complicate isotope analysis, despite its usefulness in tracking nitrate contaminants. Therefore, the reliability of source identification may be increased by combining the use of water stable isotope analysis with the isotopic composition of nitrates.

5.3. AGE DATING OF GROUNDWATER

Different rates of water flow through the hydrological cycle. The average residence time of water vapor in the atmosphere is only roughly two weeks. With circulation rates measured in hours to days, runoff in first order streams and high elevation catchments is extremely responsive to precipitation and melting. The mean residence time of precipitation in large lakes is on the order of decades to over a century, and large rivers that drain large portions of the continents integrate precipitation on a timescale of weeks to months.

The time scale at which groundwater circulates is very different. Shallow bedrock aquifers and alluvial aquifers beside drainage channels have groundwater with mean residence times ranging from months to years, and they can react quickly to recharge. Groundwaters in more confined and deeper aquifers have the potential to cycle over hundreds or thousands of years. Others might still hold onto "fossil" groundwaters, which are no longer connected to active meteorological processes, but were replenished during climate conditions that no longer exist.

Surface water flow is closely observed through the assessment of discharge rates at pivotal locations within the drainage system, serving as a vital component for flood mitigation, hydropower production, and the administration of water resources. Groundwater resources are several times larger than all surface waters, but natural flow cannot be directly measured due to their cryptic nature and slow movement. Gradient distributions and hydraulic conductivity can be used to calculate groundwater flow. Geochemical methods can also be utilized to measure it. It is crucial to ascertain the average subsurface residence time, or "age" of groundwater for a number of reasons. Planning for water resources requires consideration of renewability, which is roughly correlated with groundwater age. While using groundwater that is thousands of years old may be considered mining the resource, groundwater resources that have been identified as modern can be developed sustainably and are regularly recharged. A groundwater's age and susceptibility to contamination are related. More direct routes from the surface will allow pollutants from surface activity to infiltrate the recharge area of modern groundwaters. Older groundwaters have a higher potential for contamination attenuation because they are typically more isolated from the surface.

5.3.1. Groundwater Mean Residence Time and "Age"

Most of the time, precipitation and runoff seep into the subsurface through soils, alluvial sediments, or bedrock outcrops to replenish groundwater. It might proceed straight to the water table or slowly through an unsaturated zone. In any event, some water will pass through the pores and fissures more slowly than others, and some will pass through them more quickly. Waters that have recharged at different times are mixed together through the hydrodynamic dispersion process. In the first few centimeters, the difference in time might only be hours or days for soil waters. A given volume in the unsaturated zone and immediately below the water table may incorporate groundwaters that have been recharged over several weeks to months. Groundwater at deeper depths in the aquifer will contain components that recharge over longer time spans, ranging from months to years. The groundwater will have a mean age and a range of ages at each location.

Groundwater ages are typically categorized using a methodical approach. Water that has been recharged in the last 50 years can be identified using tritium; this type of water is referred to as "modern." The resolution of radiocarbon allows identification of "old" groundwaters recharged more than a thousand to perhaps twenty thousand years before the present, despite the complexities of subsurface carbon sources and sinks. Groundwaters classified as "ancient" or "fossil" have little to no radiocarbon left in them and are frequently disconnected from current meteorological processes. "Submodern" groundwaters are defined as those that fall between these two categories, i.e., those that are tritium-free and show no signs of radiocarbon decay.

Tritium and radiocarbon remain the most widely used and trustworthy isotope techniques for dating groundwaters, even in the face of advancements in other geochemical and isotopic tools.

5.3.2. Tritium and "Modern" Groundwaters

Either a distinctive signal that can be maintained in recently recharged waters or an input that dissipates after being removed from the meteoric source should be provided by tracers of modern meteoric water. Tritium possesses both (^3H or T). This radioisotope has a brief half-life and is created in the atmosphere naturally. It is then immediately incorporated into precipitation. It also contains a distinctive signal from nuclear weapons tests conducted in the 1960s in the atmosphere. Groundwater age can be determined by both thermonuclear peak position and decay.

Tritium, having a half-life of 12 point 32 years, is a short-lived hydrogen isotope. It is the only radioisotope that can truly date groundwater because it is directly incorporated into the water molecule (H^3HO or HTO) (Lucas and Unterweger, 2000). Tritium contamination in atmospheric water vapor during sampling is not a concern, in contrast to CFCs, ^{85}Kr, and other contemporary tracers that demand extremely stringent sampling protocols. Measuring by β-counting directly on water or on water that has been electrolytically enriched is a simple and affordable method that offers clear proof of contemporary recharge.

Thermonuclear weapon testing conducted in the upper atmosphere between May 1951 and May 1976 created sizable anthropogenic tritium reservoirs that have only lately returned to natural background levels. Since it decays so quickly from natural, pre-bomb levels, groundwater recharged before roughly 1950 is typically not able to detect it. By comparison, modern groundwaters are less than 45 years old compared to the mid-1990s. Groundwaters devoid of tritium are regarded as "submodern" or older.

5.3.2.1. Tritium Concentration Units

Knowing the concentration of tritium in meteoric waters—which are influenced by both natural and man-made sources—is necessary for using tritium in groundwater dating. The Tritium Unit, or TU, is the most commonly used unit of concentration for tritium, where:

$$\begin{aligned} 1\ \text{TU} &= 1\ \text{T atom per } 10^{18}\ \text{H atoms} \\ &= 6.7 \cdot 10^{7}\ \text{T atoms per kg } H_2O \\ &= 1.1 \cdot 10^{-16}\ \text{mol/kg } H_2O \\ &= 3.33 \cdot 10^{-10}\ \text{ppb (gfw for T is 3)} \end{aligned}$$

One of the most sensitive measurements in geochemistry is the one that enrichment methods can detect down to less than 1 TU. The measurement is carried out using either atom-counting of ^{3}He in sealed water samples that were pumped of gases and stored for a specific decay period, or scintillation counting of β – decay events. It is frequently necessary to express tritium concentrations in units of radioacitivity in order to address the issue of tritium's radiological hazard. The decay events per time unit of measurement for radioactivity is used. Since ^{226}Ra is the isotope of radium that Curie isolated from the ^{238}U decay series, 1 g of it has the same radioactivity as one curie, or Ci. This is the original unit of measurement. Now, one nuclear disintegration per second (1 Ci = 3.7×10^{10} Bq) is the bequerele, or Bq. The radioactivity of a radionuclide is equal to its change in concentration with time, $\partial N/\partial t$. Converting between units of concentration (atoms per volume or mass) and radioactivity (decay events per time) is done with the decay constant, λ, which is based on the half-life of the radionuclide:

$$\lambda = \frac{\ln 2}{t_{½}}\ (\text{time}^{-1}) \qquad \textit{decay constant}$$

$$-\frac{\partial N}{\partial t} = \lambda N\ (\text{decay / time}) \qquad \textit{radioactive decay}$$

For an equivalent concentration or mass, radionuclides with short half-lives, such as tritium, are therefore much more radioactive than those with long half-lives, such as common uranium, ^{238}U, $t½ = 4.5 \times 10^{9}$ y.

For tritium decay, radioactive decay is the rate of change in tritium concentration is a function of its concentration, N (T atoms per liter). The decay constant, λ, for tritium is:

$$\lambda = \frac{\ln 2}{t_{½}} = \frac{0.693}{12.32} = 0.05626\ \text{yr}^{-1}$$

The radioactivity is simply the product of its concentration and decay constant. The radioactivity of 1 L water with 1 TU ($N = 6.67 \times 10^{7}$ atoms T, from above) would then yield:

$$\begin{aligned} \text{radioactivity} &= \lambda N \\ &= 0.05626 \cdot 6.67 \cdot 10^{7}\ \text{disintegrations per year per liter} \\ &= 3.76 \cdot 10^{6}\ \text{disintegrations per year per liter, or} \\ &= 0.119\ \text{disintegrations per second per liter} \\ &= 0.119\ \text{Bq/L (Bq = 1 disintegration per second)} \end{aligned}$$

Water that naturally contains tritium has very low radioactivity, well below the WHO's recommended limit of 7800 Bq/L. Water would require more than 65,000 TU, a level never found in precipitation or naturally occurring groundwaters, to surpass this limit. Anthropogenic tritium fallout did not pose a radioactive risk to public health, even during its peak. Nuclear power plants, facilities that reprocess nuclear fuel, those that prepare weapons-grade nuclear material, and companies that make tritium gas (T2) for self-illuminating emergency signs and tritiated paints used in illuminating dials are currently releasing small amounts of ^{3}H into the atmosphere.

5.3.2.2. Tritium Production and Decay

Natural tritium is formed in the stratosphere where high-energy cosmic radiation produces a secondary neutron flux. Tritium is produced by collisions of these neutrons with N_2, following the reaction:

$$^{14}N + n \rightarrow \; ^{12}C + ^{3}H$$

Tritium decays with a half-life of 12.32 years to ^{3}He by release of a beta particle:

$$^{3}H \rightarrow \; ^{3}He + \beta^-$$

The newly created tritium still exhibits the same chemical characteristics as stable hydrogen (^{1}H), which allows it to react with atmospheric oxygen or hydroxide to form a water molecule. The majority of cosmogenic tritium is produced in the stratosphere and mixes into the troposphere during the spring.

Anthropogenic activities are major sources of environmental tritium because nuclear reactors and thermonuclear bombs mimic the natural neutron activation process. The first observation and prediction of this occurred during the testing of hydrogen fusion bombs in the 1950s, when a massive neutron flux was released, activating ^{14}N in the atmosphere. The explosion of the 10.4 megaton device, Mike, on November 1, 1952, marked the start of ten years of atmospheric tests of thermonuclear weapons by the United States and the Soviet Union. The tests concluded in 1962 with the spectacular release of over 140 megatons of fusion energy. Mike created a fireball that was more than three miles wide and had a neutron flux (the rate at which neutrons are released, not the total amount, naturally) that is higher than that of a supernova. Tritium precipitation increased dramatically as a result, peaking at over 5000 TU in the northern hemisphere. A massive peak was produced in the last year of the megaton tests and showed up as precipitation in the spring of 1963. This peak from 1963 was used as a benchmark in numerous hydrological studies. Precipitation ^{3}H concentrations have now stabilized at levels that are almost identical to cosmogenic pre-bomb levels.

The American tests were carried out in the Marshall Islands of the eastern Pacific Ocean at 12°N, while the Soviet tests were carried out in eastern Kazakhstan at roughly 50°N. Precipitation in the southern hemisphere was significantly lower and came a few years behind that in the northern hemisphere. In the 1970s, more testing conducted by China and France temporarily maintained levels. By the 1990s, atmospheric levels had essentially returned to background levels due to rapid decline through mixing with the oceans.

The world's meteoric ^{3}H was unaffected by the neutron flux from these relatively small and low-level fission tests, even though fission from the detonation of uranium and plutonium bombs started on July 16, 1945.

The timing and magnitude of the 1963 tritium peak were indicative of asymmetric thermonuclear tritium production between the two hemispheres. The tritium peak reached only a few hundred tritium units in the southern hemisphere, but it exceeded 5000 tritium units as far north as Ottawa (at the latitude of the Soviet tests).

There are other factors contributing to the tritium decline in precipitation than decay, which reduces tritium levels by only 5.5% annually. On the log-normal plot, radioactive decay from example initial values is represented by straight lines. Levels have decreased much more quickly through attenuation by the seas and groundwaters than through decay alone. Tritium is not produced subsurfacely as other radionuclides such as ^{129}I and ^{36}Cl are, although it can be measured. The subsurface is experiencing a neutron flux as a result of U's spontaneous fission. This can result in ^3H production through the fission of ^6Li according to:

$$^6Li + n \rightarrow {}^3H + \alpha$$

Tritium's brief half-life, however, does not provide enough time for meaningful accumulation. Though levels in uranium ore zones may be significantly higher, groundwater in U and Li rich rocks rarely has more than 0.5 TU. The majority of aquifers have negligible geogenic ^3H.

5.3.2.3. Global Distribution of Natural Tritium

Hydrogeologists are entering a new era in the use of tritium for dating groundwater. The peak of thermonuclear testing in 1963 has mostly dissipated from actively recharged aquifers, and precipitation levels are now controlled by natural production, resembling preanthropogenic levels. This shift replaces the fluctuating input from bomb testing with a more consistent input.

5.3.3. Dating Groundwaters with Tritium

In the simplest terms possible, the finding of tritium in groundwater suggests that at least a portion of contemporary recharge occurs. It is assumed that modern recharge predates the start of thermonuclear bomb experiments. This provides a 50-years time limit as of right now. Hydrogeological constraints, however, should be taken into account because the concentration of tritium will be significantly impacted by various groundwater flow and discharge scenarios. Groundwater mixing has a major impact on the range of groundwater ages, especially at the discharge zone in open boreholes and spring vents. The relative contributions from recent and pre-1950s precipitation will have a significant impact on the measured tritium in groundwater because the tritium input function from the previous 50 years varies over several orders of magnitude.

The level of tritium, as opposed to just a measurable concentration, allows for more sophisticated interpretations. It is now necessary to have some understanding of the background of tritium in local precipitation.

The following points should be considered:

- In most regions, tritium concentrations in precipitation are now back to natural production levels.
- Bomb tritium, found in groundwaters recharged prior to about 1990, is higher in the mid latitude northern hemisphere than in low latitudes, tropics and southern hemisphere.
- Coastal precipitation generally has much lower tritium concentrations than continental (interior) stations. This is true for precipitation during the thermonuclear era as well as today and is due to the lower degree of mixing with upper atmosphere vapor experienced by coastal precipitation systems.
- Groundwaters with low tritium levels (close to 1 TU) are often mixtures of modern groundwaters and submodern groundwaters (greater than at least 50 years old).
- Groundwater age from the movement of the 1963 3H peak: Sequential sampling conducted over time is a great help in determining the age of groundwater using tritium. Variations in the source term (i.e., the decreasing bomb

input) or variations in the mixing of various groundwaters can be linked to temporal variations in tritium. When tritium levels rise and then fall in a time-series measurement, the ideal scenario is when the 1963 peak moves and the groundwater flow velocity can be computed. In a similar vein, the 1963 peak can be located through spatial variations.

- Groundwater age from radioactive decay: In order to accurately calculate the age of groundwater based on the decay of a known input concentration, it is important to ensure that atmospheric levels have remained stable (such as since the 1990s when levels returned to natural levels) and that there has been minimal hydrodynamic dispersion (like plug flow). The calculation involves using the decay equation, where the measured tritium concentration (^3H_t) after a certain amount of time (t) is equal to the initial concentration (^3H_o) multiplied by the exponential decay function.

$$^3H_t = {}^3H_o\, e^{-\lambda t}$$

From above, the decay term, $\lambda = \ln 2/t½$. Using tritium's half-life, t½ = 12.32 years, this equation can be rewritten:

$$t = -17.77 \ln \frac{^3H_t}{^3H_o}$$

Graphs like this one can be used to back-extrapolate to the input function in order to determine age, since the initial tritium concentration is not easily known.

In most aquifers, plug flow—that is, no hydrodynamic dispersion—is an infrequent occurrence. However, in contemporary systems where the input function is well-known and stable, these computations could provide a first-order age approximation.

- Tritium-helium dating of groundwater: The need to know the tritium input function for calculating groundwater ages from tritium decay can be avoided by utilizing the parent-daughter decay equation. In certain situations, measuring ^3He, a byproduct of tritium decay, can result in significantly more accurate age calculations. However, the challenge lies in the fact that measuring helium isotopes is not commonly done and is only available in a limited number of laboratories.

The measured concentration of tritium, ^3He_t, can give the time, t, of decay (age) as long as the initial concentration ^3H_o is known:

$$^3H_t = {}^3H_o\, e^{-\lambda t}$$

However, given the complicated input function for tritium in precipitation knowing the initial value is not possible in most cases. On the other hand, the decay of tritium leads to an ingrowth of 3He, which can be calculated from:

$$^3He_t = {}^3H_o(1 - e^{-\lambda t})$$

By rearranging the tritium decay equation and substituting into the ^3He ingrowth equation, the input concentration of tritium ^3H_o is no longer a factor and now groundwater age, t, can be calculated from the measured tritium and ^3H_e concentrations:

$$^3He_t = {}^3H_t(e^{\lambda t} - 1)$$

The measured helium concentration, ^3He_t, is expressed in TU (1 ^3He per 10^{18} H).

It must also be corrected for atmospheric ^3He that is dissolved at the time of recharge.

The measured value of ^3He, corrected for atmospheric 3He, represents ^3He ingrown from ^3H decay, and is then used in the dating equation:

$$t = 17.77 \cdot \ln(1 + [^3He_t]/[^3H_t])$$

where $[^3He_t]/[^3H_t]$ is the concentration ratio of these two isotopes expressed in tritium units.

5.3.4. None-Routine Tools for Dating Modern Groundwater

Various isotopic methods can be utilized to determine average residence times. Changes in stable isotopes throughout the seasons can give an indication of age. The most commonly used techniques rely on the decay of radionuclides. Isotopes with longer half-lives (such as ^{14}C, ^{36}Cl, ^{39}Ar, and ^{81}Kr) can be used to date ancient groundwater. Short-lived radioisotopes (like ^3H, ^{32}Si, ^{37}Ar, ^{85}Kr, and ^{222}Rn) as well as those generated by human nuclear activities in the last forty years (^3H, ^{14}C, ^{36}Cl, and ^{85}Kr) can show if a groundwater source is modern in origin.

While ^3H and ^{14}C analysis procedures are now standard, other radioisotopes necessitate complex sampling and/or analytical methods. It is problematic to date between modern waters and paleogroundwaters during the "submodern" era. Although ^{39}Ar dating (t½ = 256 years) may be able to fill this >45 to ~1000 year range, this method requires very ideal aquifers, very large samples, complex sample preparation techniques, and specialized counting facilities. Because so few laboratories can afford it, 39Ar dating has not become a standard tool.

5.3.4.1. Thermonuclear ^{36}Cl

In addition to producing a massive peak in tritium, thermonuclear bomb testing also produced a radioactive isotope of Cl, ^{36}Cl. Early hydrogen bomb tests in marine environments produced a high neutron flux, which increased concentrations over two orders of magnitude above the abundance of hydrogen in the atmosphere. As a result, ^{36}Cl can be used to identify modern recharge, just like tritium. But unlike ^3H, its long half-life (t½ ~ 300,000 years) makes it impossible to date modern waters using ^{36}Cl decay.

Natural atmospheric production of 36Cl happens when solar radiation activates atmospheric argon, which then falls or precipitates to the earth's surface. Natural production and fallout are latitude-dependent, ranging from less than 5 near the equator and poles to nearly 20 atoms m^{-2} s^{-1} at mid-latitudes. Additionally, it is produced epigenetically when solar radiation activates the Cl, K, and Ca molecules on the land surface. The atmospheric Ar and marine Cl– were activated by the high neutron fluxes that accompanied low altitude tests in the 1950s, producing thermonuclear ^{36}Cl. Bentley et al., reconstructed glacier ice, which contains a record of this fallout. Similar to tritium, high ^{36}Cl levels in groundwater signify that recharge has taken place since this time. Concentrations have returned to more naturally occurring levels, at least by 1980, as a result of the atmospheric Cl being washed out. On the other hand, it appears that the biosphere's recycling and storage of Cl sustain a background ^{36}Cl activity that is persistently higher than that of nature.

CASE STUDY

ASSESSMENT OF AQUIFER SYSTEM IN PARTS OF SOUTH-WESTERN NIGERIA USING ISOTOPE TECHNIQUES

The present study area comprises of fast growing business and educational centres in the southern end of Kwara State in Nigeria. The towns included in the study are Offa, Omu-Aran and Osi-Opin, which have long been seats of Local Government Areas. These towns have been rapidly expanding during the last 15 years as a result of rapid industrialization and expansion of Local Government Authorities.

Offa township lies approximately 45 km SE of Ilorin, the Kwara State capital city and has remain the second largest commercial centre in the state. Its closeness to the state capital couple with the good network of roads and railway is responsible for the growing commerce and population in this area. Over the years, commercial activities in this area have steadily grown attracting more people to the urban centres and consequently increasing groundwater use.

Wastes are generated from a wide variety of industrial, commercial, agricultural and domestic activities. Omu-Aran is less than 30 kilomettrs from Offa and is less than 20 kilometres from Osi-Opin.

-Osin-Opin town comprises of more than seven communities that are linked together as one due to rapid development and good neighbourliness. These towns are connected by a major federal road linking the state to the federal capital territory in Central Nigeria. Commercial activities become tense as the communities are closely packed and densely populated.

The fast growth of population, poor town planning and no organized waste disposal systems are causing problems in supplying public services, water being one of the most affected. Public water supply is grossly inadequate in these areas and has failed to meet the growing demand for water. Therefore in order to meet the increasing water needs for domestic and industrial purposes there has been heavy dependence on groundwater.

Geological and hydrogeological background

Geologically, Offa and its neighbouring towns are underlain by Precambrian to Lower Paleozoic crystalline Basement rocks and the entire area falls within the southwestern Nigerian Basement Complex.

Radiometric dating indicates that the Basement rocks of soouthwestern Nigeria is polycyclic and has responded to several tectonic events with differing intensities from Archean to late Proterozoic (Pan African). Prominent structural trends (faulting and folding) have been observed in this region. Due to the crystalline nature of the rock types in this area, the porosity and permeability necessary for groundwater occurrence are generally lacking. However, appreciable porosity and permeability may have developed through fracturing and weathering processes. Available data have shown that borehole in Offa area are designed to tap water from the weathered zone aquifer or the jointed/ fractured basement rock aquifer.

The boreholes are either singled –screened or multiple- screened, and sometimes open wells are drilled through fractured basement rocks yielding substantial amount of water for private stakeholders. Records from available boreholes show the total borehole depth of between 25.0 to 68.2 metres while overburden thickness varies from 3 - 24 metres. Hence yields of boreholes are generally low, ranging from 0.3 to 2.53 L/sec. (with a mean value of 1.94 L/sec.).

Methods of Assessment

In the study areas, instantaneous information has been gathered from shallow regolith/ fractured bedrock aquifers as well as long-term monitoring of water levels and chemistry between 1999 and the year 2002. For the purpose of the present study several hand-dug wells and some boreholes (tapping groundwater at relatively deeper level) have been sampled for chemistry and stable isotopes. The hand-dug wells are in most cases tapping the uppermost regoliths aquifer.

Temperature, pH and electrical conductivity (EC) were measured in the field using portable (battery-operated) meters. Alkalinity test (as $CaCO_3$) was conducted in each of the wells sampled by titration methods using standard field titration kits supplied by Merck GmbH (Darmstadt, Germany). The environmental isotopes being determined in this study are the stable isotopes in water molecule – $^{18}O/^{16}O$ and $^{2}H/^{1}H$ – and the differences of these relations can supply information on the specific conditions during the formation of the water sample being analyzed. Analyses of the stable isotopes were jointly performed at the Isotope Hydrology Group, Hydrogeology Section, and Centre for Environmental Research (UFZ), Leipzig-Halle, Germany and the Stable isotopes laboratory, Earth Sciences Department, Monash University, Melbourne, Australia. Stable isotopes ratios are expressed as delta in per mil (i.e., δ) relative to VSMOW (Vienna Standard Mean Ocean Water). The isotope precision of measurement based on VSMOW is ±0.15 % for ^{18}O and ±1% for 2H.

Results

Groundwater isotopic signature in Offa area shows a range from - 19.2 to 0.8% for ä ^{2}H (with mean of - 13.7%) and - 3.6 to 3.5% for ä ^{18}O (mean: - 2.2%) as against rainwater (ä ^{2}H: 0.8 %, ä ^{18}O: -1.0 %). A plot of ä ^{2}H against ä ^{18}O show that the isotopic composition of Offa groundwater cluster closely around the Mean Meteoric Water Line (Figure 1) defined by the equation:

$$ä\ ^{2}H = 8.6\ ä\ ^{18}O + 10$$

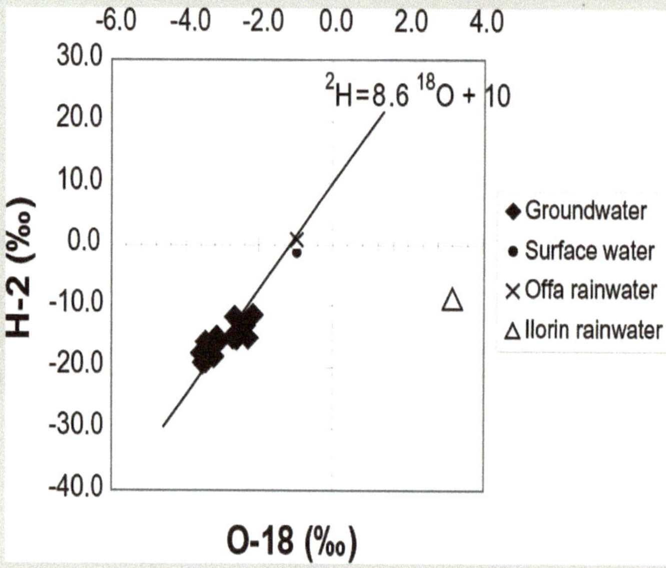

Figure *1. The isotopic composition of Offa groundwater.*

CLASS ACTIVITY

TRACING GROUNDWATER RECHARGE AND FLOW PATHS

Objective: To investigate how isotopic signatures can be used to trace groundwater recharge areas and flow paths.

Materials: Groundwater samples from various locations within a watershed, precipitation samples (rain and snow) if available, Isotope ratio mass spectrometer (or access to lab results), Maps of the study area, and graph paper or spreadsheet software for plotting

SUMMARY

- Environmental isotope hydrology is a relatively new field of investigation based on isotopic variations observed in natural waters.
- The main environmental isotopes of hydrological interest are the stable isotopes deuterium (hydrogen-2), carbon-13, oxygen-18, and the radioactive isotopes tritium (hydrogen-3) and carbon-14.
- Isotopes of hydrogen and oxygen are ideal geochemical tracers of water because their concentrations are usually not subject to change by interaction with the aquifer material.
- When evaporation of water occurs, water vapor contains at equilibrium less molecules of $H_2{}^{18}O$ or HDO in proportion to $H_2{}^{16}O$ than the liquid water from which it derives.
- Depending on the physical conditions prevailing in the air above the water and the importance of the derived kinetic isotopic effect during evaporation, residual water samples are aligned in a δD - $\delta^{18}O$ diagram on an evaporation line, with a slope lower than that of the local MWL.
- The stable isotopes of nitrogen and oxygen in nitrate ($\delta^{15}N$ and $\delta^{18}O$) are also fundamental to identifying the sources of nitrate contamination because the isotopic values are distinct from source to source.
- Hydrograph separation is the separation of streamflow components into two or more different components that contribute to the stream in a small catchment area or watershed.
- Unlike other tracers, stable water isotopes are added naturally on the scale of the watershed by precipitation (rain or snowmelt events) and, upon entering the watershed, undergo transport according to the natural movement of the body of water through the watershed.
- The stable isotope composition of water is mainly determined by isotope ratio mass spectrometry (IRMS). This technique measures the relative isotope ratios of molecular compounds by analyzing mass differences.
- Dual isotope analysis of nitrates ($\delta^{15}N$ and $\delta^{18}O$) can be a powerful tool for identifying nitrate sources and nitrate cycling mechanisms in stream water because the different sources have isotopically distinct $\delta^{15}N$ and $\delta^{18}O$ compositions.
- The analyzed nitrate $\delta^{15}N$ and $\delta^{18}O$ isotope ratios provide distinct values for each contaminant source. The value of $\delta^{15}N$ in atmospheric NO_3^- is usually in the range of −15% to +15%.
- Groundwater circulates on quite a different time scale. Alluvial aquifers along drainage channels or shallow bedrock aquifers can respond rapidly to recharge and have groundwater with mean residence times on the order of months to years.
- Surface water flow is extensively monitored by measurement of discharge rates at critical points in the drainage network, as input for flood control, hydroelectric generation and water resources management.

- Groundwater recharge occurs in most cases by infiltration of precipitation and runoff through soils, alluvial sediments or bedrock outcrop into the subsurface.
- The radioactivity of water with natural levels of tritium is low, and well below the WHO recommended limit of 7800 Bq/L.
- The presence of tritium in groundwater, at its most fundamental level of interpretation, is an indication that at least some component of modern recharge.

REVIEW QUESTIONS

1. What do you understand by isotopic fractionation during phase changes? Explain.
2. Discuss the use of water isotopes in the hydrological study of a drainage basin.
3. What is the base flow separation of a hydrograph?
4. Describe pretreatment method for nitrate isotope analysis.
5. Define the movements of nitrate from surface to stream.
6. How to date groundwaters with tritium? Discuss.

REFERENCES

1. Battaglin, W. A., Kendall, C., Chang, C. C. Y., Silva, S. R., & Campbell, D. H. (2001). Chemical and isotopic evidence of nitrogen transformation in the Mississippi River, 1997–98. *Hydrological Processes*, *15*, 1285–1300.
2. Benstead, J. P., March, J. G., Fry, B., Ewel, K. C., & Pringle, C. M. (2006). Testing isosource: Stable isotope analysis of a tropical fishery with diverse organic matter sources. *Ecology*, *87*, 326–333.
3. Böttcher, J., Strebel, O., Voerkelius, S., & Schmidt, H. L. (1990). Using isotope fractionation of nitrate-nitrogen and nitrate-oxygen for evaluation of microbial denitrification in a sandy aquifer. *Journal of Hydrology*, *114*, 413–424.
4. Durka, W., Schulze, E. D., Gebauer, G., & Voerkelius, S. (1994). Effects of forest decline on uptake and leaching of deposited nitrate determined from 15N and 18O measurements. *Nature*, *372*, 765–767.
5. Friedman, I., Benson, C., & Gleason, J. (1991). Isotopic changes during snow metamorphism. In H. P. Taylor, J. R. O'Neil, & I. R. Kaplan (Eds.), *Stable Isotope Geochemistry: A Tribute to Sam Epstein* (pp. 211–221). The Geochemical Society.
6. Heaton, T. H. E. (1986). Isotopic studies of nitrogen pollution in the hydrosphere and atmosphere: A review. *Chemical Geology (Isotope Geoscience Section)*, *59*, 87–102.
7. Hitchon, B., & Krouse, H. R. (1972). Hydrogeochemistry of surface waters of the Mackenzie River drainage basin, Canada–III. Stable isotopes of oxygen, carbon and Sulphur. *Geochimica et Cosmochimica Acta*, *36*, 1337–1357.
8. Kelly, E. F., Amundson, R. G., Marino, M. D., & DeNiro, M. J. (1991). Stable isotope ratios of carbon in phytoliths as a qualitative method of monitoring vegetation and climate change. *Quaternary Research*, *35*, 222–233.

9. Kendall, C. (1998). Tracing nitrogen sources and cycling in catchments. In C. Kendall & J. J. McDonnell (Eds.), *Isotope Tracers in Catchment Hydrology* (pp. 519–576). Elsevier.
10. Kendall, C., Campbell, D. H., Burns, D. A., Shanley, J. B., Silva, S. R., & Chang, C. C. Y. (1995). Tracing sources of nitrate in snowmelt runoff using the oxygen and nitrogen isotopic compositions of nitrate. In K. Tonnessen, M. Williams, & M. Tranter (Eds.), *Biogeochemistry of Seasonally Snow-Covered Catchments* (pp. 339–347). IAHS Publication.
11. Kim, H., Cho, S. H., Lee, D., Jung, Y. Y., Kim, Y. H., Koh, D. C., & Lee, J. (2017). Influence of pre-event water on streamflow in a granitic watershed using hydrograph separation. *Environmental Earth Sciences*, *76*, 82.
12. Lee, J., Koh, D. C., & Choo, M. K. (2014). Influences of fractionation of stable isotopic composition of rain and snowmelt on isotopic hydrograph separation. *Journal of the Korean Earth Science Society*, *35*, 97–103.
13. Lee, J., Lee, H. J., Ham, J. Y., & Kim, H. (2016). Chemical separation of the discharge generated by artificial rain-on-snow experiments in a snowpack. *Journal of the Geological Society of Korea*, *52*, 113–120.
14. Lucas, L. L., & Unterweger, M. P. (2000). Comprehensible review and critical evaluation of the half-life of tritium. *Journal of Research of the National Institute of Standards and Technology*, *105*(4), 541–549.
15. Mast, M. A., Kendall, C., Campbell, D. H., Clow, D. W., & Back, J. (1995). Determination of hydrologic pathways in an alpine-subalpine basin using isotopic and chemical tracers. In K. Tonnessen, M. Williams, & M. Tranter (Eds.), *Biogeochemistry of Seasonally Snow-Covered Catchments* (pp. 263–270). IAHS Publication.
16. Meng, Z., Yang, Y., Qin, Z., & Huang, L. (2018). Evaluating temporal and spatial variation in nitrogen sources along the lower reach of Fenhe River (Shanxi Province, China) using stable isotope and hydrochemical tracers. *Water*, *10*, 231.
17. Rahman, K., Besacier-Monbertrand, A. L., Castella, E., Lods-Crozet, B., Ilg, C., & Beguin, O. (2015). Quantification of the daily dynamics of streamflow components in a small alpine watershed in Switzerland using end member mixing analysis. *Environmental Earth Sciences*, *74*, 4927–4937.
18. Semenov, M. Y., & Zimnik, E. A. (2015). A three-component hydrograph separation based on relationship between organic and inorganic component concentrations: A case study in Eastern Siberia, Russia. *Environmental Earth Sciences*, *73*, 611–620.
19. Sommerfield, R. A., Judy, C., & Friedman, I. (1991). Isotopic changes during the formation of depth hoar in experimental snowpacks. In H. P. Taylor, J. R. O'Neil, & I. R. Kaplan (Eds.), *Stable Isotope Geochemistry: A Tribute to Sam Epstein* (pp. 205–210). The Geochemical Society.
20. Souchez, R., & Jouzel, J. (1984). On the isotopic composition in δD and δ18O of water and ice during freezing. *Journal of Glaciology*, *30*(106), 369–372.

21. Souchez, R., Petit, J. R., Tison, J. L., Jouzel, J., & Verbeke, V. (2000). Ice formation in subglacial Lake Vostok, Central Antarctica. *Earth and Planetary Science Letters*, *181*, 529–538.

22. Worden, J., Noone, D., & Bowman, K. (2007). Importance of rain evaporation and continental convection in the tropical water cycle. *Nature*, *445*, 528–532.

23. Xue, D., Botte, J., De Baets, B., Accoe, F., Nestler, A., Taylor, P., Van Cleemput, O., Berglund, M., & Boeckx, P. (2009). Present limitations and future prospects of stable isotope methods for nitrate source identification in surface and groundwater. *Water Research*, *43*, 1159–1170.

CHAPTER 6

Groundwater and Catchment Processes

LEARNING OBJECTIVES

After studying this chapter, you will be able to:

- Understand water balance equation
- Explain precipitation and evapotranspiration
- Discuss soil water and infiltration
- Know recharge estimation
- Perform hydrograph analysis
- Sxurface water – groundwater interaction

KEY TERMS FROM THIS CHAPTER

Climate variability
Conservation of mass
Evaporation
Geology
Hydrologic cycle
Irrigation scheduling
Subcatchments
Vegetation
Water balance
Collision
Drainage basin
Evapotranspiration
Humidity
Irrigation
Precipitation
Transpiration
Wastewater treatment plant
Water path types

INTRODUCTION

The physical features of a catchment or subcatchment, as well as the larger region and subregion, have an impact on the hydrological processes of that area. Within a catchment, a natural boundary or raised ridge can also divide subcatchments. Although floodwaters may cross a subcatchment boundary in comparatively flat areas, surface water generally does not flow from one subcatchment to another. Due to the underlying geology, groundwater may occasionally cross subcatchment boundaries.

6.1. WATER BALANCE EQUATION

According to the law of water balance, the amount of water entering any system or region must equal the amount leaving it, plus any changes in storage over time. A water balance equation in hydrology can be used to explain how water enters and exits a system. A system can be any of the various hydrological or water domains, including a city, an irrigation zone, a drainage basin, or a column of soil. Another name for the water balance is a water budget. One of the fundamental tasks in the science of hydrology is creating water budgets. The US Geological Survey states that: Planning and managing water resources and the environment effectively depends on having a solid understanding of water budgets and the underlying hydrologic processes. An area's water budget changes over time can be used to measure how human activity and climate variability affect water resources. Quantification of the effects of geology, soils, vegetation, and land use on the hydrologic cycle is made possible by comparing water budgets from various locations.

6.1.1. Equation for a Basin

A general water balance equation is:

$$P = Q + ET + \Delta S$$

where,

P is precipitation; Q is streamflow; ET is evapotranspiration; ΔS is the change in storage (in soil or the bedrock/groundwater).

This equation makes use of the concepts of conservation of mass in a closed system, which state that any water entering the system through precipitation needs to be converted into surface runoff (which eventually reaches the channel and exits as river discharge), transpiration, evaporation, or ground storage. The system must be closed in order for this equation to work; otherwise, it must be taken into consideration (for instance, when surface runoff contributes to a different basin). Agricultural hydrology discusses large-scale water balances. Water balances can be used to forecast potential water shortage areas and assist in managing the availability of water. Additionally, irrigation and runoff assessment (e.g., flood control, pollution control (with the RainOff model). It also finds application in the planning of subterranean drainage systems, which can be horizontal (i.e., vertical (drainage by wells) or horizontal (using pipes, tile drains, or ditches). In order to determine the necessary drainage, a hydrogeological water balance and a groundwater model (e.g., SahysMod) could be useful. A water balance graph, which typically shows levels of precipitation and evapotranspiration on a monthly scale, can be used to illustrate the water balance. A number of monthly water balance models have been created for various scenarios and objectives. Since the 1940s, models for monthly water balance have been researched.

6.1.2. Water Balance of a System

"Water must be transformed from a natural resource into one that provides services in order to be made available for its many uses and users, which requires tools and institutions." This indicates that the terms Water Resource System (WRS) and Water Use System (WUS) refer to two different categories of water systems. Water balance must be followed by a WRS, such as a lake, river, or aquifer. For instance, the amount of water entering an aquifer and the amount exiting it, plus the change in storage, must equal one another. Water storage of many WRS is declining, say, once every ten years, due to a number of factors including population growth, climate change, and poor management. This indicates that after ten years, the amount of water in a WRS dropped, i.e., during that period, inflow was lower than outflow. A water unit system (WUS) is typically a water construct of a user, such as a region, city, industry, irrigation zone, or area, rather than a physical place. The inflows and outflows are depicted in a WUS schematic. With nine Water Path Types (WPT), a WUS's storage changes very little (compared to its inflow) over the appropriate time interval, making the water balance inflow equal to outflow. (Figure 6.1).

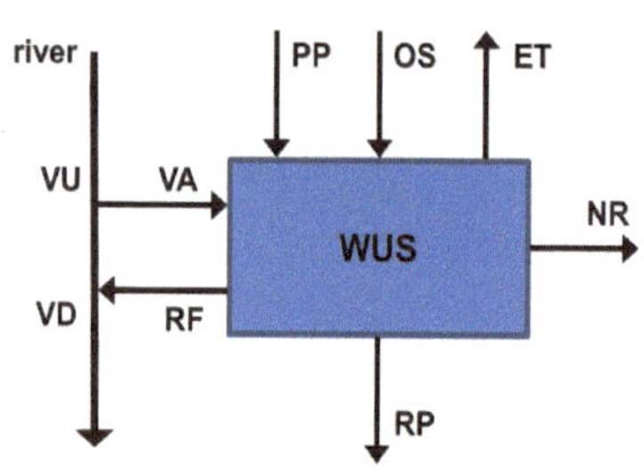

Variable	Description
ET	Evapotranspiration
NR	Non-reusable
OS	Water from other sources
PP	Total precipitation
RF	Return Flow
RP	Potential return (does not return to main source)
VA	Abstracted/applied water from main source
VD	Volume of water downstream after return flow in the main source
VU	Volume of water upstream before abstraction in the main source
V1	Volume of water at section 1 (VU or VA)
V2	Volume of water at section 2 (VD or RF)

Figure 6.1. *A typical schematic of a Water Use System (WUS) with its fixed nine Water Path Types.*

Source: By Mitral8 - Own work, CC BY-SA 4.0, https://commons.wikimedia.org/w/index.php?curid=100308799.

$$VA+OS+PP = ET+NR+RF+RP$$

Of course, an aquifer could serve as the primary source of water for a WUS in place of a river. As a simplified example, let's take a quick look at an urban water supply on an annual basis. It features very little ET and PP (WUS is a piped network), a small amount of groundwater (OS), return flow to the main source (RF) following a visit to a wastewater treatment plant, and a variety of Water Path Instances (WPI) for the RP type, including leaks and water used for green zone irrigation. Given that an urban area's annual change in storage is minimal, the water balance equation becomes.

$$VA_{riv}+OS_{gw} = NR+RF_{wwtp}+RP_{leak}+RP_{irr}$$

6.2. PRECIPITATION AND EVAPOTRANSPIRATION

Any type of water, whether liquid or solid, that descends to the earth's surface from the atmosphere is called precipitation. Precipitation includes things like snow, hail, drizzle, and rain. Rainfall is the most frequent type of precipitation in India. The process that brings water back into the atmosphere and completes the hydrologic cycle is called evapotranspiration. Evaporation and transpiration are the two components of evapotranspiration. The removal of water molecules from soil and water bodies is known as evaporation. The loss of water from plants as vapour is known as transpiration.

6.2.1. Precipitation

All liquid and solid water droplets that fall from the sky and land on the ground are called precipitation. Drizzle, rain, snow pellets, ice crystals, and hail are some of these particles. The size of a precipitation particle and a cloud particle is the primary distinction. The mass of an average raindrop is comparable to approximately one million cloud droplets. The precipitation particles can withstand the fall from the cloud to the ground because of their large size, which also gives them significant falling speeds.

Two essentially distinct processes are involved in the change from a cloud with only cloud droplets to one with a mixture of both cloud droplets and precipitation particles: the direct formation of early precipitation elements from the vapour state and their subsequent growth through aggregation and collision with cloud droplets. Ice crystals or droplets made of chemical solutions could be the first components of the precipitation.

The formation of precipitation via ice crystal growth is contingent upon the ability of cloud droplets to freeze naturally at temperatures below approximately -40°C, or -40°F. Supercooling is the process of lowering cloud droplet temperatures below the average freezing point. Ice crystals can form in supercooled clouds as a result of water vapor sublimating on specific sublimation nuclei—particles of dust in the atmosphere. Ice crystals form in naturally occurring clouds at temperatures below approximately -15°C (+5°F). The physical-chemical makeup of the sublimation nucleus greatly influences the precise temperature at which ice crystals form.

As long as the temperature inside a supercooled cloud stays below freezing, ice crystals will continue to grow. The temperature and level of vapor saturation in the surrounding air are the main factors influencing the growth rates. The water droplets are sacrificed in order for the crystals to grow. Under ideal circumstances—e.g., an ice crystal will grow to a size of approximately 0.13 millimeters (0.005 inches) in three to five minutes after formation in a large, rapidly growing cumulus cloud. This size causes the rate of sublimation growth to slow down, and the main means of further growth is by aggregation and collision with cloud droplets.

As particles that are just beginning to precipitate, small solution drops are also significant. There are a lot of tiny, soluble chemical particles in the atmosphere. The two most prevalent are sulfate-containing substances created by gaseous reactions in the atmosphere and sodium chloride that is washed up from the oceans. Due to their hygroscopic properties, these particles—known as condensation nuclei—collect water and exist as solution droplets at relative humidities greater than roughly 80%. There are frequently a lot of condensation nuclei in tropical maritime air masses. When clouds form in this type of air, they may produce several large solution droplets well before the cloud tops reach temperatures that are conducive to the formation of ice crystals.

The mechanisms of collision and coalescence account for the majority of the precipitation particle's growth, regardless of whether the initial particle is an ice crystal or a droplet formed on a condensation nucleus. The larger size of the incipient precipitation elements causes them to fall more quickly than cloud droplets. They consequently run into the droplets that are in their fall path. The relative sizes of the precipitation particle and the cloud droplets in the fall path that it actually hits, as well as the percentage of these droplets that coalesce with the particle after collision, determine the rate of growth of a precipitation particle through collision and coalescence.

6.2.2. Evapotranspiration

The term "evapotranspiration" (ET) describes the group of processes that transfer water from the Earth's surface—that is, from surfaces of open water and ice to bare soil and vegetation—into the atmosphere. It includes transpiration, which is the evaporation of water through plant leaves' stomata, or openings, as well as water evaporation, which is the direct transfer of water from soil, canopies, and water bodies to the air. Measurement of evapotranspiration is crucial for agricultural irrigation and water resource management since it plays a significant role in the local water cycle and climate. (Figure 6.2).

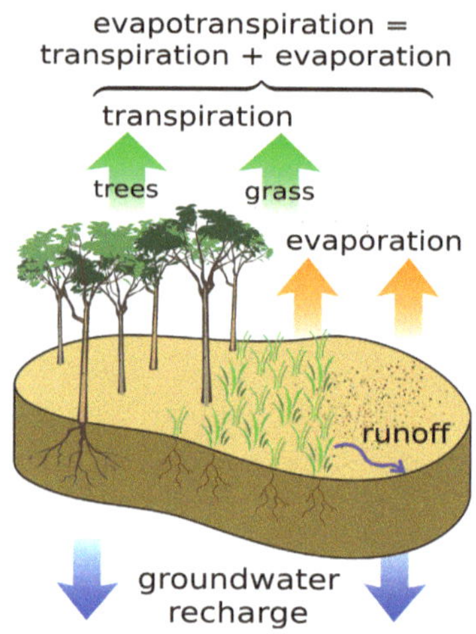

Figure 6.2. *Water cycle of the Earth's surface, showing the individual components of transpiration and evaporation that make up evapotranspiration. Other closely related processes shown are runoff and groundwater recharge.*

Source: By M. W. Toews - Own work, CC BY 4.0, https://commons.wikimedia.org/w/index.php?curid=2843655.

6.2.2.1. Definition

The process by which water is moved from the surfaces of open water and ice, bare soil, and vegetation that make up the Earth's surface to the atmosphere is known as evapotranspiration. In order to better understand crop water requirements, irrigation scheduling, and watershed management, evapotranspiration—a combination of evaporation and transpiration—is measured. The two main parts of evapotranspiration are:

- **Evaporation:** the direct flow of water from sources like soil and water bodies into the atmosphere. A few of the variables that can impact it are temperature, humidity, solar radiation, and wind speed.
- **Transpiration:** the flow of water through a plant's root systems and out into the atmosphere as water vapor. The plant's stomata are where this exit takes place. The kind of plant, the type of soil, the amount of water in the soil, the weather, and cultivation techniques can all affect the rate of transpiration.

Typically, evapotranspiration is expressed in millimeters of water i.e volume of water moved in a predetermined amount of time per unit area of the Earth's surface. Evapotranspiration is thought to return between three-fifths and three-quarters of the land's precipitation to the atmosphere on average worldwide.

Other processes that return water to the atmosphere are not generally taken into account by evapotranspiration, though some of them—such as the sublimation of snow and ice in high-elevation or high-latitude regions—can significantly contribute to atmospheric moisture even in normal circumstances.

6.2.2.2. Influencing Factors

1. *Primary Factors*

   Three main factors determine the amount of evapotranspiration in a given area: First, the amount of water present. Second, the quantity of energy found in the soil and air (e.g., heat, as determined by the earth's surface temperature); and third, the atmosphere's capacity to hold onto water (humidity). Concerning the second factor (heat and energy), rising global temperatures are a result of climate change. Evapotranspiration over land has increased as a result of global warming. One of the ways that climate change is affecting the water cycle is through increased evapotranspiration.

2. *Secondary Factors*

- **Vegetation type**

  The type of vegetation affects evapotranspiration levels. For instance, because they typically have less extensive foliage, herbaceous plants transpire less than woody plants. Additionally, because their roots can draw in more water for the plant and its leaves, plants with deep roots have the capacity to transpire water more continuously. Another illustration is that because conifer forests are evergreen, they typically have higher rates of evapotranspiration than deciduous broadleaf forests, especially during the dormant winter and early spring seasons.

- **Vegetation coverage**

  In regions with a high concentration of vegetation, transpiration makes up a greater portion of evapotranspiration than evaporation. Denser vegetation, such as forests, may therefore decrease water yield and increase evapotranspiration. Rainforests and cloud forests are the two exceptions to this rule. Within cloud forests, trees retain water in the form of fog or low clouds, which eventually trickles down to the ground. Although they frequently accumulate more water than they transpire or evaporate, these trees nevertheless contribute to evapotranspiration. When evapotranspiration (a part of which condenses and returns quickly as precipitation experienced at ground level as rain) increases humidity within the forest, rainforests yield more water than cleared, unforested land in the same climatic zone. Because of its density, the vegetation blocks out sunlight, lowers ground temperature (which lowers losses from surface evaporation), and slows down wind (which lowers the amount of moisture that escapes into the atmosphere). When the rainforest is preserved, the combined effect raises the ground water table and increases surface stream flows. As ground temperatures and wind speeds rise, vegetation cover is lost or purposefully destroyed through clearing and burning, wind reduces soil moisture, and high wind and rainfall events make soils more prone to erosion, clearing rainforests frequently results in desertification.

- **Soil and irrigation**

  Actual evapotranspiration in unirrigated areas is often no higher than precipitation, with some buffer and temporal variations based on the soil's water-holding capacity. It will typically be lower because surface runoff and percolation will lose some water. High water table regions are an exception, as capillary action may allow groundwater to rise through the soil matrix and return to the surface. Without irrigation, soil will dry out until conditions stabilize if potential evapotranspiration exceeds actual precipitation.

6.2.2.3. Measurements

1. *Direct Measurement*

   Direct measurements of evapotranspiration can be made using a pan lysimeter or scale. A lysimeter weighs a plant, the soil around it, and any water that has been added through irrigation or precipitation on a continuous basis. Next, by measuring the change in weight, the change in water storage in the soil is modeled. When applied correctly, this makes it possible to measure evapotranspiration over small areas precisely (Figure 6.3).

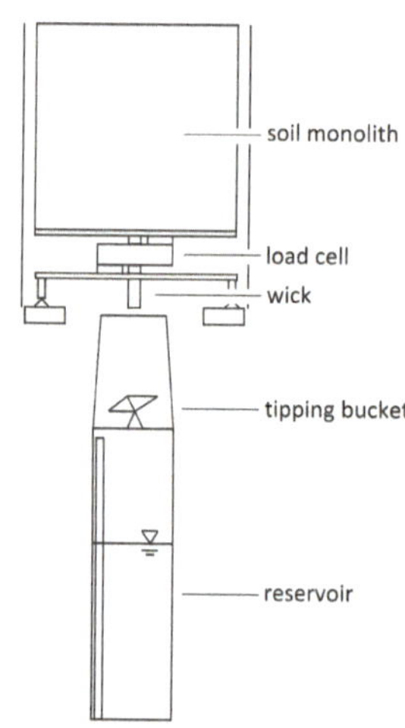

Figure 6.3. *Design for a lysimeter.*

Source: By B. R. Voortman, J. P. M. Witte - B. R. Voortman et. al. "Quantifying energy and water fluxes in dry dune ecosystems of the Netherlands ", Hydrology and Earth System Sciences doi:10.5194/hess-19-3787-2015, CC BY 3.0, https://commons.wikimedia.org/w/index.php?curid=72157572.

2. *Indirect Estimation*

Since it can be challenging or time-consuming to measure atmospheric vapor flux directly, evapotranspiration is usually estimated using one of several non-direct measurement techniques.

- **Catchment Water Balance**

The water balance equation, which links the change in water stored within the basin (S) to its input and outputs, can be used to estimate evapotranspiration for a given area.

ÄS=P-ET-Q-D

The equation links precipitation (P), which is water entering the basin, to evapotranspiration (ET), streamflow (Q), and groundwater recharge (D), which is water exiting the basin, to the change in water stored within the basin (ΔS). If values for the other variables are known, ET can be estimated by rearranging the equation:

$$ET = P - \Delta S - Q - D$$

- **Energy Balance**

An additional estimation methodology involves computing the energy balance.

$$\lambda \mathbf{E} = \mathbf{R}_n - G - H$$

where R_n is the net radiation, G is the soil heat flux, H is the sensible heat flux, and λE is the energy required to change the phase of water from liquid to gas. A scintillometer, soil heat flux plates, or radiation meters are examples of instruments that can be used to calculate the elements of the energy balance and solve for the energy required for actual evapotranspiration. Using satellite imagery, the SEBAL and METRIC algorithms find the energy balance at the Earth's surface. This makes it possible to compute evapotranspiration on a pixel-by-pixel basis, both actual and potential. An important metric for water management and irrigation effectiveness is evapotranspiration. These crucial indicators can be mapped over days, weeks, or years in both space and time using SEBAL and METRIC.

- **Estimation from Meteorological Data**

Reference ET can be computed from meteorological data, such as wind, temperature, and humidity. The Penman equation is the most popular and common formula for determining reference ET. The American Society of Civil Engineers and the Food and Agriculture Organization both advise using the Penman–Monteith variation. Although the more straightforward Blaney-Criddle equation was widely used for a long time in the West, it is less accurate in humid, wet areas. The Makkink equation, which is straightforward but needs to be calibrated to a particular place, and the Hargreaves equations are two other formulas for calculating evapotranspiration from meteorological data.

A crop coefficient and a stress coefficient must be applied in order to convert the reference evapotranspiration to the actual crop evapotranspiration. Crop coefficients, which are a common component of hydrological models, typically fluctuate throughout the year due to the seasonality of crops and the year-round variability of plant behavior. Perennial plants reach maturity over several seasons, whereas annual plants only live for a few, meaning that the way a plant responds to stress can be greatly influenced by a variety of plant characteristics.

6.2.2.4. Potential Evapotranspiration

America in the year 2100 as compared to 1980, using the aggregated findings of several climate models. The amount of water that, in the event that there was enough water available, a particular crop, soil, or ecosystem would evaporate and transpire is known as potential evapotranspiration (PET) or potential evaporation (PE). It reflects the energy needed to transpire or evaporate water as well as the energy provided by the wind to carry the water vapor up from the ground into the lower atmosphere and away from the original location. Potential evapotranspiration is expressed as a percentage of soil moisture or as a depth of water.

If we view actual evapotranspiration as the balance between the atmosphere's need for moisture from a surface and the surface's ability to provide moisture, then PET measures the amount of moisture the atmosphere demands (also known as evaporative demand). Factors such as surface and air temperatures, sunlight, and wind speed all play a role in this process. A dryland is characterized by having annual potential evaporation that surpasses annual precipitation.

At a nearby climatic station, a value for the potential evapotranspiration is typically calculated on a reference surface, typically on land with a predominance of short grass (though this may differ from station to station). The reference evapotranspiration (ET_0) is the name given to this value. When there is enough water present, actual evapotranspiration is said to equal potential evapotranspiration. Evapotranspiration can never be higher than potential evapotranspiration; however, it can be lower in the event that plants are unable to transpire maturely and readily or if there is insufficient water to evaporate. Because the full cover alfalfa reference crop has a higher ET value than the general short green grass reference, some US states use it instead of the general short green grass reference, which is 0.5 m (1.6 ft) in height.

Due to higher levels of solar radiation that provide the energy (heat) for evaporation, potential evapotranspiration is higher in the summer on clearer, less cloudy days and closer to the equator. On windy days, there is also a higher potential for evapotranspiration because more evaporation can quickly replace the moisture that has evaporated from the ground or plant surface before it precipitates.

6.3. SOIL WATER AND INFILTRATION

In semiarid regions, the potential for forage production is most limited by soil moisture. The water efficiency of irrigated and dry-land crop production systems is estimated to be 50%, and the amount of soil water available greatly influences the management choices that farmers make all year long. The amount of moisture in the soil that is suitable for plant growth makes up about 0.01% of the water that is stored worldwide. Water permeability, water-holding capacity, and infiltration are significantly influenced by the texture and structure of the soil.

6.3.1. The Key Components of Soil Texture

The composition of the soil in terms of the percentage of small, medium, and large particles (clay, silt, and sand, respectively) within a particular soil mass is referred to as its texture. For instance, sand or loamy sand is considered coarse soil, loam, silt loam, or silt is considered medium soil, and clay, silty clay, or sandy clay is considered fine soil. The arrangement of soil particles (clay, silt, and sand) into stable units known as aggregates is referred to as soil structure and is what gives soil its structure. Aggregates can form distinct, uniform patterns or they can be friable and loose. Granular structures, for instance, are loose and prone to friability, blocky structures are six-sided and may have rounded or angled sides, and platelike structures are layered and may be a sign of compaction issues.

The space between soil particles, which contains varying amounts of air and water, is referred to as soil porosity. Both soil structure and texture affect porosity. For instance, compared to a coarse soil, a fine soil has more pores, but they are smaller. In comparison to a fine soil, a coarse soil has larger particles but less porosity, or total pore space. Finer soils have the capacity to hold more water than coarse soils because small pores can hold water tighter than large ones.

6.3.2. Infiltration

The process by which water from the ground surface seeps into the soil is called infiltration. It is frequently applied in soil science and hydrology. The maximum rate of infiltration is known as the infiltration capacity. Although it can be measured in other units of distance over time if needed, it is typically measured in meters per day. As the soil moisture content of the soil's surface layers rises, the infiltration capacity falls. Runoff typically occurs when the rate of precipitation surpasses the rate of infiltration, barring a physical barrier.

The devices that can be used to measure the rates of infiltration are rainfall simulators, parameters, and infiltrometers. Adsorption, osmosis, capillary forces, and gravity are some of the causes of infiltration. Numerous attributes of the soil can also influence how quickly infiltration happens.

6.3.2.1. Factors that Affect Infiltration

1. *Precipitation:* Rainfall has various effects on infiltration. Precipitation's type, quantity, and length all matter. Compared to other precipitation events like snow or sleet, rainfall causes faster rates of infiltration. As far as quantity is concerned, more precipitation means more infiltration up until the point at which the ground becomes saturated, at which point the infiltration capacity is reached. The infiltration capacity is also affected by the length of the rainfall. Since the soil is not yet saturated, infiltration occurs quickly when the precipitation event first begins, but as time passes and the soil gets more saturated, the rate of infiltration slows. Runoff is also determined by the relationship between infiltration capacity and rainfall. Runoff will happen if the rate of rainfall exceeds the infiltration capacity.
2. *Soil Characteristics:* The infiltration capacity of soils is largely dependent on their porosity. In comparison to soils with large pore sizes, like sands, those with smaller pore sizes, like clay, have a lower infiltration capacity and slower infiltration rates. This rule does not apply in the case of dry conditions when clay is present. Large cracks in the soil may form in this situation, increasing the infiltration capacity. Compaction of the soil affects infiltration capacity as well. Reduced porosity in the soils caused by compaction lowers the soils' ability to allow infiltration. After wildfires, soils that are hydrophobic can form, which can significantly reduce or even completely stop infiltration.
3. *Soil Moisture Content:* The infiltration capacity of saturated soil has been reached, and the rate cannot rise above this point because the soil is no longer able to hold more water. Much more surface runoff results from this. A partially saturated soil can infiltrate at a moderate rate, while a fully unsaturated soil can infiltrate at the greatest rate.
4. *Organic Materials in Soils:* The infiltration capacity is increased by all organic materials present in the soil, including both plants and animals. Because the roots of vegetation grow deep into the ground, the soil becomes cracked and fractured, facilitating faster infiltration and greater capacity. Additionally, vegetation can lessen the soil's surface compaction, increasing infiltration once more. In the absence of vegetation, infiltration rates may be extremely low, resulting in excessive runoff and elevated levels of erosion. Animals that burrow in the soil produce fractures in the soil structure much like vegetation does.
5. *Land Cover:* Water cannot permeate through an impermeable surface, such as pavement, hence infiltration is not possible if the land is covered in it. There is also more runoff

as a result of this relationship. Storm drains in impermeable areas frequently empty straight into bodies of water, preventing infiltration. The land's vegetative cover affects the infiltration capacity as well. Increased precipitation interception from vegetative cover can reduce precipitation intensity, resulting in less runoff and increased interception. Elevated evapotranspiration resulting from a greater vegetation abundance can also lower the rate of infiltration. By shielding the soils from heavy precipitation events, vegetation debris like leaf cover can also speed up infiltration. The amount of litter on the ground and the basal cover of perennial grass tufts determine how quickly a certain soil infiltrates semi-arid savannas and grasslands. The infiltration rate beneath a litter cover on sandy loam soils can be nine times higher than that of bare surfaces. The existence of a soil crust or surface seal is primarily responsible for the low rate of infiltration in bare areas. Water quickly seeps through a tuft's base and is directed toward the roots of the tufts themselves.

6. *Slope:* When the slope of the land is higher runoff occurs more readily which leads to lower infiltration rates.

6.3.2.2. Process

Only when there is space for more water at the soil's surface can the infiltration process proceed. The porosity of the soil and the speed at which water that has already permeated the soil can move away from the surface determine how much more water is able to be absorbed by the soil. The infiltration capacity is the fastest rate at which water can permeate soil under specific conditions. When water arrival at the soil surface is less than the capacity for infiltration, it can occasionally be analyzed using mathematical models called hydrology transport models, which take runoff, infiltration, and channel flow into account to forecast river flow rates and stream water quality.

6.3.2.3. Infiltration in Wastewater Collection

Systems for collecting wastewater are made up of lift stations, junctions, and lines that transport sewage to a wastewater treatment facility. Stormwater inflow/infiltration frequently happens when these pipes are damaged by fracturing, bursting, or tree root incursion. This situation may result in the overflow of a sanitary sewer or the release of raw sewage into the surrounding area.

6.3.2.4. Infiltration Calculation Methods

A part of the general mass balance hydrologic budget is infiltration. The amount and/or rate at which water percolates into the soil can be estimated in a number of ways. The rigorous standard that solves Richards' equation numerically fully couples surface water to groundwater through non-homogeneous soil. A more recent approach, the Finite water-content vadose zone flow method solution of the Soil Moisture Velocity Equation, permits 1-D groundwater and surface water coupling in homogeneous soil layers and is connected to the Richards equation. For a single rainfall event, there are some good approximation methods to solve the infiltration flux in the case of uniform initial soil water content and deep, well-drained soil. The methods of Green and Ampt (1911), Parlange et al., (1982). There are numerous other empirical techniques in addition to these, like the SCS method and Horton's method. that are essentially exercises in curve fitting.

6.3.2.5. General Hydrologic Budget

The overall hydrologic budget, including all elements, concerning infiltration factor F. The infiltration question can be answered with basic algebra given all the other variables and infiltration as the only unknown.

$$F = BI + P - E - T - ET - S - I_A - R - B_O$$

where, F is infiltration, which can be measured as a volume or length; BI is the boundary input, which is essentially the output watershed from adjacent, directly connected impervious areas; B_O is the boundary output, which is also related to surface runoff, R, depending on where one chooses to define the exit point or points for the boundary output; P is precipitation; E is evaporation; T is transpiration; ET is evapotranspiration; S is the storage through either retention or detention areas; I_A is the initial abstraction, which is the short-term surface storage such as puddles or even possibly detention ponds depending on size; R is surface runoff.

The only thing to keep in mind when using this method is that doubles are common, so you have to be careful about which variables to include and which to leave out. When the evaporation (E), transpiration (T), and evapotranspiration (ET) are included in the equation, it is a simple example of double counting variables. ET has part of E and part of T included in it. Not only raw precipitation but also interceptions must be taken into consideration.

6.3.2.6. Richards' Equation (1931)

A partial differential equation with highly nonlinear coefficients, Richards' equation is the standard rigorous method for computing infiltration into soils. The Richards equation can occasionally have issues with mass conservation, is computationally costly, and is not always guaranteed to converge.

6.3.2.7. Finite Water-content Vadose Zone Flow Method

The partial differential equation of Richards (1931) is approximated using this method, which downplays the importance of soil water diffusion. This was determined by utilizing specific forms of the soil constitutive relations to compare the solution of the advection-like term of the soil moisture velocity equation against exact analytical solutions of infiltration. The approximation's impact on the calculated infiltration flux was found to be negligible due to the small diffusive flux. Additionally, the finite water-content vadose zone flow method was found to be a valid solution to the set of three ordinary differential equations, with guaranteed mass conservation and convergence. It is predicated on the ideas that soil is uniform within layers and that flow only happens in a vertical direction (one dimension).

6.3.2.8. Green and Ampt

Green and Ampt are the two men from whom the name was derived. When estimating infiltration, the Green-Ampt method takes into consideration a lot of factors that other methods—like Darcy's law—do not. It depends on the time, porosity, hydraulic conductivity, and soil suction head.

$$\int_0^{F(t)} \frac{F}{F + \psi\,\Delta\theta}\,dF = \int_0^t K\,dt$$

where, φ is wetting front soil suction head (L); θ is water content (-); K is hydraulic conductivity (L/T); F(t) is the cumulative depth of infiltration (L).

Once integrated, one can easily choose to solve for either volume of infiltration or instantaneous infiltration rate:

$$F(t) = Kt + \psi\,\Delta\theta \ln\left[1 + \frac{F(t)}{\psi\,\Delta\theta}\right].$$

By solving for F(t), one can quickly find the volume using this model. The variable that needs to be solved for, though, is part of the equation itself, so the variable in question needs to be set to converge on zero or another suitable constant. A good first guess for F is the larger value of either Kt or $\sqrt{2\psi\,\Delta\theta Kt}$. These values can be obtained by solving the model of the zeroth and second orders, respectively, with a log substituted for its Taylor-Expansion around one. When applying this formula, the only thing to keep in mind is that it is necessary to assume that h_0, or the water head—the depth of ponded water above the surface—is very small. The instantaneous infiltration rate at the time, t, F was measured can then be found by substituting F into the corresponding infiltration rate equation below using the infiltration volume from this equation.

$$f(t) = K\left[\frac{\psi\,\Delta\theta}{F(t)} + 1\right].$$

6.3.2.9. Horton's Equation

Named after the same Robert E. Horton mentioned above, another useful tool for determining ground infiltration volumes or rates is Horton's equation. According to an empirical formula, infiltration begins at a constant rate (f_0) and decreases exponentially over time (t). The rate of infiltration will eventually level off to the rate f_c when the soil saturation level reaches a particular value.

$$f_t = f_c + (f_0 - f_c)e^{-kt}$$

where, f_t is the infiltration rate at time *t*; f_0 is the initial infiltration rate or maximum infiltration rate; f_c is the constant or equilibrium infiltration rate after the soil has been saturated or the minimum infiltration rate; k is the decay constant specific to the soil.

Another way to apply Horton's equation is the following. It can be utilized to calculate the overall amount of infiltration, F, at a given time t.

$$F_t = f_c t + \frac{(f_0 - f_c)}{k}(1 - e^{-kt})$$

6.3.2.10. Kostiakov Equation

An empirical equation that is named after its creator, Kostiakov, makes the assumption that the intake rate decreases with time in accordance with a power function.

$$f(t) = akt^{a-1}$$

where a and k are empirical parameters.

This expression's primary drawback is its dependence on the zero final intake rate. Instead, the infiltration rate typically approaches a finite steady value, which occasionally may happen quickly. This is fixed by the Kostiakov-Lewis variant, sometimes referred to as the "Modified Kostiakov" equation, which includes a steady intake term in the original formula.

$$f(t) = akt^{a-1} + f_0$$

in integrated form, the cumulative volume is expressed as:

$$F(t) = kt^a + f_0 t$$

where

f_0 approximates but does not necessarily equate to the final infiltration rate of the soil.

6.3.2.11. Darcy's Law

A condensed form of Darcy's law is applied in this infiltration technique. Many would contend that using this method is inappropriate because it is too simple. Compare it to the previously mentioned Green and Ampt (1911) solution. Assuming that the infiltration gradient occurs over an arbitrary length L, this method is similar to Green and Ampt but lacks the cumulative infiltration depth. As a result, it is incomplete. In this model the ponded water is assumed to be equal to h_0 and the head of dry soil that exists below the depth of the wetting front soil suction head is assumed to be equal to $-\varphi - L$.

$$f = K\left[\frac{h_0 - (-\psi - L)}{L}\right]$$

where, φ is wetting front soil suction head; h_0 is the depth of ponded water above the ground surface; K is the hydraulic conductivity; L is the vague total depth of subsurface ground in question. This vague definition explains why this method should be avoided.

or

$$f = K\left[\frac{L + S_f + h_0}{L}\right]$$

where, f Infiltration rate f (mm hour$^{-1)}$); K is the hydraulic conductivity (mm hour$^{-1)}$); L is the vague total depth of subsurface ground in question (mm). This vague definition explains why this method should be avoided; S_f is wetting front soil suction head $(-\varphi)$ = $(-\varphi f)$ (mm); h_0 is the depth of ponded water above the ground surface (mm).

6.4. RECHARGE ESTIMATION

A vital part of any watershed's water balance is groundwater recharge. Nonetheless, a variety of techniques, varying greatly in complexity and expense, have been employed to estimate recharge because direct measurement is practically impossible (Lerner et al., 1990; Scanlon et al., 2002). The most accurate recharge estimates are usually achieved by practicing hydrologists through the use of relatively simple methods that require only readily available hydrologic data.

Recharge is usually calculated through an examination of stream flow records, ground-water levels, or the water balance for a watershed in the humid eastern United States, where the majority of streams are growing and the water table is relatively shallow. While acknowledging that it is likely less than the amount replenishing the ground-water system, base flow has occasionally been utilized as an approximation of recharge (Daniel, 1996; Holtschlag, 1997; Szilagyi et al., 2003). Recharge should be estimated using a variety of techniques, with the outcomes compared, according to a standard recommendation (Nimmo et al., 2003; Healy and Cooke, 2002). Although good-quality data are typically not available to make estimates from multiple methods, this is a prudent approach none the less. However, there are two hydrologic research sites in east-central Pennsylvania where long-term climate, surface water, ground water, and unsaturated zone monitoring allows comparison of various approaches to estimating ground-water recharge with available data. The websites are run by the U.S. Department of Agriculture, Agricultural Research Service (ARS), as a component of the Pasture Systems and Watershed Management Research Unit Research Watershed. These ARS sites provide long-term, continuous hydrologic records that are typical of the humid continental climate of the northeastern United States. In addition, they contain streamflow data from gages in nested watersheds and measurements of unsaturated-zone drainage from gravity-drainage lysimeters—a dataset that is rarely available. In collaboration with the ARS, this study was carried out as part of the U.S.

Ground-Water Resources Program, United States Geological Survey (USGS) (Grannemann, 2001). It was one of many research projects aimed at improving our knowledge of techniques for calculating recharge in the humid eastern United States.

There are several techniques for estimating the natural and artificial recharge of the aquifer; the choice of technique relies on the data that is available, the topography and geography of the area, the necessary spatial and temporal scale, and the accuracy of the results produced by the various techniques. As stated by Scanlon et al., (2002) methods based on groundwater data provide estimates of actual recharge, those based on surface water and unsaturated zone data provide estimates of potential recharge. He recommended using multiple techniques to increase the reliability of the results because of the uncertainties associated with each approach.

6.4.1. Water Table Fluctuation (WTF) Method

The fundamental premise of the WTF method is that recharge water reaching the water table is the only reason for the increase in groundwater level in an unconfined aquifer. The recharge is computed as

$$R = S_y\ dh/dt = S_y\ \Delta h/\Delta t$$

where, R = rate of recharge (LT^{-1}); Sy = specific yield, ($M^0L^0T^0$); Dh = Δh = water table rise, (L); Dt = Δt = time within which rise dh takes place, (T).

The calculated value of R can be multiplied by the surface area of the aquifer to determine the recharge rate in terms of volume per unit time.

Water arrives in an area and takes some time to redistribute to other components such as base flow, groundwater evaporation, and net sub-surface flow. The WTF method can be used to estimate changes in subsurface storage over longer time periods, such as seasonal or annual intervals. Healy et al., (2002) stated that numerous studies have employed the WTF method for groundwater recharge estimation since it was first used in the 1920s.

Since no assumptions are made regarding the mechanism by which water reaches groundwater, the method is fairly straightforward. The technique also has certain drawbacks. Only unconfined aquifers can be used with the water table fluctuation method, which is unable to account for a steady rate of recharge. This means that water levels won't change and the WTF method won't predict any recharge if the rate of recharge from a given area is equal to the rate of drainage. The computation of particular yield values presents additional challenges.

This approach for estimating groundwater recharge has been tested by numerous researchers. Allison et al., (1990) estimated artificial recharge in southern Australia using the water table fluctuation method. After the native vegetation was cleared, they noticed a steady increase in groundwater levels of 0.1 m/year. With a specific yield of 0.2, this translates to a 20 mm/year increase in recharge. This value was discovered to be in line with the recharge calculated using different, independent techniques. Healy and Cook (2002) provided in-depth analyses of the groundwater recharge estimation techniques that rely on groundwater level data. They came to the conclusion that the most popular approach for estimating groundwater recharge might be the WTF method, which makes use of specific yield and changes in water table level over time.

6.4.2. Water Budget Method

The water budget equation serves as the foundation for the water budget methods. A basin's water budget can be expressed as

$$P + Q_{on} = ET + Q_{off} + \Delta S$$

where, P = precipitation (and may also include irrigation) (mm/day); Q_{on} and Q_{off} = water flow onto and off the site (surface flow, interflow and groundwater flow) (mm/day); ET = evapotranspiration (mm/day) and ΔS = change in storage (mm/day). Schict and Walton (1961) developed the following budget equation for recharge estimation based on the water balance equation mentioned above:

$$R = \Delta S^{gw} + ET^{gw} + (Q^{gw}_{off} - Q^{gw}_{on}) + Q^{bf}$$

where, R = recharge; ΔS^{gw} = change in subsurface storage; Q^{bf} = base flow; ET^{gw} = evaporation from groundwater; and $Q^{gw}_{off} - Q^{gw}_{on}$ = net surface flow from the basin.

All other parameters in the model above, with the exception of R, are measurable or estimable. This approach is applicable to a broad variety of temporal and spatial scales. However, a significant drawback of this strategy is that the precision with which other elements of the water balance equation are measured determines how accurate the recharge estimates will be (Scanlon et al., (2002).

6.4.3. Darcy's Law

According to Darcy's law, the fluid flux, such as recharge in an aquifer system, can be determined by knowing both the head gradients and hydraulic conductivities. This law is used to calculate recharge (R) in the saturated zone using the following equation:

$$R = -\frac{K(\theta)dH}{dz} = -K(\theta)\frac{d}{dz}(h+z) = -K(\theta)\left(\frac{dh}{dz}+1\right)$$

where, $K(\theta)$ = hydraulic conductivity at the ambient water content; H = total head; h = metric potential head; z = horizontal distance between the two points where hydraulic head is measured.

The vertical total head gradient and the unsaturated hydraulic conductivity at the ambient soil-water content must be measured or estimated in order to apply Darcy's law.

Numerous studies conducted in arid and semiarid environments have used this methodology.

The value of the metric potential head can be assumed to be 1 in regions of unform porous media where a thick unsaturated zone exists. The unit-gradient assumption sets recharge equal to the hydraulic conductivity at ambient water content and eliminates the need to measure the metric pressure gradient.

6.4.4. Empirical Relationships

Seasonal groundwater balance studies can also be used to develop empirical relationships between rainfall and groundwater recharge. One such attempt was made for the Upper Ganga Canal command area by Kumar and Seethapathi (2000). By fitting the estimated values of rainfall recharge and the corresponding values of rainfall during the monsoon season using non-linear regression techniques, an empirical relationship for the estimation of ground water recharge was proposed. The equation as illustrates the relationship between recharge and rainfall.

$$R = 0.63\ (P - 15.28)^{0.76}$$

where, R = recharge (m); P = precipitation (m).

6.4.5. Groundwater Models

There is still a significant degree of uncertainty in field recharge measurements, and there is a lot of research being done on the topic (Sanford, 2002). In addition to the range of methods employed for field measurements, researchers have estimated recharge using groundwater models. Based on the geologic characteristics and rate of recharge, models can also be used to forecast the distribution of recharge in temporal and spatial scales.

Groundwater system studies make extensive use of models for groundwater flow and contaminant transport.

The rate and direction of groundwater flow through subsurface aquifers and confining units are determined by groundwater flow models. Simulations are the names given to these computations. The hydro-geologic features of the site must be thoroughly understood in order to simulate groundwater flow.

In order to accurately predict groundwater flow directions and the movement of contaminants, the model must be successfully calibrated and verified. Two significant problems with groundwater models have been brought to light by Sanford (2002). Since groundwater recharge is a basic feature of the majority of groundwater models, it is important to evaluate how recharge is depicted in the models and how it is calculated using groundwater models. Using groundwater models is incredibly beneficial. Groundwater modeling exercises have improved the analysis suggested by the artificial recharge scheme.

6.4.6. Tracer Techniques

In recent times, the use of heat or chemical isotopic tracers has become increasingly significant in determining groundwater recharge. Analyzing the concentration of environmental tracers that indicate groundwater age has become a popular method in this field. Isotopic tracers are proving to be a valuable tool in groundwater research. According to Coplen (1993), there has been significant technological advancement in the application of isotopic analyses to groundwater hydrology, which is helping to interpret and define groundwater flow paths, ages, recharge areas, leakage, and interactions with surface water.

Datta (1999) looked into the occurrence and recharge of groundwater in Delhi's National Capital Territory (NCT) using the signatures of ^{18}O isotopes. These signatures showed that the aquifer in the region is not a homogeneous system in terms of lateral extent and that the groundwater in the Delhi area's wells is a mixture of various proportions of different water sources.

Owing to the significant uncertainties associated with measuring each method's individual parameters, numerous researchers (Healy and Cook, 2002, Scanlon et al., 2002) have indicated that using a variety of estimation techniques is very helpful in obtaining results that are somewhat trustworthy. Three separate techniques were employed by McCartney and Houghton (1998) to calculate the groundwater recharge on Jersey's Channel Island. Chloride balance (a), stream base flow analysis (b), and rainfall-recharge-runoff simulation (c) are these. The results from all three approaches were fairly consistent, showing that 16–19% of the average annual rainfall is the long-term recharge. Additionally, modeling results suggest that in 5 out of 28 years, groundwater abstraction may have exceeded recharge.

A vital part of any watershed's water balance is groundwater recharge. Nonetheless, a variety of techniques have been employed to estimate recharge because it is practically hard to measure directly. In certain situations, base flow has been utilized as an approximation of recharge. Although it's generally advised that recharge be estimated using a variety of techniques and the results compared, in practice this may be challenging due to methodological differences. Understanding the various recharge mechanisms and their significance in the study area is crucial for estimating natural ground water recharge. The study's goals, the data that are currently available, and the opportunities to obtain more data should all be taken into consideration when choosing a methodology. The economy also plays a significant role. Estimates, however, typically contain significant errors. There is currently no one complete estimation method that stands out from the variety of approaches that produce trustworthy outcomes. As a result, using multiple methods based on different input data is preferable.

6.5. STREAM GAUGING TECHNIQUES

Hydrologists and environmental scientists use a stream gauge, also known as a streamgage or stream gauging station, to test and observe terrestrial bodies of water. In addition to biota and water quality observations, hydrometric measurements of water level, surface elevation ("stage"), and/or volumetric discharge (flow) are typically made. Gauging station locations are frequently indicated on topographical maps. A central data logging facility may receive telemetry data from some highly automated gauging stations.

6.5.1. Measurement Equipment

Currently, automated direct stream discharge measurement is challenging. From a mathematical perspective, quantifying stream discharge involves calculating the volumetric flow rate, which is typically a flux integral and necessitates numerous cross-sectional velocity measurements. One or more proxy variables, such as surrogate measurements, can be used to generate discharge values in lieu of directly measuring stream discharge. As the surrogate, a stage measurement—that is, the elevation of the water surface—is employed most of the time. Variable downstream channel conditions have a significant impact on low gradient (or shallow-sloped) streams.

A second stream gauge would be put in for these streams, and the difference in the gauges' slopes would be used to determine the water's surface slope. Combined with the stage measurement, this value would help determine the streamflow discharge more precisely. At locations with a stable cross-sectional area, the use of water velocity as a trustworthy stand-in for streamflow discharge has also been made possible by advancements in velocity sensor accuracy. These sensors measure the velocity at a specific point in the stream and are permanently installed in it. When a stage measurement is the only available surrogate, a rating curve needs to be created. The functional relationship between stage and discharge is known as a rating curve. The cross-sectional area of the channel is ascertained by taking numerous, discrete measurements of the streamflow discharge using a velocimeter and other tools to measure the channel geometry. In order to measure the hydrologic extremes (droughts and floods) and make a discharge measurement, the technicians and hydrologists in charge of creating the rating curve make frequent trips to the site and adhere to specific guidelines known as standard operating procedures (SOPs). (Figure 6.4). The volumetric streamflow discharge can be ascertained by using the rating curve in conjunction with stage measurements once it has been established. The amount of water that passes by the stream gauge is then evaluated using this record, which is helpful for numerous hydrological tasks.

Figure 6.4. *December 12, 2001 photo of the USGS streamflow-gaging station at Huey Creek, McMurdo Dry Valleys, Antarctica.*

Source: By Jerrod Wheeler (submitted to USGS by Kirk Miller, Wyoming) - USGS Surface Water Photo Gallery, item 10 10, Public Domain, https://commons.wikimedia.org/w/index.php?curid=9016574.

An index velocity determination is carried out in those cases where a velocity measurement is also utilized as a surrogate. In this analysis, the flow velocity at a specific point in the stream cross section is measured using a velocity sensor, which is typically either magnetic or acoustic. Once more, the technician or hydrologist measures the streamflow discharge discretely at different points in time. By dividing streamflow discharge by cross-sectional area, the mean velocity of the cross section is found for each discrete determination of streamflow discharge. The mean velocity and the index velocity from the permanently mounted meter are used to create a rating curve that resembles the one used for stage-discharge calculations. A second rating curve that connects the stream's stage to its cross-sectional area is created. The automatically collected stage yields an estimate of the cross-sectional area using these two ratings, and the automatically collected index velocity yields an estimate of the cross-section mean velocity. The product of the estimated cross section area and the estimated mean velocity of the streamflow is used to calculate the streamflow discharge.

To increase the accuracy of the rating table by enhancing the dependability of using water level as a substitute for flow, a range of hydraulic structures and primary devices are employed, such as:

- Weirs
    - V-notch,
    - broad-crested,
    - sharp-crested and
    - combination weirs
- Flumes
    - Parshall flume

Other equipment commonly used at permanent stream gauge include:

- Cableways - for suspending a hydrographer and current meter over a river to make high flow measurement
- Stilling well - to provide a calm water level that can be measured by a sensor
    - Water level gauges:
- Staff (head) gauges - for a visual indication of water depth
- Water pressure measuring device (Bubbler) - to measure water level via pressure (typically done directly in-stream without a stilling well)
- Stage encoder - a potentiometer with a wheel and pulley system connected to a float in a stilling well to provide an electronic reading of the water level
- Simple ultrasonic devices - to measure water level in a stilling well or directly in a canal.

- Electromagnetic gauges

In the absence of a recognized stream gauge, a current meter or an Acoustic Doppler current profiler can be used to measure the discharge of a stream or canal. The "float method," which involves observing a floating object, like a piece of wood or an orange peel, floating down a stream, is an unofficial technique that is not appropriate for any official or scientific purpose but can be helpful.

6.5.2. Methods for Steam Gauging

The most widely utilized techniques for stream gauging are: 1. Method of Area-Velocity 2. Area Slope Approach 3. Structures for measuring discharge.

6.5.2.1. Area-Velocity Method

It should be noticed that a stream's depth of flow varies in a section across the direction of the main flow. Similar to this, the flow velocity varies throughout the stream's cross-sectional area. Therefore, the most effective way to measure stream flow would be to first ascertain the velocity variation at a channel section, then integrate (i.e., sum) the products of the flow cross-section's velocity in a specific small elemental area with that area. To estimate the stream flow, other, less complex measurements are made because, in most cases, this may not be feasible. For instance, a flow section can be split into an appropriate number of vertical strips, each with a measured average depth. A current meter is used in each of these strips to measure the velocity at 0.2 h and 0.8 h depths below the free surface. Here, h represents the strip's flow depth.

The average flow velocity in any given strip can be roughly calculated by averaging these two values. By multiplying the average velocity and strip area for each individual strip by the sum, the total discharge in the stream can be found, i.e.,

$$Q = \sum_{i=1}^{N} Qi = \sum_{i=1}^{N} B_i h_i v_i$$

where N strips make up the flow cross-section and B_i, h_i, v_i, and Q_i stand for width, average depth, average velocity, and discharge of the ith strip. An alternative method of determining the average velocity in a strip is to measure the velocity solely at 0.6 h depth below the free surface, then take the value and apply it to the average velocity in the strip. An approximate average velocity can also be obtained by measuring the surface velocity within the strip and multiplying it by 0.9.

6.5.2.2. Area-Slope Method

Determining peak flow conditions and measuring velocity within that duration is typically challenging during floods. Installing two gauges at different distances apart that have the same datum is comparatively easier. These gauges read the water's surface levels during floods. Occasionally, flood level markers on the banks can also serve this function. By calculating the area of flow cross-section at the two gauge locations and using the river cross-section, one can calculate the average value of area A. Likewise, it is also possible to obtain the hydraulic radius R.

The gauge level readings and the separation between the two gauge stations can also be used to calculate the water surface slope S. By using the Manning's equation and assigning a suitable value to the Manning's n, one can predict the discharge Q in the stream,

$$Q = \frac{1}{n} AR^{2/3} S^{1/2}$$

6.5.2.3. Discharge Measuring Structures

The most commonly used structure for the purpose of discharge measurement in a river is a weir which is, generally, either triangular (for relatively small discharges) or rectangular (for relatively large discharges) in shape.

These weirs are made to resemble a barrier built over a river. The discharge Q flowing over a weir is related to the head over the weir, i.e., H and the weir characteristics. For instance, the discharge Q of a triangular weir with notch angle ⊠ is expressed as.

$$Q = \frac{8}{15} C_d \sqrt{2g} H^{ü} \quad \tan \theta / 2$$

Likewise, for rectangular sharp-crested weir

$$Q = \frac{2}{3} C_d B \sqrt{2g} H^{3/2}$$

in which C_d is the Coefficient of discharge, B the length of weir and g is the acceleration due to gravity.

6.6. HYDROGRAPH ANALYSIS

Analyzing hydrographs allows one to see how water levels vary over time. Rivers, reservoirs, and other bodies of water can have their water flow understood and managed with the help of this information. Engineers and planners can use hydrographs to better understand where and how to use water resources.

There are various varieties of hydrographs, and each has unique properties. The most popular kind is the stage-discharge hydrograph, which displays the volume of water released from a reservoir or river as well as the change in water level over time. The recession curve and the detention-time curve are two more forms of hydrographs. A recession curve illustrates the change in volume or rate of discharge over time as a result of evaporation and other factors, whereas a detention-time curve illustrates how long it takes for water to move through a reservoir.

6.6.1. Analysis of Flood Hydrograph

Determine the factors influencing the peak discharge and volume passing through a specific reach by analyzing flood hydrographs. This can assist engineers in creating structures like culverts and bridges that can better withstand large flows and divert water from flood-prone areas. Flood control techniques are frequently used to manage the amount of water entering reservoirs and to stop floods in rivers and streams.

6.6.2. Meaning of Flood Hydrograph

An illustration of the volume of water flowing through a river or stream during a flood is called a flood hydrograph. It displays the maximum volume of water that leaves one place and the maximum discharge rate at which water enters another. Analysis of flood hydrographs can be used to pinpoint high-risk areas and assess the magnitude of a flood event.

6.6.3. Analysis of Unit Hydrographs

A unit hydrograph (UH) shows the runoff from a drainage basin over time in a graphical format. It can assist in determining and managing the water level in reservoirs, rivers, and other bodies of water. The rising limb and the falling limb are the two components that make up the UH. The rising limb symbolizes the gradual rise in water level, while the falling limb denotes the gradual fall in water level. The drainage characteristics of the area dictate how long each part (rising and falling) lasts.

Several factors can influence a unit hydrograph, including:

- The size and shape of the drainage basin
- The amount and type of precipitation
- The elevation of the drainage basin
- The surface roughness of the land
- The rate of groundwater recharge
- The rate of evapotranspiration

By comprehending these elements, we can more accurately forecast changes in water levels over time and implement control measures as needed. For instance, we can release water from the reservoir to reduce flooding if we are aware that a significant storm is approaching.

6.6.4. Applications of Hydrographs

There are several uses for hydrographs in the analysis of water levels. A hydrograph can be used to quickly and easily determine the peak flow time and magnitude, which is crucial information for designing culverts or bridges that are based on river channels. Engineers use flood frequency curves, also known as recurrence intervals, which we can determine from hydrographs in order to design levees, detention ponds, and other structures that are intended to protect people and infrastructure from flooding. Hydrographs can also be analyzed to learn more about the functioning of a watershed and improve the management of available water supplies. Understanding the volume of water flowing through a watershed and its rate of flow is one of the most crucial parts of hydrology. Hydrographs are useful in understanding both the velocity and the amount of discharge (flow), which can be used to identify areas that will be inundated by flooding.

Mathematical systems known as hydrological models are employed to forecast hydrographs by utilizing historical data. Hydrologists can forecast future water levels and how they will fluctuate due to a variety of factors, including urbanization and climate change, with the aid of these models.

6.7. SURFACE WATER – GROUNDWATER INTERACTION

Connectivity between surface and groundwater pertains to the direction and volume of water flow between above-ground and below-ground resources. The climate, geology, and topography are some of the factors that can alter the magnitude and direction of these flows. Evaluating surface water-groundwater interactions is a challenging and intricate process. To evaluate the degree of connectivity and comprehend the mechanisms governing the transfer of water from surface to subsurface storages, a variety of instruments are accessible.

Groundwater extraction has increased significantly in Australia in recent years due to rising water demands and a decrease in surface water availability. Allocating the same water twice is a risk associated with the independent management of surface water and groundwater.

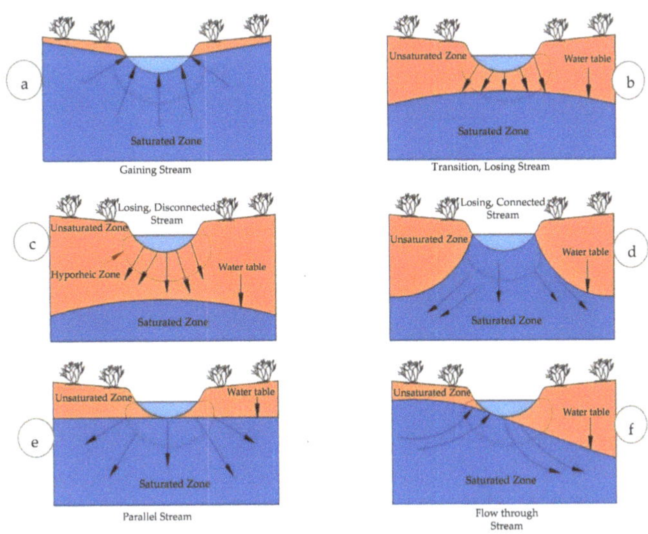

Figure 6.5. *Different stream-water and groundwater interaction scenarios, (a) gaining stream, (b) transition-losing stream, (c) losing-disconnected stream, (d) losing-connected stream, (e) parallel stream, (f) flow-through stream. The arrows denote the directions of fluid flow.*

Source: Dolon Banerjee, Sayantan Ganguly, Groundwater–Surface Water Interaction, 18 Apr 2023, 10.3390/w15081552.

During the period of low flow, fresh groundwater discharge into a stream is essential for aquatic ecosystems and surface water users. The amount of baseflow that is pumped into the stream from an aquifer next to a river can vary significantly. Conversely, increased groundwater discharge may be detrimental to surface water quality if the groundwater is contaminated or salty. Comprehending the degree of connectivity between surface water and groundwater is essential for the sustainable administration of the entire water resource. (Figure 6.5).

There are various scenarios and methods in which surface water features and aquifers can interact. Depending on the season, a river may lose water to its aquifer or receive groundwater from it along its entire course.

A stream reach, lake or wetland can either be:

- Gaining Stream - receiving inflow of groundwater
- Losing Stream - losing water to the groundwater system by leakage to the aquifer.

CASE STUDY

INFILTRATION BASIN, SAO PAULO

- **SuDS used**
  - Infiltration Plan
  - Water quality and quantity management
- **Benefits**
  - Encourage the regulation of low-impact local planning and development;
  - Encourage the prevention of soil and groundwater pollution;
  - Preserve the areas of vegetation cover so that there is an increase of the resistance in the flow, reducing the speed of flow and increasing the time of course of the roof and the roads to the drainage system;
  - Conduct runoff to ensure clean water enters the natural drainage system;
  - Reduce costs of construction and maintenance of drainage infrastructure.

Location

Infiltration basin at Federal University of Sao Carlos, Sao Paulo, Brazil

Description

The University Federal of Sao Carlos campus is sectorized in North Area and South Area and in its northern area there is a space destined for the expansion of the university. In the same locality there is a weather station that performs the monitoring of the rains. Because of this two characteristics, was delimited a microbasin for study with 4,65ha, where a set of Best Management Practices (BMPs) were implemented to promote a sustainable urban development and to reduce the urbanization impact of the campus (Figure 1).

For defined the possibilities implantation for BMPs at microbasin, were verified the following characteristics: the physical aspects, including topography, the permanent water supply, the existence of a permanent exudate and the level of the water table; the urban and infrastructure aspects, with the space availability and the existence of underground infrastructure; and the environmental and sanitary aspects, where were verified the pollution and sanitarian risk beyond the costs of implantation. This way was defined two infiltration plan, as it accommodate big volumes and need large areas, and which in this project will be connected through of a lawn canal.Until this moments, were executed the lawn canal, with a extension of 80m and deep of 0,50m and one infiltration plan with a ellipse arc form, with de semi axis bigger of 25,85m and the semi axis smaller of 19,90m, supporting a total volume of 111,34m^3 (Figure 2).

Figure 1.

Source: https://www.susdrain.org/case-studies/pdfs/suds_awards/010_18_03_30_susdrain_suds_awards_infiltration_basin_sao_paulo.pdf.

Figure 2.

Source: https://www.susdrain.org/case-studies/pdfs/suds_awards/010_18_03_30_susdrain_suds_awards_infiltration_basin_sao_paulo.pdf.

How it Works

At first, tests were carried out that verified a high degree of surface compaction of the soil, which prevented the infiltration of rainwater. Double ring tests were performed, which confirmed the infiltration rates practically null before the construction of the plane. Thus, the recovery of the infiltration capacity was predicted by lowering the surface of the land by 0,30 m for water storage and decomposition of the soil with scarification of 0,50m.

Figure 3.

Source: https://www.susdrain.org/case-studies/pdfs/suds_awards/010_18_03_30_susdrain_suds_awards_infiltration_basin_sao_paulo.pdf.

The first stage of the construction of the infiltration plan was the location of the project. To achieve the desired levels of design, it was necessary, with an excavator, to lower the ground by 0,30m. After the terrain cut, a topographic survey was carried out to verify the required quotas. Then, the soil was subsoiled up to 0,50m, with subsequent manual adjustment to avoid compacting the soil, because the subsoil increased the terrain level and caused the loss of project quotas. The connection of the building with the plan was made by a canal. After these steps the grass was planted and a bottom drain pipe was installed as an alternative drainage drain.

Figure 4.

Source: https://www.susdrain.org/case-studies/pdfs/suds_awards/010_18_03_30_susdrain_suds_awards_infiltration_basin_sao_paulo.pdf.

The operation of the Infiltration Plan takes place following the following steps: step 1 - the water precipitated on the roof is directed by the pluvial building installation to a channel; step 2 - Connection of the rain installation with the plane and measurement of the inlet flow; step 3 - the water is then directed to a channel, filled with gravel, with the function of distributing it and avoiding the preferential path in the plane; step 4 - After passing through the channel of gravel (0,40m by 0,30m), the water is distributed to the plan, where it will be stored temporarily until its infiltration into the ground; step 5 - if necessary, the excess volume of the infiltration plane capacity is exhausted by the ridge of the bus. In relation to the lawn channel, the rainwater coming from the roof of the Medicine Department building, with 1.501,5m², is drained by a vertical tubulation and then directed to the lawn channel (Figures 3, 4 and 5).

Figure 5.

Source: https://www.susdrain.org/case-studies/pdfs/suds_awards/010_18_03_30_susdrain_suds_awards_infiltration_basin_sao_paulo.pdf.

Specific Project Details

The infiltration plan was dimensioned to receive the water coming from the roof of the building of the Physiotherapy Department and the expansion area located between the building and the plan, totaling an area of 3.001,30m². The method adopted for the design of the structure was that of the envelope curve, using the rainfall equation of the city of Sao Carlos. The payback period used was 10 years. The sizing was done with the constant saturated K rate of 125mm/h of site at 100m. After recovery procedure, made during construction, infiltration rates were compared by means of triplicate double ring assay results. The design rain has an intensity of 41,1mm/h, duration of 70 minutes and water height of 48,0mm. The calculated storage volume was 106,5 m³ and the available volume after construction is 111,34 m³. In the lawn channel five solids discharges were simulated, with duplicate tests for each event, changing the input concentration, totalizing ten tests. For this purpose, a reservoir of 5m³ of useful volume was used. The initial level (t = 0) inside the reservoir was kept constant for each initial test condition, corresponding to a 1,34m blade, which refers to a volume of 4,2 m³. At the outlet of the reservoir was installed a short PVC pipe of 75mm, connected to a sphere register and to a 90° elbow. The flow rates were calibrated at the Hydraulics Laboratory of UFSCar, Sao Carlos - SP. The soil used to simulate particulate matter present in the rainwater was collected from an area near the lawn channel to represent the particulate material present in the rainwater. Subsequent to the collection, the soil was separated sieve with a diameter equivalent to 0,074 mm, weighed and stored in plastic bags, for later introduction into the unloading tank. The suspension was constantly shaken to prevent background deposit. No deposits were found on any component of the unloading system. The remaining residual concentration of particulate material was evaluated at different points along the extension of the grass channel by installing collectors at different positions along the length of the channel The results indicate that there is a front of removal of particulate material with a tendency to increase upstream efficiency. The higher the initial concentration of particulate material (Co), the greater the extension of the path required for its removal, with a maximum observed limit of around 80 m, for concentrations above 196 mg.L-1. The results explain that there was a greater dispersion of the experimental data for Co of 65 mg.L-1, when compared to the other investigated situations. It is believed that the lower concentration is more susceptible to experimental errors, being more influenced by transport of material deposited in the past.

Maintenance & Operation

The plan was implemented in 2012. Since then, further research has been carried out in the area. Because it is basically a grassy area, its maintenance requires simple care such as cleaning leaves and removing debris loaded with water.

Monitoring and Evaluation

The monitoring of precipitations, tributaries and water levels in the infiltration plan were performed between December 16, 2013 to March 18, 2014. During this period, thirty-two rain events were recorded, of which seven had water level recorded by the level sensor inside of the plan. It was verified that the event with the highest value of Tr recorded between the trials (6.1 years) had a maximum water depth of 0,24 m, not exceeding the maximum depth of 0,59 m.

Benefits and Achievements

The implementation of the infiltration plan allowed:

- Encourage the regulation of low-impact local planning and development;
- Encourage the prevention of soil and groundwater pollution;
- Preserve the areas of vegetation cover so that there is an increase of the resistance in the flow,
- reducing the speed of flow and increasing the time of course of the roof and the roads to the drainage system;
- Conduct runoff to ensure clean water enters the natural drainage system;
- Reduce costs of construction and maintenance of drainage infrastructure.

Lessons Learnt

Working in a public area of study has the advantage of being able to innovate and propose solutions wich in an extremely urbanized area, such as urban centers, we would have space difficulties. However, these characteristics also indicate that the choice of structure should be thought directly with the type of customer, in addition to the construction issues already mentioned. One of the challenges faced was the lack of knowledge of this type of drainage by campus managers and executing companies.

Interaction with Local Authority

In relation to the authorities with whom we worked along the process, the greatest difficulty was the issue of ignorance, since the local authorities did not obtain knowledge of this type of drainage and felt distrusted in relation to the efficiency of the techniques. Nevertheless, after meetings about the existing research and the possibilities of intervention on campus, we were given authorization to research, implement and execute the project in the area of the microbasin.

CLASS ACTIVITY

MODELLING GROUNDWATER: PERCOLATION AND COLLECTION

Materials Needed:

- Clear plastic cups
- Soil
- Gravel
- Sand
- Plastic wrap
- Rubber bands
- Blue food coloring
- Plastic pipettes or droppers

Activity Instructions:

With a groundwater percolation model exercise, let's delve into the world beneath our feet. Give each student a transparent plastic cup, and instruct them to fill it with layers of sand, gravel, and soil. A portion should be left at the top for the water. To create a barrier, cover the top of the cup with plastic wrap and fasten it with a rubber band. Now, transfer the blue food coloring-tinted water onto the plastic wrap. Students can mimic rainfall by using plastic pipettes or droppers. Observe the water seeping through the layers of soil, symbolizing the movement of groundwater. In addition to showing percolation, this practical exercise helps to solidify the idea that groundwater accumulates beneath the surface of the Earth.

SUMMARY

- The law of water balance states that the inflows to any water system or area is equal to its outflows plus change in storage during a time interval. In hydrology, a water balance equation can be used to describe the flow of water in and out of a system. A system can be one of several hydrological or water domains, such as a column of soil, a drainage basin, an irrigation area or a city.
- Precipitation is any form of solid or liquid water that falls from the atmosphere to the earth's surface. Rain, drizzle, hail and snow are examples of precipitation.
- Evapotranspiration is the process which returns water to the atmosphere and thus completes the hydrologic cycle. Evapotranspiration is the process which returns water to the atmosphere and thus completes the hydrologic cycle. Evapotranspiration consists of two parts, Evaporation and Transpiration. Evaporation is the loss of water molecules from soil masses and water bodies. Transpiration is the loss of water from plants in the form of vapour.
- Soil texture and structure greatly influence water infiltration, permeability, and water-holding capacity.
- Soil texture refers to the composition of the soil in terms of the proportion of small, medium, and large particles (clay, silt, and sand, respectively) in a specific soil mass. For example, a coarse soil is a sand or loamy sand, a medium soil is a loam, silt loam, or silt, and a fine soil is a sandy clay, silty clay, or clay.
- Infiltration is the process by which water on the ground surface enters the soil. It is commonly used in both hydrology and soil sciences. The infiltration capacity is defined as the maximum rate of infiltration. It is most often measured in meters per day but can also be measured in other units of distance over time if necessary.
- Ground-water recharge is a fundamental component in the water balance of any watershed.
- In the application of WTF method the basic assumption is that, the rise in the groundwater level in unconfined aquifer is only due to recharge water arriving at the water table.
- The water budget methods are those that are based on water budget equation. The water budget of a basin can be stated as $P + Q_{on} = ET + Q_{off} + \Delta S$
- Darcy's law states that fluid flux; such as recharge in an aquifer system can be calculated if both the head gradients and hydraulic conductivities are known.
- A stream gauge, streamgage or stream gauging station is a location used by hydrologists or environmental scientists to monitor and test terrestrial bodies of water. Hydrometric measurements of water level surface elevation ("stage") and/or volumetric discharge (flow) are generally taken and observations of biota and water quality may also be made.
- Hydrograph analysis is the process of studying how water levels change over time. This information can be used to understand and control water flow in rivers, reservoirs, and other bodies of water. By understanding hydrographs,

engineers and planners can make better decisions about where and how to use water resources.

- Groundwater-surface water connectivity refers to the direction and magnitude of flow between water resources located above and below ground.

REVIEW QUESTIONS

1. What is precipitation? Explain with suitable example.
2. What is transpiration? What affects transpiration?
3. How does transpiration affect groundwater?
4. What are the methods used to determine infiltration?
5. What is the groundwater estimation method?
6. What are the techniques used in stream flow measurement?
7. What is surface water groundwater connectivity?

REFERENCES

1. Allison, G. B., Cook, P. G., Barnett, S. R., Walker, J. R., Jolly, I. D., & Hughes, M. W. (1990). Land clearance and river salinization in the western Murray basin, Australia. *Journal of Hydrology*, *9*, 1–20.
2. Brun, P., Zimmermann, N. E., Hari, C., & Pellissier, L., & Karger, D. N. (preprint). Global climate-related predictors at kilometer resolution for the past and future. *Earth System Science Data Discussions*. https://doi.org/10.5194/essd-2022-212.
3. Coplen, T. B. (1993). Uses of environmental isotopes. In W. A. Alley (Ed.), *Regional Ground Water Quality* (Chap. 10, pp. 227–254). Van Nostrand Reinhold.
4. Daniel, C. C., III. (1996). Ground-water recharge to the regolith-fractured crystalline rock aquifer system, Orange County, North Carolina. *U.S. Geological Survey Water-Resources Investigations Report* 96–4220, 59 p.
5. Datta, P. S. (1999). Groundwater situation in New Delhi: Red alert. Nuclear Research Laboratory, IARI, New Delhi-12.
6. Grannemann, N. G. (2001). U.S. Geological Survey groundwater resources program, 2001. *U.S. Geological Survey Fact Sheet* 056-01, 2 p.
7. Green, W. H., & Ampt, G. A. (1911). Studies on soil physics. *The Journal of Agricultural Science*, *4*, 1–24. https://doi.org/10.1017/S0021859600001441.
8. Haie, N. (2020). *Transparent Water Management Theory: Sefficiency in Sequity*. Springer.
9. Healy, R. W., & Cooke, P. G. (2002). Using groundwater levels to estimate recharge. *Hydrogeology Journal*, *10*, 91–109.
10. Healy, R. W., Winter, T. C., LaBaugh, J. W., & Franke, O. L. (2007). *Water Budgets: Foundations for Effective Water-Resources and Environmental Management*. U.S. Geological Survey.

11. Holtschlag, D. J. (1997). A generalized estimate of ground-water recharge rates in the Lower Peninsula of Michigan. *U.S. Geological Survey Water-Supply Paper* 2437, 37 p.
12. Kumar, C. P., & Seethapathi, P. V. (2000). Assessment of natural ground water recharge in upper Ganga canal command area.
13. Lerner, D. N., Issar, A. S., & Simmers, I. (1990). Groundwater recharge—A guide to understanding and estimating natural recharge. *International Association of Hydrogeologists, International Contributions to Hydrogeology*, *8*, 147 p.
14. McCartney, M. P., & Houghton-Carr, H. A. (1998). An assessment of groundwater recharge on the Channel Island of Jersey. *Journal of the Institution of Water and Environmental Management*, *12*(6), 445–451.
15. Nimmo, J. R., Stonestrom, D., & Healy, R. W. (2003). Aquifer recharge. In B. A. Stewart & T. A. Howell (Eds.), *Encyclopedia of Water Science* (pp. 1–4). Marcel Dekker.
16. Ogden, F. L., Allen, M., Zhu, J., Lai, W., Seo, M., Douglas, C. C., & Talbot, C. A. (2017). The soil moisture velocity equation. *Journal of Advances in Modern Earth Systems*, *9*(2), 1473–1487. https://doi.org/10.1002/2017MS000931.
17. Ogden, F. L., Lai, W., Steinke, R. C., Zhu, J., Talbot, C. A., & Wilson, J. L. (2015). A new general 1-D vadose zone flow solution method. *Water Resources Research*, *51*(6), 4282–4300. https://doi.org/10.1002/2015WR017126.
18. Parlange, J. Y., Lisle, I., Braddock, R. D., & Smith, R. E. (1982). The three-parameter infiltration equation. *Soil Science*, *133*(6), 337. https://doi.org/10.1097/00010694-198206000-00001.
19. Richards, L. A. (1931). Capillary conduction of liquids through porous mediums. *Physics*, *1*(5), 318–333. https://doi.org/10.1063/1.1745010.
20. Sanford, W. (2002). Recharge and groundwater models: An overview. *Hydrology Journal*, *10*, 110–120.
21. Scanlon, B. R., Healy, R. W., & Cook, P. G. (2002). Choosing appropriate techniques for quantifying groundwater recharge. *Hydrology Journal*, *10*, 18–39.
22. Sutcliffe, J. V. (2004). *Hydrology: A Question of Balance*. International Association of Hydrological Sciences (IAHS).
23. Szilagyi, J., Harvey, F. E., & Ayers, J. F. (2003). Regional estimation of base recharge to ground water using water balance and a base-flow index. *Ground Water*, *41*(4), 504–513.
24. Viessman, W., & Lewis, G. L. (1996). *Introduction to Hydrology* (4th ed.). HarperCollins College Publishers.

CHAPTER

7 Groundwater Investigation Techniques

LEARNING OBJECTIVES

After studying this chapter, you will be able to:

- Evaluate the measurement and interpretation of groundwater level data
- Discuss on precipitation and evapotranspiration
- Learn about the recharge estimation
- Examine the removal of cuttings from under the bit

KEY TERMS FROM THIS CHAPTER

Antisplash grid
Barometric pressure
Evapotranspiration
Fertilizers
Groundwater
Industrial fracking
Infiltration
Municipal water supplies
Pesticides
Autographic gauges
Brine pits
Extraction wells
Garbage
Hydrogeology
Industrial waste
Isotropic aquifer
Oil field
Saltwater intrusion

INTRODUCTION

The water found in the pore spaces of rocks, soil, and rock formation fissures beneath the surface of the Earth is known as groundwater. Groundwater makes up about 30% of all freshwater that is readily available worldwide. When a unit of rock or an unconsolidated deposit yields a useful amount of water, it is referred to as an aquifer. The water table is the depth at which the pore spaces in soil or the cracks and voids in rock are totally saturated with water. Groundwater receives its replenishment from the surface; it can naturally seep and spring out of the ground, creating oases and wetlands. Groundwater is also frequently extracted through the construction and maintenance of extraction wells for use in industry, agriculture, and cities. Hydrogeology, also known as groundwater hydrology, is the study of groundwater distribution and movement. Though technically speaking, groundwater can also include soil moisture, permafrost (frozen soil), immobile water in very low permeability bedrock, and deep geothermal or oil formation water. Generally, groundwater is thought of as water that flows through shallow aquifers. It is theorized that groundwater lubricates surfaces, which may have an impact on fault movement. Water is probably present in a large portion of Earth's subsurface, where it may occasionally mix with other fluids.

Compared to surface water, groundwater is frequently less expensive, more practical, and less prone to pollution. It is therefore frequently utilized for public water supplies. For instance, groundwater is the main source of usable water storage in the US, and California is the state that extracts the most groundwater annually out of all the states. More water is contained in underground reservoirs than in any surface reservoir or lake in the United States, not even the Great Lakes. Groundwater is the only source of water for many cities. It is the main source of water for more than 2 billion people worldwide. Groundwater use by humans contributes to environmental issues. For instance, contamination in groundwater is more difficult to remove and less obvious than that in rivers and lakes. Waste that is improperly disposed of on land most frequently leads to groundwater pollution. The main sources are leaking underground oil storage tanks and pipelines, sewage sludge, septic systems, industrial waste lagoons, excessive fertilizers and pesticides used in agriculture, tailings and process wastewater from mines, industrial fracking, and industrial and household chemicals and garbage landfills. Furthermore, unsustainable groundwater extraction can result in land subsidence and sinking cities like Bangkok, as well as elevation loss similar to the several meters lost in the Central Valley of California. Groundwater is also vulnerable to saltwater intrusion in coastal areas. Sea level rise and other consequences of climate change, especially those on the water cycle, exacerbate these problems. The act of human groundwater pumping has caused a 31-inch shift in Earth's axial tilt.

7.1. MEASUREMENT AND INTERPRETATION OF GROUNDWATER LEVEL DATA

Groundwater level data collection and measurement are essential components of hydrogeology. An aquifer unit's groundwater level data can be utilized for a number of tasks, such as creating a flow net, drawing a hydrograph, and mapping the potentiometric surface to ascertain the direction of groundwater flow. In order to forecast how an aquifer will behave in the future, the process of developing and testing a numerical groundwater flow model also depends heavily on hydraulic head values.

7.1.1. Water Level Measurement

A submersible pressure transducer placed slightly below the lowest anticipated groundwater level or a water level dipper are both used to measure the groundwater level in the field. The depth from a practical measurement datum, such as the well top or borehole flange, to the groundwater level position is the necessary field measurement. The elevation of the groundwater level (or groundwater head, h) can be measured as the height above or below a local base level, usually mean sea level, if the elevation of the measurement datum is known through leveling techniques.

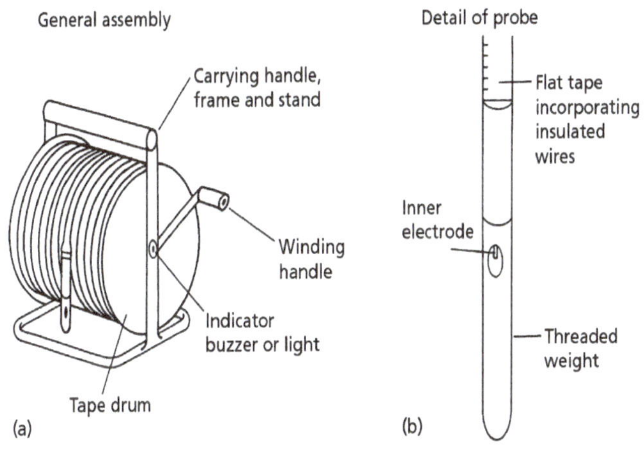

Figure 7.1. *Design of a water level dipper showing (a) the general assembly and (b) detail of the probe.*

Source: https://www.geokniga.org/bookfiles/geokniga-hydrogeologyprinciplesandpractice.pdf.

A pair of electrodes are fastened to the end of a length of twin-core cable that has been graduated in centimeters and meters and wound onto a drum to create a field water level dipper (Figure 7.1). A circuit is completed when the electrodes come into contact with the water's surface, activating either a buzzer or a light, or both. The precision of measuring water levels is ±0.005 m.

A solid-state pressure sensor housed in a submersible housing made of stainless steel makes up a pressure transducer. The transducer is equipped with a waterproof cable that links the water pressure sensor to a data logger or other monitoring device. This allows for the acquisition of a temporal record of fluctuations in groundwater levels (Figure 7.5). With the use of a pressure transducer, the measurement of fluid pressure, P (= $\rho g \psi$), recorded at a depth elevation, z, is converted to a value of groundwater head h.

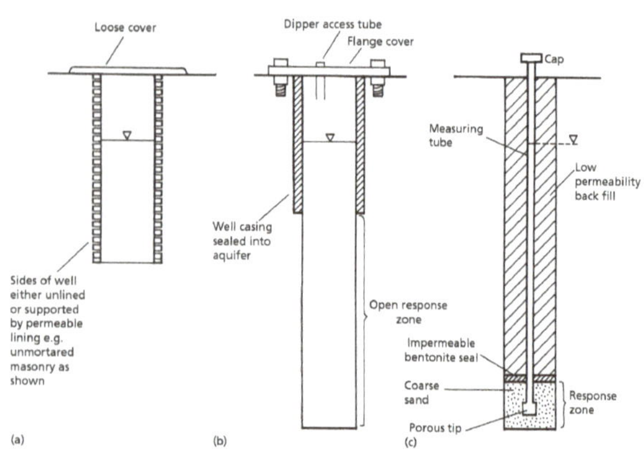

Figure 7.2. General designs of (a) well, (b) observation borehole and (c) piezometer for the measurement of groundwater level.

Source: https://www.geokniga.org/bookfiles/geokniga-hydrogeologyprinciplesandpractice.pdf.

Either a well (Figure 7.2a) or a surface water level sensor or an intentionally constructed observation borehole (Figure 7.2b). Wells are usually unconfined aquifers that reach the top of the local water table. They are lined with unmortared bricks and have a shallow depth. Depending on the strength of the aquifer rock, observation boreholes can be either uncased (open) or cased. They are drilled using a percussion or rotary rig (Brassington, 1998). Groundwater levels in confined aquifers (where a potentiometric surface exists) and unconfined aquifers (where a water table exists) are recorded by observation boreholes. A unique kind of installation called a piezometer (Figure 7.2c) is intended to give an aquifer's hydraulic head measurement at a specific depth. In order to determine the vertical component of groundwater flow—either upward in a discharge zone or downward in a recharge area—a bundle of piezometers nested within a single borehole installation can provide data on hydraulic heads at multiple depths in an aquifer.

7.1.2. Borehole Hydrographs and Barometric Efficiency

Hydrographs from wells or boreholes usually show data gathered on a monthly basis and document variations in groundwater levels. As depicted in Figure 7.3a to illustrate the position of monthly groundwater levels measured in relation to historical minima and maxima as well as the long-term average, more data can be displayed on a hydrograph. Entire records are priceless. It is possible to distinguish between artificial and climatic effects, such as the recurrent occurrence of wet and dry years and the over-exploitation of groundwater resources that gradually lowers groundwater levels (see Figure 7.3b). By using a hydrograph record that was not impacted by the pumping test, groundwater level drawdown data collected during pumping tests can also be adjusted for background trends in the regional potentiometric surface.

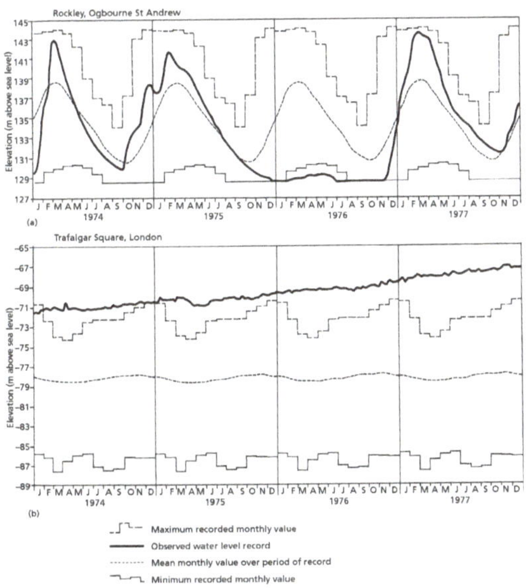

Figure 7.3. *Borehole hydrograph records (1974–1977) for the Chalk aquifer in southern England: (a) Rockley, Ogbourne St Andrew (NGR 1655 7174; datum 146.6 m above sea level; records from 1933 to 1973 used to produce the maximum, minimum and average values); (b) Trafalgar Square, London (NRG TQ 2996 8051; datum 12.6 m above sea level; records from 1953 to 1973 used to produce the maximum, minimum and average values). The record for Rockley shows the effect of drought in the dry year 1976 followed by the wet year 1977 when the groundwater level recovered. The record for Trafalgar Square shows the effect of rising groundwater levels in the London Basin following the reduction in abstraction rates after over-exploitation of the Chalk aquifer during the nineteenth and first half of the twentieth centuries. The rate of increase in groundwater level shown in this section of the borehole hydrograph is 1.2 m a*$^{-1}$.

Source: https://www.geokniga.org/bookfiles/geokniga-hydrogeologyprinciplesandpractice.pdf.

Changes in atmospheric pressure can result in significant variations in the water levels in wells and boreholes within confined aquifers. Water levels are seen to decrease as barometric pressure rises. This phenomenon is related to a shift in the stress field applied to the aquifer and is also thought to be a temporary result of external loading from passing trains, construction blasting, and earthquakes (Jacob, 1940). Applying the effective stress principle and taking into account the scenario depicted in Figure At position X at the top of a confined aquifer, Figure 7.4a, the stress equilibrium is given by:

$$\sigma_T + P_A = \sigma_e + P_w \quad (1)$$

where P_A is atmospheric pressure, σ_T is the total stress created by the weight of the overlying aquitard, σ_e is the effective stress acting on the aquifer material and P_w is the fluid pressure in the aquifer. The fluid pressure creates a pressure head, ψ, that is measured in the well penetrating the aquifer. At position Y in the well, the balance of pressures is:

$$P_A + \gamma\psi = P_w \quad (2)$$

where γ is the specific weight of water. If, as shown in Figure 7.4b, the atmospheric pressure is increased by an amount dP_A, the change in the stress field at position X is given by:

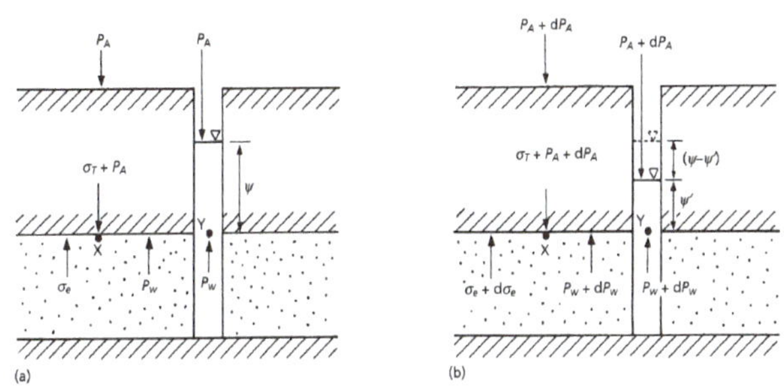

Figure 7.4. *Diagram showing the change in aquifer stress conditions caused by an increase in atmospheric pressure from PA in (a) to PA + dPA in (b) resulting in a decrease in groundwater level ($\psi - \psi'$) in a confined aquifer.*

Source: https://www.geokniga.org/bookfiles/geokniga-hydrogeologyprinciplesandpractice.pdf.

$$dP_A = d\sigma_e + dP_w \qquad (3)$$

Now, it can be seen that the change in dP_A is greater than the change in dP_w such that at position Y in the well, the new balance of pressures is:

$$P_A + dP_A + \gamma\psi' = P_w + dP_w \qquad (4)$$

which, on substitution of Equation 2 in Equation 4, gives:

$$dP_A - dP_w = \gamma(\psi - \psi') \qquad (5)$$

Since $dP_A - dP_w$ is greater than zero, then $\psi - \psi'$ is also greater than zero, proving that an increase in atmospheric pressure leads to a decrease in water level (Figure 7.5). In a horizontal, confined aquifer the change in pressure head, $d\psi = \psi - \psi'$ in Equation 5, is equivalent to the change in hydraulic head, dh, and so provides a definition of barometric efficiency, B, expressed as:

$$B = \frac{\gamma dh}{dP_A} \qquad (6)$$

Constrained aquifers typically have a barometric efficiency between 0 and 75 (Todd, 1980). Expressions connecting the barometric efficiency of a confined aquifer to water and aquifer characteristics, such as the storage coefficient, were further developed by Jacob (1940). The water level in an observation well remains constant in unconfined aquifers because variations in air pressure are directly transferred to the water table in both the aquifer and the well.

However, pressure variations can affect air bubbles trapped in pores beneath the water table, which can result in fluctuations smaller than those seen in confined aquifers.

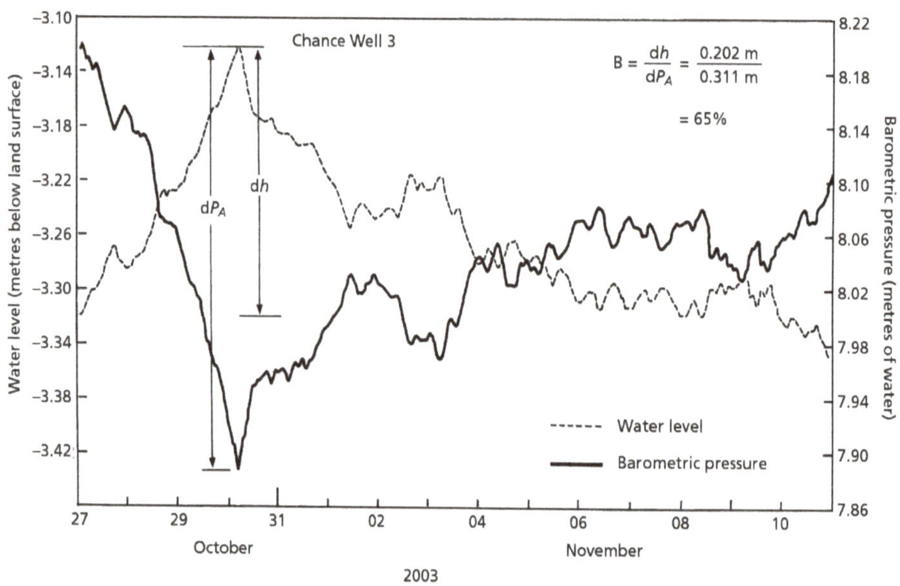

Figure. 7.5. *Relationship between groundwater level fluctuation and atmospheric pressure in Chance Well 3 (CW-3), Long Valley Caldera, California (well depth 267 m; altitude 2155.90 m above sea level; latitude 37°38′49″, longitude 118°51′30″). At this site, water level and barometric pressure are monitored using a pressure transducer and barometer, respectively. These data are recorded on a data logger every 30 minutes and the data transmitted by satellite to the United States National Water Information System database.*

Source: https://www.geokniga.org/bookfiles/geokniga-hydrogeologyprinciplesandpractice.pdf.

7.1.3. Construction of Groundwater Level Contour Maps

At least three observation points are needed, as indicated in Figure 7.6, in order to create a map of the groundwater level, which will show the potentiometric surface and the direction of groundwater flow. The process involves first plotting the field groundwater levels' location on a scale plan and then relating them to a common datum (sea level or a map datum, for convenience). The next step involves drawing lines between three measurements of the groundwater level and dividing them into a number of short, equal lengths according to the elevation difference at each end of the line (in the example, each division on line AB and BC is 0.2 m, while on line AC each division is 0.1 m). After joining points on each line that have the same elevation, the next step is to choose a contour interval that corresponds to the overall variation in the water levels in the mapped area (in this case, 0.5 m). Once one or two important contour lines can be mapped, the same process is carried out for additional pairs of field observation points. By interpolating between the field values at this point, the remaining contour lines can be drawn.

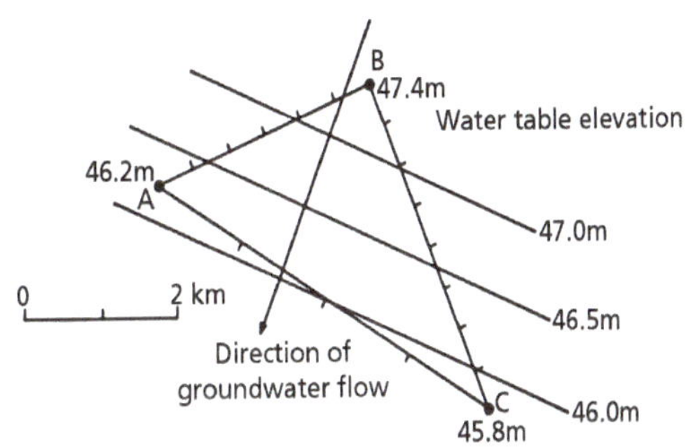

Figure 7.6. *Graphical construction method for determining the direction of groundwater flow from three groundwater level measurements.*

Source: https://www.geokniga.org/bookfiles/geokniga-hydrogeologyprinciplesandpractice.pdf.

Knowledge of a region's general topography, records of the elevations of springs known to discharge from an aquifer, and the elevations of gaining streams and rivers can all be used to complete a potentiometric surface map (see Figure 7.8a) that flow over the aquifer outcrop, as these points show where the water table is intercepted by the ground surface. It is customary to depict the coastline as a groundwater contour with an elevation equal to sea level (0 m) for unconfined aquifers that border the sea. Large surface water bodies located inland can be assumed to be similar (Brassington, 1998). Drawing the groundwater flow direction in an isotropic aquifer in the direction of decreasing hydraulic head at right angles to the contour lines on the potentiometric surface is possible. This presupposes that the material in the aquifer is isotropic. The flow lines in anisotropic materials, such as fractured or fissured aquifers, will be angled with respect to the potentiometric contour lines. An example of a finished potentiometric surface map showing the London Basin's Chalk aquifer. To compute variations in the amount of water stored in an aquifer and evaluate the local effects of groundwater recharge and abstraction, potentiometric surface maps at low and high groundwater levels can be constructed (Brassington, 1998).

7.2. PRECIPITATION AND EVAPOTRANSPIRATION

Groundwater recharge and streamflow generation are significantly impacted by near-surface hydrological processes like evapotranspiration, infiltration, and precipitation. Rainfall is the primary source of water input for a catchment; however, other factors such as soil type, climate, geology, and land use that affect catchment runoff properties must also be considered in order to ensure that the water is available to support river flows, replenish aquifer storage, and maintain water supplies.

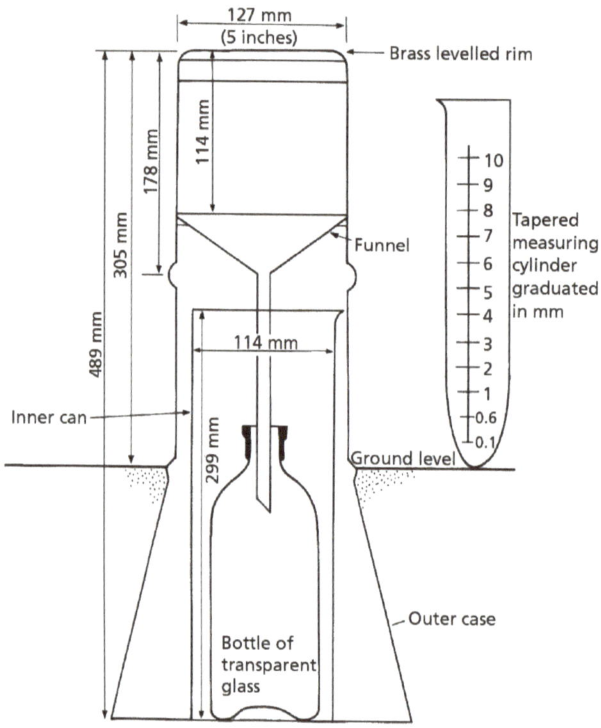

Figure 7.7. *The copper 5-inch standard rain gauge used by the UK Meteorological Office. The rain gauge consists of a 5-inch (127-mm) diameter funnel with a sharp rim, the spout of the funnel being inserted into a glass collecting jar. The jar is in an inner copper can and the two are contained in the main body of the gauge, the lower part of which is sunk into the ground with the rim 12 inches (305 mm) above the surrounding short grass or gravel, this height being chosen so that no rain splashes from the surroundings into the funnel. The funnel has a narrow spout in order to reduce evaporation loss. Normally, the gauge is sited such that its distance from any obstructions (trees, houses, etc.) is at least four times the height of the obstruction.*

Source: https://www.geokniga.org/bookfiles/geokniga-hydrogeologyprinciplesandpractice.pdf.

7.2.1. Precipitation Measurement

Though it can also come in the form of hail, sleet, snow, fog, or dew, precipitation mostly falls as rain. Figure 7.7 depicts the layout of the common rain gauge used in the United Kingdom with a 127 mm (5 inches) diameter cylinder. Standard rain gauges in the United States and Canada measure 8 or 9 inches (203 or 229 mm) in diameter. Every day, at 0900 hours in the UK, for instance, standard rain gauges are read. High winds can impact the rain gauge's catch in exposed areas, so it's generally agreed that using a rain gauge with its rim at ground level will yield more accurate readings. A ground-level installation necessitates housing the gauge in a pit and enclosing it with an antisplash grid, despite the higher cost. Standard rain gauges can also be used to measure snowfall, though their accuracy may be compromised by turbulence around the gauge's rim. The snow is captured, melted, and the corresponding volume of water is measured. In remote, infrequently visited locations, recording gauges, also known as autographic gauges, are helpful because they can automatically measure or weigh precipitation. The tipping-bucket rain gauge serves as an illustration, wherein the frequency with which a small bucket with a known volume fills and tips is noted. A reed switch on each tip triggers an electrical pulse that is sent to a logger. When rainfall fills a chamber in the tilting siphon autographic gauge, a chart pen that floats upward and tips when the chamber is full returns the pen to the bottom of the chart. The benefit of recording gauges is that they can measure both the total amount of rainfall and its intensity. However, they are more costly and prone to error (very low rainfall amounts are not recorded).

Table 7.1. *Density of Rain Gauges Required in a Hill Area*

Catchment area (km^2)	Number of gauges
4	6
20	10
80	20
160	30

Source: https://www.geokniga.org/bookfiles/geokniga-hydrogeologyprinciplesandpractice.pdf.

Where precipitation gradients are indicated, a greater number of rain gauges is needed to provide a trustworthy estimate of catchment rainfall. Given that massive thunderstorm systems may only cover an area of roughly 20 km^2, a minimum density of 1 gauge per 25 km^2 is advised. Higher rain gauge densities are required in the initial years of measurement in hilly terrain, where orographic effects may result in significant and sustained variations in rainfall over short distances (Table 7.1). While annual totals show only a slight gradient, daily rainfall in tropical regions varies greatly in space. The rain gauge densities in Table 7.1 will be excessive in these areas, so obtaining homogeneous long-term records at a small number of trustworthy sites should take precedence.

A technique of averaging the individual gauge measurements must be used in order to evaluate a representative value of rainfall over large areas. The simplest approach is to compute the arithmetic mean of the total amounts noted by all of the local rain gauges. This method produces results that are acceptable if the rainfall gradients are small and the point distribution is uniform. A different approach is to draw lines between neighboring gauges, divide the lines with perpendiculars, and then join the perpendiculars to form a boundary enclosing the gauge using Thiessen polygons to define each rain gauge's zone of influence (Figure 7.8a).

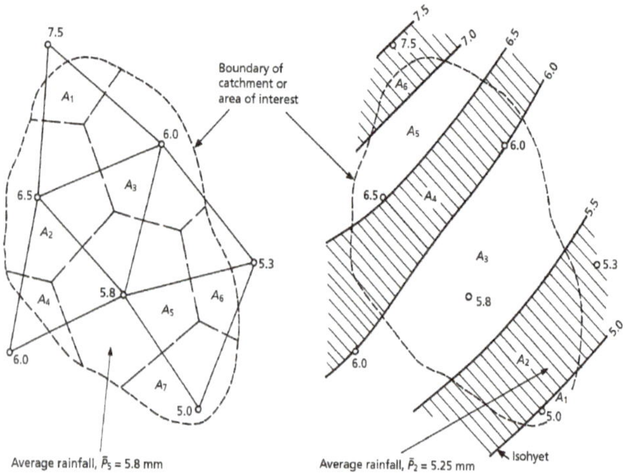

Figure 7.8. *Calculation of area-weighted average rainfall amount using the methods of (a) Thiessen polygons and (b) isohyets.*

Source: https://www.geokniga.org/bookfiles/geokniga-hydrogeologyprinciplesandpractice.pdf.

It is considered that the average amount of rainfall in the boundary area is equal to that of the enclosed gauge. This method can also be modified by cutting the height difference between neighboring gauges in half to determine the perpendiculars; however, the outcome is not significantly altered by this altitude-corrected method. As depicted in Figure 7.8b, an additional technique is to create contours, or isohyets, with equal rainfall depths. The areas in between each isohyet are then measured and given an average rainfall amount.

7.2.2. Evapotranspiration Measurement and Estimation

Within the hydrological cycle, water evaporation from open water surfaces, like lakes, reservoirs, and rivers, happens at a rate that is determined by the temperature of the water as well as the humidity and temperature of the air layer above the water's surface. The rate at which vapour is removed from the water's surface is also influenced by wind speed. An evaporation pan or tank can be used to measure evaporation directly. Pans are placed above ground on a little plinth, while tanks are buried in the ground. The British standard tank is 0.61 meters deep and measures 1.83 meters by 1.83 meters in size. Made of unpainted, galvanized metal, the United States National Weather Service (NWS) Class A pan has a diameter of 1.22 meters and a depth of 0.25 meters. It is widely recognized as an international standard. Compared to pans that lose more vapor due to heat from their exposed walls and shallow water, evaporation tanks typically provide a more accurate measurement of actual evaporation. It is possible to use pan coefficients, or correction factors, to accurately measure "true evaporation.". The range of pan coefficients for the US NWS Class A pan and the British standard tank is 0.60–0.80, and for both, 0.93–1.07 (WMO 1994).

Water is lost from vegetated surfaces through transpiration through plant stomata and evaporation from bare soil. The combination of these effects is referred to as evapotranspiration, which is an important process for catchment water balances and frequently the main source of water loss from a catchment. The amount of water that would be extracted from a vegetated surface if there was enough water in the soil to meet the water demand is known as potential evapotranspiration, or PE. The actual evapotranspiration rate will be lower than PE if the soil is not saturated, as would occur after a rainfall event or in an area with a high water table, such as a groundwater discharge area. A lysimeter is a large container that is placed outside and holds a monolith of soil and plants. It is used to measure evapotranspiration directly. For the lysimeter, evapotranspiration is calculated using the balance of irrigation and precipitation inputs, changes in soil moisture content, and water loss through soil drainage. Alternatively, catchment-scale evapotranspiration losses can be computed using field estimates of soil moisture content in conjunction with data from groundwater monitoring, river discharge, and precipitation.

PE can be computed theoretically using meteorological data and is dependent on the atmosphere's evaporative capacity. The most widely used techniques for estimating PE are those of Penman (1948) and Penman–Monteith (Monteith 1965, 1985), which are energy-budget approaches requiring additional meteorological data, and Blaney and Criddle (1962) and Thornthwaite (1948), which are based on empirical correlations between evapotranspiration and climatic factors. Two conditions must be satisfied for continuous evaporation to occur, according to the Penman formula for estimating evaporation from meteorological data. First, radiation must be present to provide the latent heat of vaporization. Secondly, vapour must be removed using an aerodynamic mechanism (wind and humidity) once it has been produced. The following is the Penman formula for evaporation from open water, or E_o, in milliliters per day^{-1}.

$$E_o = \frac{\frac{\Delta}{\gamma}H + f(u)(e_{sat} - e_{act})}{\frac{\Delta}{\gamma} + 1} \qquad (7)$$

where H = net radiation balance in mm of water equivalent, Δ = rate of change of saturated vapour pressure with temperature, ⊠ = psychrometric or hygrometric constant (different values depending on the temperature units and the method of ventilation (aspiration) of the wet and dry bulb thermometers), f(u) = aerodynamic coefficient (function of wind speed), e_{sat} = saturated vapour pressure of air at temperature, t, e_{act} = actual vapour pressure of air at temperature, t.

Penman's formula for evaporation (eq. 7) has been adapted to calculate potential evapotranspiration by the application of empirically derived factors. Penman introduced the empirical formula:

$$\text{PE} = fE_o \qquad (8)$$

where f is a seasonal correction factor that includes the effects of differing solar insolation intensity, day length, plant stomatal response and geometry. For example, the evaporation rate from a freshly wetted bare soil is about 90% of that from an open water surface exposed to the same weather conditions. For a grassed surface in temperate latitudes the value of PE is, on average, about 75% of the open water evaporation rate. A more process-based approach to calculating PE, following Penman's method and extended by experimental work, is given by the Penman–Monteith formula (eq. 9) that incorporates canopy stomatal and aerodynamic resistance effects, to calculate evapotranspiration rate in mm day^{-1} as follows:

$$\text{ET} = \frac{\Delta H + \frac{\rho c_p(e_{sat} + e_{act})}{r_{sfc}}}{\lambda\left\{\Delta + \frac{\gamma(r_{sfc} + r_{aero})}{r_{aero}}\right\}} \qquad (9)$$

where, further to the symbols applied in Equation 7, ρ = density of water, λ = latent heat of vaporization, c_p = specific heat capacity of water, r_{sfc} = surface resistance, r_{aero} = aerodynamic roughness. The Penman–Monteith formula is used as the basis for the national computerized system, morecs, the United Kingdom Meteorological Office Rainfall and Evaporation Calculation System (Thompson et al., 1981). morecs provides an areal-based estimate of evapotranspiration which supplements the Penman approach with simulation of soil water flux and a consideration of local vegetation cover. The model is based on a two-layer soil and provides estimates of areal precipitation (P), PE, actual evapotranspiration (AE), soil moisture deficit (SMD) and hydrologically effective rainfall (P – AE – SMD) for 40 × 40 km grid squares on a weekly basis.

7.3. RECHARGE ESTIMATION

The quantity of surface water that percolates through the unsaturated zone or directly contacts the permanent water table through the riparian zone is known as groundwater recharge. According to Rushton and Ward (1979), groundwater recharge is the amount that is eventually available for both abstraction and sustaining the baseflow component of rivers. In a catchment water balance, it is crucial to calculate effective rainfall, the amount of rainfall that remains after evapotranspiration, and how this hydrological excess water is divided between surface water and groundwater. Lerner et al., go into great detail about how to estimate the amount of water that is recharged directly to the water table and indirectly through fissures and fractures in hard rock or limestone terrain, which can occur as lateral runoff to an aquifer at the edge of a confining layer or as localized infiltration beneath water-filled surface depressions. (1990). Three methods are covered in the following sections and are available for calculating direct recharge: direct measurement over areas up to 100 m^2 using lysimeters; Darcian approaches to calculate flow in the unsaturated zone above the water table; hydrograph analysis from boreholes and streams; empirical techniques that simplify recharge as a function of rainfall quantity; methods for estimating soil water budgets at the field or catchment scale; and use of applied or environmental tracers to track the saturated movement of water in the unsaturated zone.

All throughout the nation, groundwater levels are dropping because human withdrawals are outpacing aquifers' natural recharge rate. Using artificial groundwater recharge is one way to manage falling water levels. Planned recharge, also known as artificial groundwater recharge, is the process of manually replenishing underground aquifers. The extra water that is present above the earth's surface is designed to seep through the ground and store itself for periods when it is scarce. Artificial groundwater recharge serves as a unique means of addressing the groundwater deficit in addition to a number of other issues.

The process of increasing the amount of water that enters an aquifer under human control is known as artificial recharge. Groundwater can be artificially replenished, for instance, by adding irrigation furrows or sprinkler systems, rerouting water across the land surface through canals, infiltration basins, or ponds, or by simply injecting water directly into the subsurface through injection wells. Water resource managers and scientists around the world use a technique called aquifer storage and recovery to store water. Basically, it's pumping water into aquifers via wells that are completed and then using that water to draw from them later on when it's dry. The recovery of water from these wells has significant advantages for urban, agricultural, and environmental applications. Through the use of appropriate civil construction techniques, the artificial recharge to ground water aims to augment ground water reservoir by changing the natural movement of surface water.

Artificial recharge techniques normally address to following issues -

a. To enhance the sustainable yield in areas where over-development has depleted the aquifer.
b. Conservation and storage of excess surface water for future requirements, since these requirements often changes within a season or a period.
c. To improve the quality of existing ground water through dilution.
d. To remove bacteriological and other impurities from sewage and waste water so that water is suitable for re-use.

The basic purpose of artificial recharge of ground water is to restore supplies from aquifers depleted due to excessive ground water development.

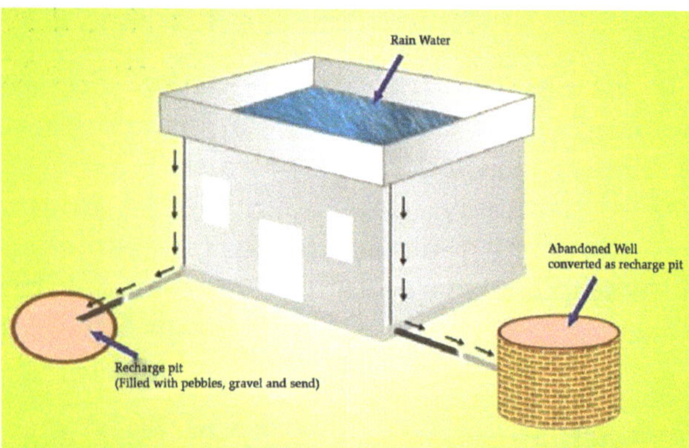

Figure 7.9. *Recharge Structures for groundwater recharge.*

Source: https://media.licdn.com/dms/image/C5612AQG_UnN1Q4bR8w/article-cover_image-shrink_600_2000/0/1520211113242?e=2147483647&v=beta&t=2IndtSZataTu8m9N9ZF7Z38nrplHzD7V3ptZ492I3HA.

7.3.1. Borehole Hydrograph Method

When combined with stream hydrograph separation, the borehole hydrograph method offers a practical way to figure out how much effective rainfall is split between groundwater discharge and surface water runoff during a recharge season. As seen in Figure 7.10, fluctuations in borehole hydrographs indicate modifications in aquifer storage. The net recharge is calculated by multiplying the amplitude of the water level change, Δh, by the aquifer storage coefficient. The addition of groundwater outflows (spring flow and base flow, determined by hydrograph separation) to net recharge equals the total recharge. The method's application is restricted by the requirement for a good dispersion of observation boreholes in the catchment of interest. It can be helpful in the preparation of an initial catchment water balance or in supporting regional groundwater flow modeling.

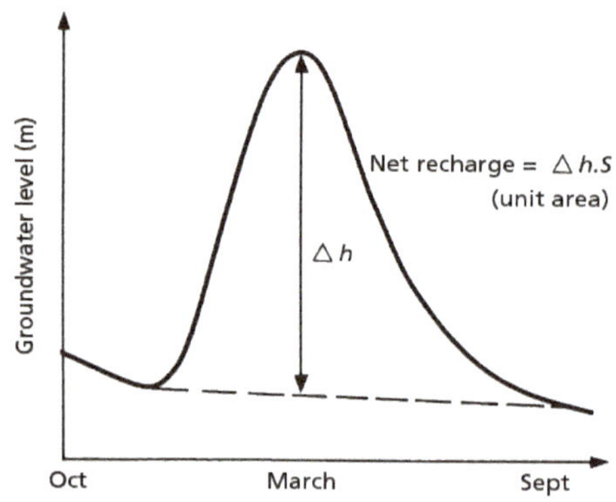

Figure 7.10. *Borehole hydrograph record representing the change in aquifer storage during a single recharge period. The net recharge is equal to the product of the amplitude of groundwater level rise (Δh) and the aquifer storage coefficient, S.*

Source: https://www.geokniga.org/bookfiles/geokniga-hydrogeologyprinciplesandpractice.pdf.

7.3.2. Soil Water Budget Method

Penman and Grindley's studies (Penman 1948, 1949; Grindley 1967, 1969) serve as the foundation for the traditional method of estimating recharge using a soil water budgeting approach. In concept, the method is straightforward. Rainfall replenishes the soil moisture store with water, while evapotranspiration removes it. When the soil is fully hydrated, surplus precipitation, or the hydrological excess, or HXS, is directed toward surface water and groundwater for recharge. The conceptual quantity of soil moisture deficit (SMD), which measures the volume of water needed to bring the soil back to field capacity, is zero. Calculating actual evapotranspiration (AE) is the most challenging part of the process. Potential evapotranspiration (PE) is a quantity that is generally defined as the maximum rate of evapotranspiration over short-rooted vegetation with an infinite water supply under current meteorological conditions.

A root constant (RC), a function of soil and vegetation characteristics, and a measure of readily available water within the root range regulate how much potential and actual evapotranspiration rates diverge. This budgeting procedure converts PE to AE. Table 7.2 provides representative values of RC expressed as an equivalent water depth.

Table 7.2. *Monthly Root Constant (RC) Values for the Penman–Grindley Method*

Month	Crop type													
	1	2	3	4	5	6	7	8	9	10	11	12	13	14
Jan. and Feb.	25	25	25	25	25	25	25	25	25	25	56	76	13	203
Mar.	56	56	56	25	25	56	25	25	25	25	56	76	13	203
Apr.	76	76	76	76	56	56	56	25	25	25	56	76	13	203
May	97	97	97	56	56	56	56	56	25	25	56	76	13	203
Jun. and Jul.	140	140	140	76	76	25	56	56	56	25	56	76	13	203
Aug.	140	140	25	97	97	25	25	56	56	25	56	76	13	203
Sept.	140	25	25	97	25	25	25	25	56	25	56	76	13	203
Oct.	25	25	25	97	25	25	25	25	56	25	56	76	13	203
Nov. and Dec.	25	25	25	25	25	25	25	25	25	25	56	76	13	203

Values in millimeters.

Source: https://www.geokniga.org/bookfiles/geokniga-hydrogeologyprinciplesandpractice.pdf.

There is disagreement over how much PE and AE diverge, and several models have been put forth to illustrate how plant transpiration decreases as soil moisture content decreases. Figure 7.9 displays a drying curve example, showing the reduction in AE as wilting occurs in plants and ultimately reaches dieoff. In Figure 7.11., the intricacy of the wilting process is merely depicted as a one-step function, where the AE remains equal to the full PE until an SMD of RC + 0.33RC is attained. After that, the AE sharply drops to a tenth of the PE rate. Until die-off, there is no more AE decline visible. One way to account for AE during the drying phase is to follow the development of SMD by using a daily, weekly, or monthly budgeting approach (Figure 7.11).

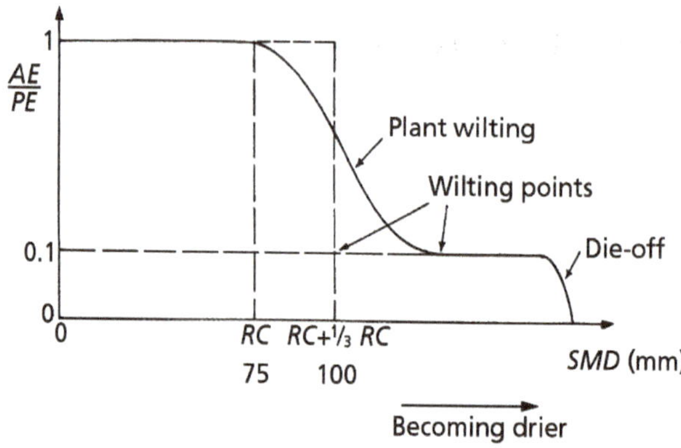

Figure 7.11. *Drying curve for a vegetated soil of short-rooted grass with a root constant (RC) of 75 mm showing the decline in the ratio of actual to potential evapotranspiration (AE/PE) for increasing values of soil moisture deficit (SMD).*

Source: https://www.geokniga.org/bookfiles/geokniga-hydrogeologyprinciplesandpractice.pdf.

Table 7.3, which provides an example of the method, shows the calculation that was started on April 1st, assuming that the soil was at field capacity and that the SMD = 0. What happens when the step function depicted in Fig 7.9. Reducing the rate at which the SMD develops with a cap effectively reached for values above a SMD = 100 mm. At the die-off point, no more SMD decrease is anticipated. Groundwater recharge for the period of interest (291 mm in the example shown in Table 7.4) can be estimated using the cumulative hydrological excess that is predicted to occur for those times when the SMD = 0 at the conclusion of the accounting procedure and in the absence of surface runoff.

Table 7.3. Monthly root constant (RC) values for the Penman–Grindley method

Month	Crop type													
	1	2	3	4	5	6	7	8	9	10	11	12	13	14
Jan. and Feb.	25	25	25	25	25	25	25	25	25	25	56	76	13	203
Mar.	56	56	56	25	25	56	25	25	25	25	56	76	13	203
Apr.	76	76	76	76	56	56	56	25	25	25	56	76	13	203
May	97	97	97	56	56	56	56	56	25	25	56	76	13	203
Jun. and Jul.	140	140	140	76	76	25	56	56	56	25	56	76	13	203
Aug.	140	140	25	97	97	25	25	56	56	25	56	76	13	203
Sept.	140	25	25	97	25	25	25	25	56	25	56	76	13	203
Oct.	25	25	25	97	25	25	25	25	56	25	56	76	13	203
Nov. and Dec.	25	25	25	25	25	25	25	25	25	25	56	76	13	203

Source: https://www.geokniga.org/bookfiles/geokniga-hydrogeologyprinciplesandpractice.pdf.

Table 7.4. *Example of the Penman–Grindley Soil Water Budget Method to Calculate Values of Actual Evapotranspiration (AE), Soil Moisture Deficit (SMD) and Recharge (Hydrological Excess, HXS) Using Precipitation (P), and Potential Evapotranspiration (PE) Data*

Month	P (mm)	PE (mm)	P – PE (mm)	AE (mm)	ΔSMD (mm)	SMD (mm)	HXS (mm)
Apr.	20	48	−28	48	28	28	0
May	12	56	−44	56	44	72	0
Jun.	24	72	−48	(24 + [100 − 72] + 0.1[20]) = 54	(28 + 2) = 30	102	0
Jul.	9	68	−59	(9 + 0.1[59]) = 15	6	108	0
Aug.	31	42	−11	(31 + 0.1[11]) = 32	1	109	0
Sep.	60	28	32	28	−32	77	0
Oct.	75	20	55	20	−55	22	0
Nov.	106	10	96	10	−22	0	74
Dec.	94	5	89	5	0	0	89
Jan.	69	5	64	5	0	0	64
Feb.	40	18	22	18	0	0	22
Mar.	72	30	42	30	0	0	42
Apr.	18	50	−32	50	32	32	0
Total recharge							291

Source: https://www.geokniga.org/bookfiles/geokniga-hydrogeologyprinciplesandpractice.pdf.

The calculation assumes that the permeable soil is covered by a short-rooted grass with a root constant (RC) of 75 mm and that the soil is at field capacity on 1 April (SMD = 0).

The traditional soil water budgeting approach states that the presence of an SMD prevents recharge. Rushton and Ward's (1979) critique demonstrated that the recharge amount determined by the Penman-Grindley method frequently underestimates the amount determined by other methods that rely on hydrograph analysis, lysimeters, and tracers. The Penman-Grindley method, in particular, seems to underestimate recharge in the summer and early fall. Rushton and Ward (1979) discovered that the time-step of the accounting procedure, the estimate of PE, the choice of root constant, the functional relationship between AE and PE, and the date of harvesting produced an error in calculated recharge of up to 15% based on a sensitivity analysis of recharge calculations for an area of the Chalk aquifer in Lincolnshire, England. Rushton and Ward (1979) allowed a direct component of recharge, conceptualized as a bypass flow component via Chalk fissures, equivalent to 15% of actual precipitation over 5 mm plus 15% of the effective precipitation, to account for an underestimate in their recharge calculation compared with known outflows and borehole hydrograph records. The remaining recharge was calculated using the traditional Penman–Grindley method. It is important to remember that this method does not provide the proper daily recharge. This monthly distribution of recharge is considered acceptable when used as an input to a regional groundwater flow model on a monthly basis.

7.3.3. Chloride Budget Method

Since the aforementioned soil water budgeting techniques were created for temperate climates, they are less applicable in semi-arid and arid regions, where they frequently yield zero values and underestimate recharge. Using a conservative tracer species, like chloride, is an alternative geochemical method to estimate the amount of recharge and, under favorable conditions, the history of recharge. In order to optimize water use and guarantee sustainable crop production, soil water budgeting techniques are essential for efficient water management in agriculture. With these techniques, the equilibrium between water inputs (rainfall, irrigation) and outputs (evaporation, transpiration, drainage) in a soil system is calculated. The objective is to sustain an ideal soil moisture content that satisfies crop needs without causing water waste or degrading the soil. The soil water balance equation, which is $\Delta S = P + I - ET - D - R$, is one of the fundamental methods in soil water budgeting. In this equation, P stands for precipitation, I for irrigation, ET for evapotranspiration, D for drainage, and R for runoff. This equation facilitates decision-making regarding irrigation scheduling and water management techniques by helping to understand how various factors affect soil moisture levels.

Accurate measurement or estimation of each component is necessary for the successful application of this method. Usually, weather stations or rain gauges provide precipitation data. Farmers manage the irrigation inputs, and careful observation is necessary to prevent over- or under-irrigation. Crop coefficients and meteorological data are commonly used to estimate evapotranspiration, which comprises both transpiration from plants and evaporation from the soil surface. Empirical models like the Penman-Monteith equation or instruments like lysimeters, which calculate the volume of water lost through evapotranspiration, are frequently employed for this purpose. Although they are more difficult to quantify directly, runoff and drainage are crucial for avoiding soil erosion and waterlogging. These are frequently calculated using hydrological simulations or by analyzing the features of the landscape and the properties of the soil. In order to fine-tune the water balance, methods like using soil moisture sensors can provide real-time data on soil water content. Geographic information systems and remote sensing are also used in advanced soil water budgeting techniques (GIS). Drones and satellite imagery are two examples of remote sensing technologies that provide extensive and high-resolution data on land surface conditions, vegetation cover, and soil moisture. These technologies enable more accurate irrigation planning and spatial analysis of water distribution when combined with GIS.

Moreover, soil water budgeting and crop growth models are combined to simulate the effects of varying water management practices on crop yield and water efficiency. These models, like AquaCrop or DSSAT, forecast the results of various irrigation schedules and practices by taking into account a number of variables, such as crop characteristics, weather, and soil type. Precision agriculture, whose goal is to apply water sparingly so that every plant receives the proper amount at the right time, depends on the integration of technology into soil water budgeting. This minimizes the effects on the environment, such as soil erosion and nutrient leaching, while simultaneously increasing crop productivity and conserving water resources.

7.3.4. How Are Groundwater Resources Recharged?

For humans, groundwater is vitally important. It is a significant freshwater source that is actively utilized in many different industries, including agriculture. Mineral groundwater is used in medicine to treat a variety of human ailments, while the thermal energy of groundwater is used to heat buildings and create electricity. The water cycle gradually replenishes groundwater storage. However, an artificial groundwater recharge system can be used to expedite the recharging process due to the resource's ever-increasing demand.

7.3.4.1. Natural Groundwater Recharge System

In the natural groundwater recharge system, surface water infiltrates the subsurface, seeps down to the water table through the unsaturated zone, and adds to the groundwater reservoir (Figure 7.12).

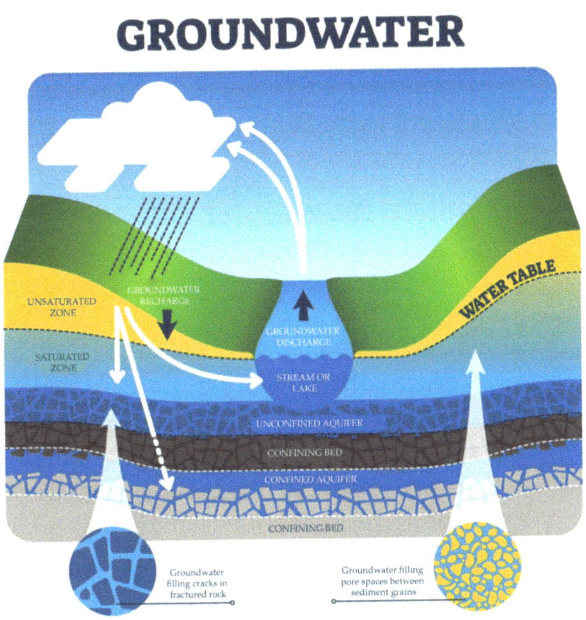

Figure 7.12. *Benefits of artificial groundwater recharge system.*

Source: https://media.licdn.com/dms/image/D5622AQEYLpn9fLQutw/feedshare-shrink_2048_1536/0/1702444199873?e=2147483647&v=beta&t=tb0Qc4WQ2Edx2xsR0QoUhKKiqnLsj-KNJAKd5zyCAR0.

7.3.4.2. Artificial Groundwater Recharge System

Artificial groundwater recharge is a set of engineering measures to increase groundwater recharge and maintain aquifers' operational reserve.

Groundwater can be artificially replenished by diverting surface water via ponds, infiltration basins, or injecting water directly into the subsurface via injection wells.

1. Benefits of an Artificial Groundwater Recharge System

   Overuse of groundwater can lead to problems that can be resolved by artificial recharge of groundwater. By increasing groundwater levels, it contributes to long-term groundwater management. Artificial groundwater recharge is being promoted as a way to manage water resources in various countries, with the goal of addressing water scarcity. Surface water sources can run for longer periods of time because of increased groundwater storage brought about by artificial recharge. In some places, seasonal streams can even develop a perennial flow. Utilizing an artificial groundwater recharge system does not require building massive buildings, like dams, to store vast amounts of water.

   It doesn't have any detrimental effects on society, like displacing people or taking up valuable agricultural land. Natural or artificial disasters cannot affect the water that is stored using an artificial recharge system. An increase in vegetative cover will produce more soil moisture when artificial recharge techniques raise groundwater availability. This could reduce soil erosion and improve the flora and fauna overall by attracting migratory birds and other wildlife. Wetland ecosystems benefit from artificial recharge because it keeps saltwater out of the area and stores water for later use. Additionally, it can improve the salinity of the water in saline aquifers, making them suitable for farming and cattle ranches close to the coast.

An indirect advantage for the environment is that less energy will be needed to pump the same amount of water as groundwater levels rise as a result of artificial recharge. Depending on the recharge process and the amount of energy needed, this could lead to net reductions in energy requirements. Aquifer thermal storage for matching cooling and heating demand has been greatly aided by artificial recharge. This technique has garnered attention recently because it offers significant electricity and fossil fuel savings. Future studies on artificial recharge control are planned in order to maximize energy storage system performance and boost efficiency, especially for large-scale applications. The unique economic advantages of artificial recharging are often associated with agriculture. The possibility of increasing the total area under irrigation owing to increased water availability or raising crop yields per hectare owing to increased land productivity are two frequently mentioned advantages. Shallow drilled wells and hand pumps are the main sources of drinking water in rural developing nations. Water from these sources can be filtered by an artificial recharge system, which uses biological and physical processes to get rid of pollutants, pathogens, and suspended particles. The continuous operation of such sources can be ensured even by artificial recharge activities that elevate water levels locally. Though they might not be enormous or represent a significant increase in net water availability, the volumes involved in these situations could be noteworthy locally. Preserving the water supply in nearby village wells can have additional advantages, as it can significantly cut down on the time and effort required to gather water for household purposes. This can impact everything from general hygiene and community health levels to time available for other productive endeavors.

2. *Future Outlook of Groundwater Recharge Systems*

Protecting groundwater resources is essential due to their limited availability and susceptibility to pollution. Currently, it is a top priority in addressing global environmental challenges. Artificial recharge systems, which are often simple and community-driven, have the potential to effectively support various physical and socioeconomic environments. However, there is a lack of rigorous evaluation of their effectiveness, leading to anecdotal claims of their benefits. To successfully manage groundwater using artificial recharge, it is important to establish clear criteria for determining where and how it can best address water issues.

7.3.5. Methods of Groundwater Recharge

1. **Spreading Basins:** Using this technique, water is flooded on the surface of basins that have been dug out of the existing topography. Highly permeable soils are appropriate for effective recharge, and it's important to keep a layer of water on top of the highly permeable soil. The infiltration rate, percolation rate, and horizontal water movement capacity are the three variables that determine how much water enters the aquifer when direct discharge is used. However, clogging at the aquifer's surface happens due to a variety of factors, including the growth of algae, the dispersion of soil and colloidal swelling caused by water, microbial activity, and the deposition of particles carried by the water in suspension or solution. The best places for spreading basin recharge are those with layers beneath the surface of the ground and the aquifer, as well as areas with clear water that can be recharged.
2. **Recharge Pits and Shafts:** Situations that allow surface flooding techniques for synthetic replenishment are comparatively uncommon. Low permeability lenses are frequently found between the water table and the land surface. In such a scenario, artificial recharge mechanisms like shafts and pits might be useful for reaching the dewatered aquifer. It has been observed that as the side slope of the pits increased, so did the rate of recharge. In order to maintain the high recharge rates, unfiltered runoff water leaves a thin layer of sediments on the sides and bottom of the pits. Shafts can have a square, rectangle, or circular cross section, and porous materials can be used as the backfill. The end of the excavation might be above the water table. Recharge rates in pits and shafts may eventually drop as a result of microbial activity-induced plugging and the buildup of fine-grained materials. (Figure 7.13).

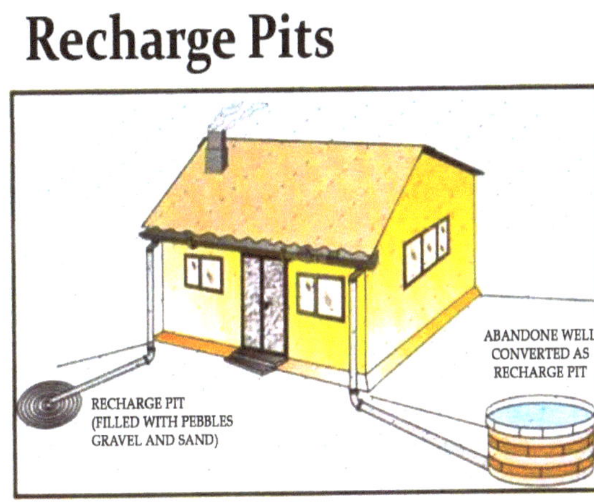

Figure 7.13. *Recharge pits*

Source: https://www.chaitanyaproducts.com/blog/wp-content/uploads/2018/10/Recharge-Pits-%E2%80%93-A-Widely-Used-Artificial-Method-for-Recharging-Ground-Water.jpg.

3. **Ditches:** A ditch is defined as a long, narrow trench where the depth is greater than the width at the bottom. A ditch system is made to fit the specific geological and topographic conditions of the location. A sequence of trenches extending down the topographic slope could be part of the design for a ditch and flooding recharge project. The ditches may end in a collection ditch that is intended to remove any water that does not seep in, preventing ponding

and minimizing the buildup of fine particles.

4. **Recharge Wells:** The deep-water bearing strata are directly replenished through the use of injection or recharge wells. In the event that the earth materials are not well-consolidated, a screen can be inserted into the well's injection zone. Recharge wells could be excavated through the material covering the aquifer. Only in regions where there is a substantial impermeable layer between the soil's surface and the aquifer that needs to be replenished are recharge wells appropriate. They also have advantages in places where there is a shortage of land. With this method, a comparatively high rate of recharge can be achieved. The recharge well's water level may build up excessively if the well screen or aquifer becomes clogged. (Figure 7.14).

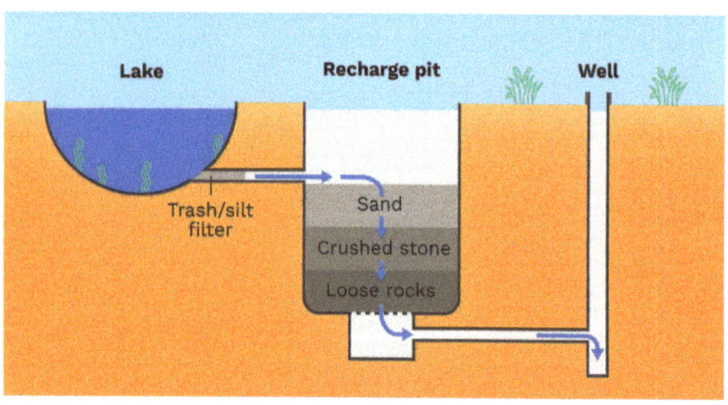

Figure 7.14. *Recharge or injection wells.*

Source: https://dialogue.earth/content/uploads/2022/11/Water-diagram-1.png.

5. **Harvesting in Cistern from Hill Sides:** This technique involves building small drains along the contours of hilly areas, with the goal of collecting runoff in these drains in a cistern at the foot of a mountain or hill. This water is of good quality and is used for both drinking and irrigation.

6. **Subsurface Dams:** From a higher pressure head to a lower one, groundwater flows. This will be beneficial in areas of the semi-arid zone, particularly in the upper reaches where there is a high groundwater velocity. More surface water can be used indirectly by using more ground water in the upper reaches, which will decrease inflow into the lower reaches of the supply. Either artificial sand storage dams or subsurface dams made of natural aquifer materials are used to store groundwater.

7. **Farm Ponds:** Traditional rainwater harvesting structures are these ones. Small storage buildings known as "farm ponds" are used for both irrigation and drinking water collection. Farm ponds can be divided into three types based on how they were built and which topographies they are best suited for. These include excavated farm ponds, which are ideal for flat terrain, embankment ponds, which are best suited for hilly and uneven terrain, and excavated cum embankment type ponds. Rainfall, land topography, soil type, texture, permeability, water-holding capacity, land-use pattern, etc. are some of the factors that influence the choice of where to put farm ponds.

8. **Historical Large Well across Streamlet:** If any historical wells are located near the streamlet, then allow the water into the well from streamlet by connecting drains. In this case the historical wells act

as a recharge well so that ground water can be improved.

9. **Check Dams:** For the purpose of collecting rainwater, check dams are little obstacles erected to reverse the direction of a shallow river or stream. A tiny catchment area behind the building is where the little dams store extra water flow during monsoon season. The impounded water enters the ground due to pressures generated in the catchment area. Restocking neighboring wells and ground water reserves is the main environmental advantage. Reducing a stream's gradient and velocity to manage erosion is the most typical use for check dams. (Figure 7.15).

Figure 7.15. *Check dams for water recharge.*

Source: https://dtnext-prod.s3.ap-south-1.amazonaws.com/imported/import/Articles/2020/May/20200506073438368_Dug-wells-indicate-lifestyle-change-in-Jawadhu-Hills_SECVPF.gif.

7.3.6. Issues to Consider

To guarantee the aquifer's suitability and the best possible design for the necessary surface installations, thorough feasibility studies must be carried out prior to implementation. When choosing a location, it's also important to evaluate legal, economic, and environmental aspects against other possibilities. Both public acceptance and the distance to the water use point are part of this.

- Advantages
  - No large structures (dams) needed to store large volumes of water
  - Stored water is relatively well protected from evaporation and pollutants
  - Schemes can be implemented incrementally, keeping initial capital investment relatively low
  - Depleted aquifers can be restored and saltwater intrusion can be controlled by applying artificial recharge
  - Purification and improvement of water quality by infiltration and percolation of surface water through the soil
  - Minimal land use for water storage required
  - Cost-effective implementation and environmental friendliness in comparison with surface storage and dams
- **Challenges**
  - Detailed feasibility studies required before implementation of large schemes
  - Continuous maintenance and qualified staff required in order to avoid clogging of injection wells and infiltration ponds, and prevent contamination of groundwater

  - Regulatory constraints: compliance of artificial recharge with the country's water and environmental legislation
  - Recovery of stored water requires wells and pumping (energy demand)
  - Undesirable chemical reactions can mobilize harmful substances
- **Pits and shafts**
  - In areas where impervious layer is encountered at shallow depth the pits and shaft are suitable structure for artificial recharge.
  - These structures are cost effective to recharge the aquifer directly.
  - The advantage of shaft or pit structure is that they do not require large piece of land like percolation tank and other spreading methods.
  - There is no loss of water in form of soil moisture and evaporation like methods of spreading.
- **Dug wells**
  - In alluvial as well as hard rock areas there are thousands of dug wells have either gone dried due to considerable decline of water level.
  - These dug wells can be used as recharge structure water and surplus water from canal etc. can be diverted into this structure to directly recharge the dried aquifer.
  - The water for recharge should be guided through a pipe to the bottom of well to avoid entrapment of bubbles in the aquifer.

7.4. REMOVAL OF CUTTINGS FROM UNDER THE BIT

In order to remove cuttings from the wellbore's bottom satisfactorily, the drilling fluid must use all of its hydraulic power or force to strike the hole's bottom. The pressure loss in the circulating system needs to be understood, regardless of the criteria that are applied. There may be laminar flow in some areas of the well and turbulent flow in other areas. One set of equations is needed to calculate pressure loss in laminar flow, and a different set is needed to calculate pressure loss in turbulent flow. The amount of laminar or turbulent flow in the system affects the pressure loss as well as the fluid's temperature; for nonaqueous drilling fluid (NADF), both temperature and pressure are important factors. The precise components of the drilling fluid affect these variables.

7.4.1. Cuttings Removal Process

In the rotary drilling process the rock surface is crushed by the drilling bit. In order to provide further crushing and progress, the cuttings must be removed instantly and efficiently from the rock surface. For this purposes, two forms of energy are brought from the surface to the rock surface, illustrated in Figure 7.16 (left).

- Mechanic energy
- Hydraulic energy

Energy should be applied as efficiently as possible once it reaches the rock surface. Hydraulic energy is used to remove cuttings, while mechanical energy is used to crush rocks. Both forms of energy are crucial for crushing and removing rock when drilling in soft and medium hard formations. Increased hydraulic energy, measured in hydraulic horse power, jet impact force, or jet velocity through the bit nozzles, has been found to significantly increase drilling rate in a number of studies. There is disagreement over which hydraulic quantity affects drilling rate the most.

The methods of hydraulic program design can be divided into two groups:

a. Determining the level of bottom hole cleaning energy required in order to balance the mechanical energy level.
b. Maximise criterion of estimation, e.g., jet impact force.

Group 1 methods necessitate field testing. These tests are based on the hypothesis that, under suitable bottom hole cleaning conditions, bit weight and drilling rate are proportionate. The minimal amount of bottom hole cleaning energy needed to prevent bit balling at that level of WOB is identified when the proportionality curve, as seen in Figure 7.16, starts to veer from its upward trend.

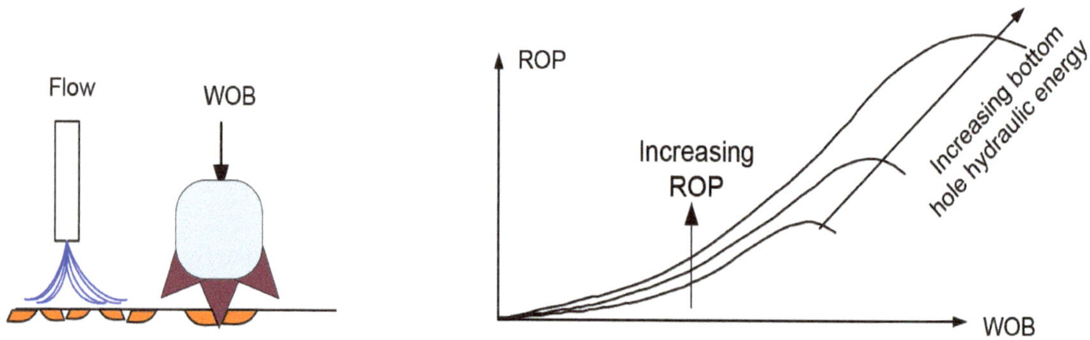

Figure 7.16. *ROP vs. WOB for different bit hydraulic energy levels. The points where the curves are turning and pointing downwards are called the bit balling point.*

Source: https://www.google.co.in/books/edition/Drilling_Fluid_Engineering/y7BwaMGyAK0C?hl=en&gbpv=1&dq=drilling+fluid+engineering&printsec=frontcover.

For methods in group 2, optimum hydraulics is the proper balance of the hydraulic elements (flow rate, bit nozzles diameter) which satisfy some criterion of estimation (the objective function) e.g., maximizing the jet impact force. In this book we select the hydraulic quantity q/d_{nozzle} as a common quantity representing the above mentioned bit hydraulic objective function. From experiments and observations demonstrated through Figure 7.14, we suggest a mathematical model that relates ROP and the hydraulic objective function:

$$ROP = A \cdot (q / d_{nozzle})^{a_8} \quad (10)$$

Figure 7.16 demonstrates rather clearly, that there is a general increasing trend of ROP when hydraulic energy is increased. Constant A in eq. (10) includes the effect of all other parameters (others than q, d_{nozzle}, and a_8) having an influence on ROP. All these parameters are assumed to be kept constant during the experimental phase and assumed to have little influence on the hydraulic parameters. Our task is to utilize the hydraulic energy of the drilling fluid to achieve the best possible cleaning below the bit so that ROP is maximized.

7.4.2. Boundary Conditions of the Drilling Process

First, we have the limitations of the pump itself. Each pump is defined with a maximum recommended pressure, p_{max}, and a maximum flow rate.

7.4.2.1. Friction Loss Increases with Depth

Due to friction in the ever extending wellbore, the pump would at some stage reach its maximum available pressure during transmission of hydraulic effect through the circulation system. Previously it has been shown that the pump pressure is a mirror image of the pressure losses in the circulating system. For hydraulic optimization purposes, we now split the friction losses in two parts; bit pressure losses, Δp_{bit}, and remaining pressure losses in the circulation system, Δp_{loss}, also called parasitic pressure losses. This is expressed through Eq. (10):

$$p_{pump} = \Delta p_{bit} + \Delta p_{loss} \tag{10}$$

Figure 7.17 displays both mentioned pressure losses, when they are embedded on to the pump characteristics. When the flow rate increases, for a given well length, the pump pressure will be more and more consumed as friction, Δp_{loss}, and less is left for the bit.

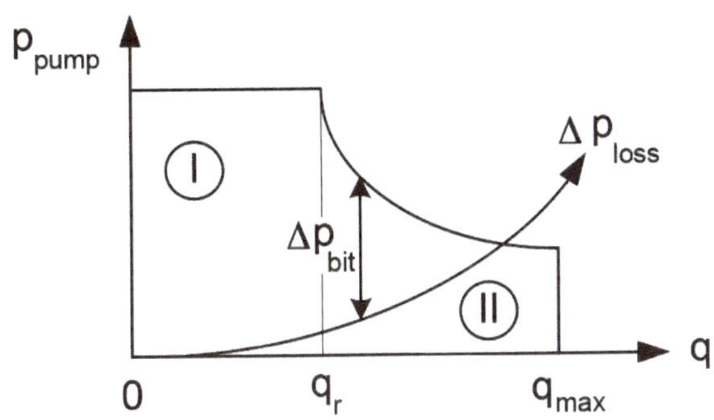

Figure 7.17. *Hydraulic friction as a function of flow rate, superimposed on the pump characteristics.*

Source: https://www.google.co.in/books/edition/Drilling_Fluid_Engineering/y7BwaMGyAK0C?hl=en&gbpv=1&dq=drilling+fluid+engineering&printsec=frontcover.

For hydraulic optimization purposes, we apply a simpler expression, reduced to variables important for hydraulic optimization:

$$\Delta p_{loss} = K_1 \cdot D \cdot q^m \tag{11}$$

Parasitic pressure loss varies linearly with depth and is an exponential function of pump rate.

The constants K_1 and m can be found by solving Eq. (11). Since there are two unknowns, two data sets are needed, at two different pump rates, rate 1 and rate 2. Taking the logarithm of Eq. (11) for two rates, we obtain:

$$m = \frac{\log \Delta p_{loss2} - \log \Delta p_{loss1}}{\log q_2 - \log q_1} \tag{12}$$

When m is known, K_1 is fond through:

$$K_1 = \frac{p_{loss2}}{D \cdot q_2^m} \tag{13}$$

Now a practical example of finding m and K_1 from simple rig experiments is summarized in Table 7.5.

Table 7.5. *Rig Test for Determining Hydraulic Parameters*

Pump rate	Pump pressure	Estimated bit pressure loss
$q_1 = 0.0135 m^3/s$	$p_1 = 100 \cdot 10^5\ Pa$	$\Delta p_{bit_1} = 11 \cdot 10^5\ Pa$
$q_2 = 0.027\ m^3/s$	$p_2 = 320 \cdot 10^5\ Pa$	$\Delta p_{bit_2} = 44 \cdot 10^5\ Pa$

Source: https://www.google.co.in/books/edition/Drilling_Fluid_Engineering/y7BwaMGyAK0C?hl=en&gbpv=1&dq=drilling+fluid+engineering&printsec=frontcover.

Observe that in Table 7.5 the bit pressure loss is estimated. Was applied. Now Δp_{loss} can be found from Eq. (11):

$$\Delta p_{loss1} = p_1 - \Delta p_{bit_1} = 89 \cdot 10^5$$

$$\Delta p_{loss2} = p_2 - \Delta p_{bit_2} = 276 \cdot 10^5$$

m and K_1 are found from Eq. (13) and (16) to be:

$$m = \frac{\log\left(\Delta p_{loss2} / \Delta p_{loss1_1}\right)}{\log\left(q_2/q_1\right)} = 1.63$$

$$K_1 = \frac{p_{loss1}}{D \cdot q_1{}^m} = 1.99 * 10^6 \tag{14}$$

Table 7.6 presents some typical calculated values for m and K_1, based on the same field tests as described through Table 7.5.

Table 7.6. *Pressure Loss Constants in Two Well Sections*

Bit no.	d_b	Drilling depth	Nozzle dia.(3)	ρ_m	μ_{pl}/Y_p	n	K	Δp_{bit}	Δp_{loss}	m	K_1
	in	m	32.in	kg/m^3	cP/lb/100ft^2		Pas^{-n}	bar	bar		10^6
9	12 1/4	950	14	1100	13 / 13	.58	.40	151.7	55.8	1.79	1.46
11	12 1/4	1700	16	1200	13 / 11	.62	.38	100.4	105.3	1.78	1.45
14	12 1/4	1840	14	1210	12 / 12	.58	.32	122.2	76.7	1.77	1.28
16	12 1/4	2100	13	1210	12 / 12	.58	.30	139.8	61.4	1.76	1.01
24	8 1/2	2560	11	1200	16 / 14	.62	.27	168.1	43.6	1.72	.78
28	8 1/2	2663	13	1320	15 / 10	.68	.18	132.1	73.2	1.76	1.08
44	8 1/2	2800	11	1290	18 / 9	.74	.13	150.7	55.6	1.68	.92
47	8 1/2	2916	12	1280	16 / 9	.71	.13	137.4	73.5	1.72	1.09
51	8 1/2	3014	12	1280	15 / 8	.72	.13	125.6	81.9	1.71	1.24
53	8 1/2	3060	12	1280	13 / 10	.65	.15	144.9	75.3	1.74	1.11

Source: https://www.google.co.in/books/edition/Drilling_Fluid_Engineering/y7BwaMGyAK0C?hl=en&gbpv=1&dq=drilling+fluid+engineering&printsec=frontcover.

Recorded data in left section, estimated to the right. Table 7.6 shows that this well used a very high number of bits. It was 1985, predating the significant advancement in steel quality selection, and frictional bearings were used in the 1990s. Table 7.6 also shows that bit nozzle replacements occurred on an equal basis. These days, it's possible that they'll drill a whole wellbore section using just one rock bit! The exponent m will get close to 2.0 if the flow is mostly turbulent in the annulus and drill string. In the full flow system, m will get close to the rheological power law exponent n when the flow is fully laminar. Since laminar pressure (in the annulus) is subordinated to turbulent pressure drop (in the drill string), m is typically somewhere in the middle, if slightly closer to 2.0.

7.4.3. Annular Flow Velocity Limitations

Two limits must be maintained for the flow velocity in the annulus of vertical wells. The upper one is typically made to prevent the development of turbulence around the drill collars. Turbulent flow is often permitted in the BHA annulus of horizontal wells since it greatly aids in cutting removal from this area of the annulus. If this is accepted, it is determined by ECD. The lower boundary is established by cuttings concentration. If it rises above 0.04, or 4 volume percent, issues relating to the buildup of cuttings—such as struck pipe—show an increasing pattern. As opposed to vertical wells, inclined wells have different criteria for annular flow velocities, which will lead to a different hydraulic program. The lowest recommended flow rate is known as the required flow rate, or qr dot. Probably the best place to start for inclined wells is with the optimal hydraulic parameter estimate that is derived for vertical wells.

7.4.4. General on Optimization

Due to cleaning requirements, flow rate often precede pressure in importance, and the use of maximum available flow rate and RPM as specified in the table for each relevant section is recommended. It must be ensured that these parameters are not in conflict with limitations for top side equipment or downhole tools and also that they do not lead to formation fracture pressure (Table 7.7).

Table 7.7. *Recommended Whole Cleaning Parameters*

	17 1/2" section	12 1/4" section	8 1/2" section
Flow rate (LPM)*	4000–5000	3000–3500	2000–2500
Drill string rotation (RPM)	140–180	130–175	120–170
Circulating Bottoms Up	1–3	2–4	3–4

Source: https://www.google.co.in/books/edition/Drilling_Fluid_Engineering/y7BwaMGyAK0C?hl=en&gbpv=1&dq=drilling+fluid+engineering&printsec=frontcover.

$v_{max} < v_{turb}$ (around drill collars)

$v_{min} \geq q_r$ (cuttings concentration in largest annulus ≤ 0.04)

7.4.5. Optimizing Rop, Liner by Liner

As indicated by the heading, the hydraulic program is dependent on first selecting the most suitable liner of the piston pump, before optimizing. It is the annular flow rate limitation, q_r, which determines the selection of liners, as indicated in Figure 7.18. The smallest possible liner should be selected since this will result in the highest possible pressure available. The hydraulic variables are given by Eq. (1.). By requiring pump pressure in Eq. (11) to be maximum and combining it and) with Eq. (12) and multiplying by q2 we obtain:

$$p_{max} \cdot q^2 = K_1 D q^{m+2} + 1.11 \frac{1}{2} \rho \cdot \frac{q^2 \cdot q^2}{\left(\frac{\pi}{4} d_e^2\right)^2} \tag{15}$$

Here the equivalent single nozzle diameter d_e is defined through total nozzle area. Assuming there are three nozzles in the bit, the equivalent bit nozzle becomes:

$$d_e^2 = \left(d_1^2 + d_2^2 + d_3^2\right) \tag{16}$$

If the tree diameters are of identical size:

$$d_e = d_1 \sqrt{3} \tag{17}$$

Rearranging Eq. (15) we obtain:

$$K_{bit} \left(\frac{q}{d_e}\right)^4 = p_{max} \cdot q^2 - K_1 \cdot D q^{m+2} \tag{18}$$

The constant $K_{bit} = \cdot 1.11 \cdot \frac{1}{2} \rho \Big/ \left(\frac{\pi}{4}\right)^2$ Solving Eq. (18) with respect to q/de yields:

$$\frac{q}{d_e} = \left[\frac{1}{K_{bit}} \left(p_{max} \cdot q^2 - K_1 \cdot D q^{m+2}\right)\right]^{1/4} \tag{19}$$

Combining Eq. (19) and Eq. (1.) and deriving with respect to the two variables, flow rate and nozzle diameter, and equating it to zero, Eq. (20) defines the optimum of variable q:

$$\frac{\partial R_p}{\partial q} = \frac{\partial}{\partial q} A \left[\frac{1}{K_{bit}} \left(p_{max} \cdot q^2 - K_1 D q^{m+2}\right)\right]^{\frac{a_8}{4}} = 0 \tag{20}$$

Deriving with respect to the variables de gives the following answer; minimize d_e. Moreover, since minimizing d_e is not feasible, only the q-derivation represents an applicable solution of Eq. (20). In Eq. (20) the constant p_{max} is the maximum recommended pressure for the selected liner. The only meaningful solution yields:

$$2 p_{max} \cdot q - (m+2) K_1 D \cdot q^{m+1} = 0 \tag{21}$$

which, when solved for q results in the optimal solution:

$$\Delta p_{bit,opt} = \frac{m}{m+2} p_{max} \tag{22}$$

Combining Eqs. (21) and (11), and solving for Δp_{bit} we obtain an additional optimal expression:

$$\Delta p_{bit,opt} = \frac{m}{m+2} p_{max} \tag{23}$$

The equivalent nozzle diameter d_e is sought:

$$d_e = \left[K_{bit} \cdot \frac{q^2}{(\Delta p_b)} \right]^{1/4} \tag{24}$$

Now the job is done, we have determined optimal parameters for a selected liner. The above procedure and solutions can also be solved graphically. Δpl_{oss} from Eq. (12) becomes a straight line in a log-log plot. Δp_{loss} represents the losses of the available pump pressure. When $\Delta p_{bit,opt}$ is secured, through the horizontal line, the sum of the two, $\Delta p_{loss} + \Delta p_{bit} = p_{pump}$, is given by the point where the two lines meet.

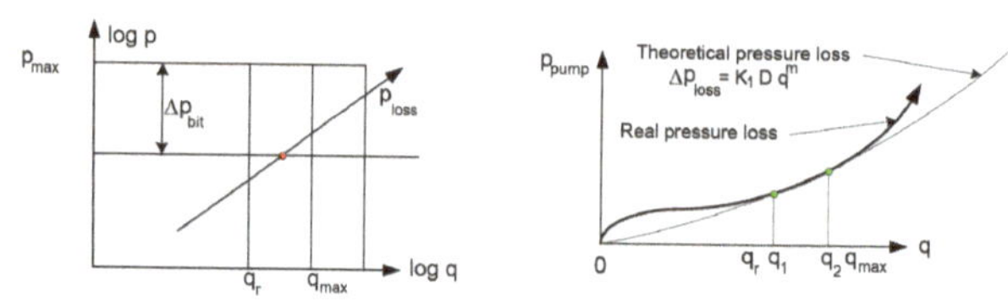

Figure 7.18. *Graphical determination of optimal flow rate.*

Source: https://www.google.co.in/books/edition/Drilling_Fluid_Engineering/y7BwaMGyAK0C?hl=en&gbpv=1&dq=drilling+fluid+engineering&printsec=frontcover.

A flow sheet of how to routinely determine the optimal bit nozzle and optimal flow rate is presented in Figure 7.19. Within a selected liner, the optimal flow rate will be a function of depth in accordance with Eq. (22), while nozzle size will stay constant in accordance with Eq. (23). With high quality drilling bits and pumps, it is probably sufficient to change liner and bit every time a new section is started. This program would suggest which liner, which nozzles and which flow rate is optimal at the selected depth.

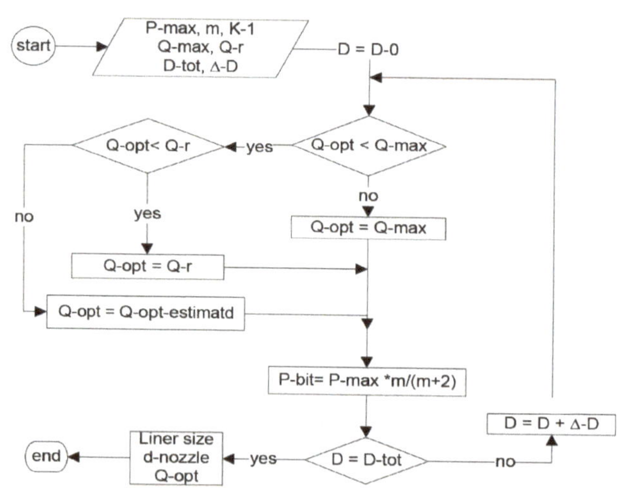

Figure 7.19. *Flow diagram of how to compute optimal flow rates and nozzle diameter in a vertical well when pump liner has been selected.*

Source: https://www.google.co.in/books/edition/Drilling_Fluid_Engineering/y7BwaMGyAK0C?hl=en&gbpv=1&dq=drilling+fluid+engineering&printsec=frontcover.

7.4.6. Optimizing the Complete Well

All the constants must be known before the well is drilled. One positive side effect obtained through this technique, is that it will be possible to suggest at what depth the pump liners can be changed. A complete hydraulic program would suggest which liner, which nozzles and which optimal flow rate should be chosen for every section. First we need to define two pump operating ranges, as already suggested in Figure 7.15. In operating range I, defined by the smallest liner, the pump pressure can be held constant. In order to utilize the pumps full capability, and thereby optimizing ROP, the pump pressure is kept at its maximum pressure. In range II, defined by the remaining liners, the pump power, E_{pump}.

Range I: $p_{max} = const$

Range II: $E_{max} = const$

The optimal solution in range II is found by combining Eq. (10) and (19):

$$ROP = A\left(\frac{q}{d_e}\right)^{a_8} = \frac{1}{K_{bit}}\left(p_{max}q^2 - K_1 \cdot D \cdot q^{m+2}\right)^{a_8/4} \quad (25)$$

By substituting $p_{max} \cdot q$ by E_{max} (which is constant over pump range II) in Eq. (25) we obtain:

$$A\left(\frac{q}{d_e}\right)^{a_8} = \frac{1}{K_{bit}}\left(E_{max} \cdot q - K_1 D q^{m+2}\right)^{a_8/4} \quad (26)$$

By differentiating Eq. (26) with respect to q, we obtain one meaningful solution:

$$\frac{\partial ROP}{\partial q} = E_{max} - (m+2)K_1 \cdot D \cdot q^{m+1} = 0 \quad (27)$$

which, when solved for q yields:

$$q_{opt_I} = \left[\frac{E_{max}}{K_1 \cdot D\,(m+2)}\right]^{\frac{1}{m+1}} \quad (28)$$

Combining the optimal solution in Eq. (27) with Eq. (11) and solving for Δp_{bit} we obtain

$$(\Delta p_{bit})_{opt_I} = \frac{m+1}{m+2} \cdot p_{max} \quad (29)$$

ROP, which is proportional to the cleaning function, can be plotted as a function of pump rate, as shown in Figure 7.20.

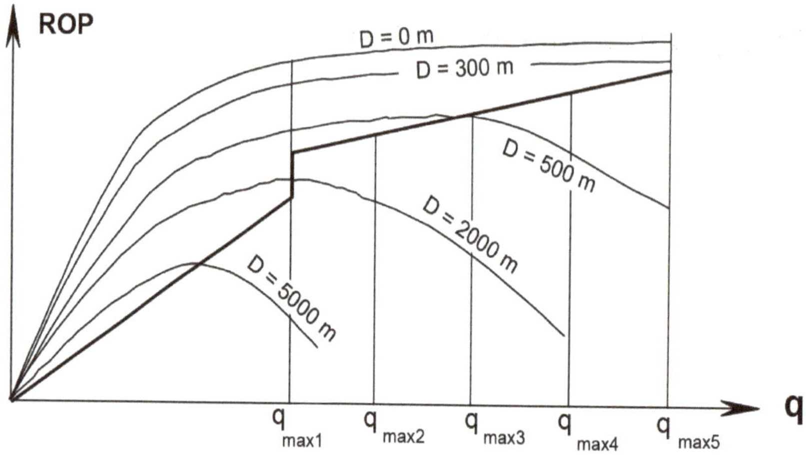

Figure 7.20. *ROP (eqn. (10) in range I and Eq. (17) in range II) vs. pump rate and depth.*

Source: https://www.google.co.in/books/edition/Drilling_Fluid_Engineering/y7BwaMGyAK0C?hl=en&gbpv=1&dq=drilling+fluid+engineering&printsec=frontcover.

We are now able to decide when to change liner in order to fully utilize the mud pump. By studying Figure 7.21 we can find graphically at which depth the pump liner has to be changed. A pc-program to monitor pump range and respective optimal solution is suggested in Figure 7.21.

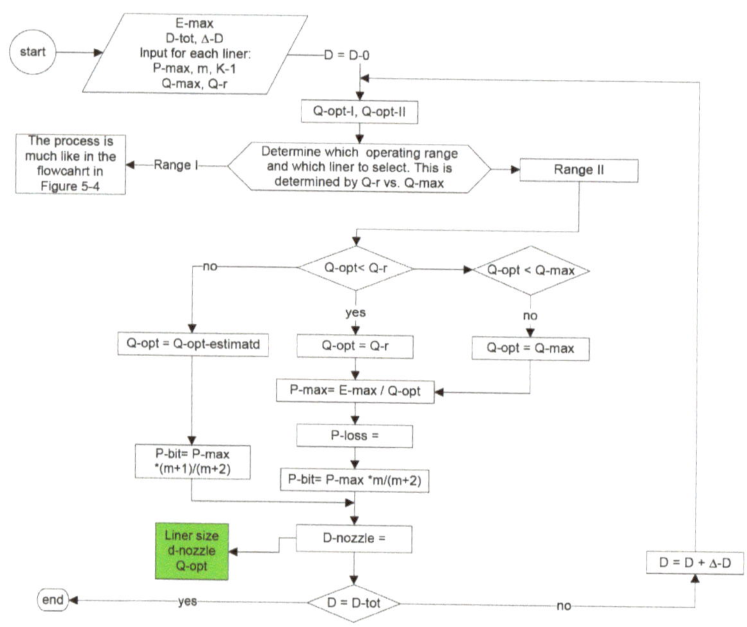

Figure 7.21. *A simplified flow sheet to compute optimum hydraulic parameters in pump operating range II.*

Source: https://www.google.co.in/books/edition/Drilling_Fluid_Engineering/y7BwaMGyAK0C?hl=en&gbpv=1&dq=drilling+fluid+engineering&printsec=frontcover.

CASE STUDY

MONITORING THE HYDROLOGY OF THE EVERGLADES IN SOUTH FLORIDA

The Everglades in South Florida, a unique and expansive wetland ecosystem, is the focus of intense hydrological monitoring efforts due to its ecological significance and the challenges posed by human activity and climate change. The monitoring of the Everglades' hydrology involves comprehensive and multifaceted approaches to understand and manage its complex water dynamics, which are crucial for maintaining the health of this diverse habitat. Hydrological monitoring in the Everglades is essential for several reasons. Firstly, the Everglades is a critical habitat for numerous plant and animal species, many of which are endangered. The region's water levels and quality directly influence the health and distribution of these species. Secondly, the Everglades serves as a vital water source for South Florida, supporting both the natural environment and human populations. Thus, maintaining the balance of water flow and storage is crucial for ecological and human needs. The hydrology of the Everglades is characterized by a slow-moving river system that depends on seasonal rainfall and the natural topography of the region. Historically, the area experienced natural water flow from Lake Okeechobee through the Everglades down to Florida Bay. However, human activities, including urban development, agriculture, and water management infrastructure like canals and levees, have significantly altered this natural flow. These changes have disrupted the natural hydrological cycles, leading to problems such as reduced water quality, habitat degradation, and increased frequency of extreme events like droughts and floods. To address these challenges, a variety of hydrological monitoring techniques are employed. These include remote sensing technologies, ground-based observations, and the use of hydrological models. Remote sensing, through satellites and aerial surveys, provides large-scale and continuous data on water levels, land cover, and vegetation health. This technology is crucial for tracking changes over vast areas and identifying trends that might not be apparent through ground-based methods alone. Ground-based observations complement remote sensing by providing detailed, localized data. This includes the use of stream gauges, weather stations, and groundwater monitoring wells. Stream gauges measure the flow and level of water in rivers and canals, while weather stations provide data on precipitation, temperature, and other climatic variables that influence hydrological processes. Groundwater wells help in understanding the dynamics of aquifers, which are critical for sustaining water flow during dry periods.

Hydrological models integrate data from remote sensing and ground-based observations to simulate the behavior of the Everglades' water system under various scenarios. These models are essential for predicting the impacts of different management strategies and climate change on the region's hydrology. They help in understanding how water flows through the landscape, how it interacts with vegetation and soils, and how changes in one part of the system might affect the whole.

One of the primary objectives of hydrological monitoring in the Everglades is to restore more natural water flow patterns. This involves efforts like the Comprehensive Everglades Restoration Plan (CERP), which aims to reestablish the historic flow of water to the region. CERP includes projects such as constructing reservoirs, removing or modifying levees and canals, and restoring wetlands. Monitoring data are crucial for the planning and implementation of these projects, providing the information needed to assess their effectiveness and make adjustments as necessary. Water quality monitoring is another critical aspect of hydrological studies in the Everglades. Nutrient pollution, particularly from agricultural runoff, has been a significant issue, leading to problems like algal blooms and the degradation of habitats. Monitoring efforts track the levels of nutrients, sediments, and other pollutants in the water, helping to identify sources of contamination and evaluate the success of mitigation efforts. Climate change poses additional challenges for the hydrology of the Everglades. Rising temperatures and changes in precipitation patterns can alter the timing and quantity of water entering the system. Sea level rise is another significant concern, as it can lead to saltwater intrusion into freshwater habitats, affecting the delicate balance of the ecosystem. Monitoring data are vital for understanding these impacts and developing strategies to adapt to changing conditions. The involvement of various stakeholders is essential for effective hydrological monitoring and management in the Everglades. This includes federal and state agencies, local governments, environmental organizations, and the public. Collaborative efforts ensure that diverse perspectives are considered and that management strategies are comprehensive and effective. Public education and outreach are also important, helping to build support for restoration efforts and encouraging practices that protect the Everglades.

CLASS ACTIVITY

UNDERSTANDING THE HYDROGEOLOGY PROCESS

- **Objective:** Students will learn about the hydrogeology process, including the water cycle, groundwater movement, aquifers, and the impact of human activities on groundwater.
- **Materials:** Large clear plastic containers or aquariums, Sand, gravel, and clay, Small pebbles or stones, Water, Plastic cups with small holes in the bottom, Straws, Worksheets with diagrams of the water cycle and groundwater flow, Markers and pens.
- **Introduction:** Begin with a brief lecture on the basics of hydrogeology, explaining key concepts such as the water cycle, groundwater, aquifers, and the impact of human activities on groundwater. Use visual aids such as diagrams of the water cycle and groundwater flow to help illustrate these concepts.

SUMMARY

- Groundwater is the water present beneath Earth's surface in rock and soil pore spaces and in the fractures of rock formations. About 30% of all readily available freshwater in the world is groundwater. A unit of rock or an unconsolidated deposit is called an aquifer when it can yield a usable quantity of water.
- Groundwater is often cheaper, more convenient and less vulnerable to pollution than surface water. Therefore, it is commonly used for public water supplies.
- Groundwater level data for an aquifer unit can be used for several purposes including plotting a hydrograph, determining the direction of groundwater flow by constructing a map of the potentiometric surface, and in completing a flow net.
- Measurement of groundwater level in the field is undertaken using either a water level dipper or by use of a submersible pressure transducer positioned just below the lowest expected groundwater level.
- Large fluctuations in water levels in wells and boreholes in confined aquifers can be caused by changes in atmospheric pressure. With increasing barometric pressure, water levels are noticed to decrease.
- Near-surface hydrological processes such as precipitation, evapotranspiration and infiltration have a profound influence on stream flow generation and groundwater recharge.
- In the hydrological cycle, evaporation of water occurs from free water surfaces such as lakes, reservoirs and rivers with the rate dependent on factors such as water temperature and the temperature and humidity of the layer of air above the water surface.

REVIEW QUESTIONS

1. How is water level measurement performed in hydrological studies?
2. What are borehole hydrographs and barometric efficiency, and why are they important?
3. How are evapotranspiration measurements and estimations carried out in hydrology?
4. What is the borehole hydrograph method, and how does it work?
5. What are the boundary conditions of the drilling process, and why do they matter?

REFERENCES

1. Abraham, M., & Mohan, S. (2015). Effectiveness of artificial recharge structures in enhancing groundwater storage: A case study. *Indian Journal of Science and Technology*, *8*(20), 1–10. https://doi.org/10.17485/ijst/2015/v8i20/81596.

2. Bhalerao, S. A., & Kelkar, T. S. (2013). Artificial recharge of groundwater: A novel technique for replenishment of an aquifer with water from the land surface. *International Journal of Geology, Earth and Environmental Sciences*, *3*, 165–183.
3. Blaney, H. F., & Criddle, W. D. (1962). Determining consumptive use and irrigation water requirements. *United States Department of Agriculture Technical Bulletin*, *1275*.
4. Brassington, R. (1998). *Field Hydrogeology* (2nd ed.). John Wiley.
5. Choi, M. R., & Kim, G. B. (2022). Effects of hybrid-type artificial groundwater recharge and underground barrier in a small basin. *Water*, *14*(12), 1849. https://doi.org/10.3390/w14121849.
6. Grindley, J. (1969). The calculation of actual evaporation and soil moisture deficits over specified catchment areas. *Hydrological Memoir*, *38*. Meteorological Office, Bracknell.
7. Hartmann, A. (2021). The hydrology of groundwater systems—From recharge to discharge. *Encyclopedia of Inland Waters* (Second Edition). https://doi.org/10.1016/B978-0-12-819166-8.00097-9.
8. Jacob, C. E. (1940). On the flow of water in an elastic artesian aquifer. *Transactions of the American Geophysical Union*, *22*, 574–586.
9. Lerner, D. N., Issar, A. S., & Simmers, I. (1990). *Groundwater Recharge: A Guide to Understanding and Estimating Natural Recharge* (Vol. 8). International Contributions to Hydrogeology. Verlag Heinz Heise.
10. Monteith, J. L. (1965). Evaporation and the environment. *Proceedings of the Symposium of the Society for Experimental Biology*, *19*, 205–234.
11. Penman, H. L. (1948). Natural evaporation from open water, bare soil and grass. *Proceedings of the Royal Society of London, Series A*, *193*, 120–145.
12. Rushton, K. R., & Ward, C. (1979). The estimation of groundwater recharge. *Journal of Hydrology*, *41*, 345–361.
13. Shi, X., Jiang, S., Xu, H., Jiang, F., He, Z., & Wu, J. (2016). The effects of artificial recharge of groundwater on controlling land subsidence and its influence on groundwater quality and aquifer energy storage in Shanghai, China. *Environmental Earth Sciences*, 75(3), 1–18. https://doi.org/10.1007/s12665-015-5019-x.
14. Thornthwaite, C. W. (1948). An approach towards a rational classification of climate. *Geographical Review*, *38*, 55–94.
15. Todd, D. K. (1980). *Groundwater Hydrology* (2nd ed.). John Wiley.
16. Veeranna, J., & Jeet, P. (2020). Groundwater recharge technology for water resource management: A case study. In *Groundwater Management and Resources*. IntechOpen. https://doi.org/10.5772/intechopen.93946.

CHAPTER

8

Groundwater Quality and Contaminant Hydrogeology

LEARNING OBJECTIVES

After studying this chapter, you will be able to:

- Discuss the groundwater and contaminant hydrology
- Focus on sources and consequences of groundwater contamination

KEY TERMS FROM THIS CHAPTER

Chromatography
Groundwater
Nanoparticles
Nitrogen contaminants
Reduction
Flow velocity
Hydrology
Nitrates
Organic liquids

INTRODUCTION

Within environmental science, groundwater quality and contaminant hydrogeology are crucial disciplines that concentrate on the evaluation and control of water quality in subterranean aquifers. Anthropogenic and natural sources can contaminate groundwater, which is an essential resource for industry, agriculture, and drinking water. Significant risks to human health and ecosystems are posed by contaminants like heavy metals, nitrates, pesticides, and industrial chemicals.

It is necessary to analyze physical, chemical, and biological parameters in order to comprehend groundwater quality. Temperature and turbidity are examples of physical parameters; pH, salinity, and concentrations of dissolved minerals and pollutants are examples of chemical parameters. Microbial populations and the existence of pathogens are examples of biological parameters. These parameters are detected and quantified through regular monitoring and the use of sophisticated analytical techniques like mass spectrometry and chromatography, which together provide a thorough evaluation of groundwater quality.

The subdiscipline of contaminant hydrogeology focuses on the behavior and movement of pollutants in groundwater systems. To comprehend how pollutants move through aquifers, it combines concepts from environmental engineering, hydrogeology, and geochemistry. Pollutant concentration and spread are influenced by a number of factors, including aquifer porosity, groundwater flow velocity, and chemical interactions between contaminants and soil or rock materials. For example, a contaminant's mobility and persistence in groundwater are determined by its solubility and adsorption characteristics.

There are many different sources of contaminated groundwater. Fertilizers and pesticides from agricultural operations contribute by leaching, and heavy metals and organic solvents from industrial processes can also be introduced. Significant sources of contamination also include septic systems, leaking underground storage tanks, and inappropriate waste disposal. Groundwater quality is also impacted by naturally occurring sources, such as the dissolution of minerals or the existence of naturally occurring radioactive materials. Groundwater contamination mitigation calls for a multidisciplinary strategy. Remediation methods like pump-and-treat, in situ bioremediation, and permeable reactive barriers, as well as pollution prevention measures like implementing best management practices in industry and agriculture, are examples of strategies. Cutting-edge modeling techniques aid in the prediction of contaminant transport and direct the remediation process.

Groundwater management must be done well in order to protect this vital resource. In order to ensure sustainable use and protection, it includes not only monitoring and remediation but also regulatory frameworks and public education. Sustaining high groundwater quality and developing contaminant hydrogeology are critical for environmental resilience and health in the face of mounting demands on water resources and pollution pressures.

8.1. GROUNDWATER AND CONTAMINANT HYDROLOGY

Numerous studies on the movement and fate of pollutants in soils and groundwater are conducted under the umbrella of groundwater and contaminant hydrology. The center's purview comprises: 1) the creation of novel methods for site characterization and sampling; and 2) additional enhanced methods for groundwater remediation.

The study of processes that impact pollution of both surface and ground waters is known as contaminant hydrology. It makes use of chemistry and hydrology concepts. Research on contaminant hydrology and water quality aims to comprehend the impact of soil characteristics and hydrologic processes on the contamination of ground and surface waters and to create plans to lessen these effects. Research is conducted on both established and newly discovered contaminants at all scales, from the soil pore to the basin scale. The fate and transport of jet fuel leaks from oil depots, the production of water injection in shallow wells from oil and gas exploration field concession areas, the effects of tile drainage on sediment and nutrient pollution on rivers, the pathogen losses from manure application, the veterinary pharmaceuticals from land-applied manure, the fate and transport of pharmaceutical industry disposal wastes in unlined evaporation ponds, sediment-turbidity relationships, water quality modeling, and TMDL and paired watershed studies are among the studies of groundwater and contaminant hydrology.

Some research institutes use a variety of techniques, including field testing, laboratory experiments, and theoretical and numerical groundwater flow and transport investigations, to address the needs of both domestic and international communities for subsurface contaminant characterization and remediation.

The chemical evolution of highly alkaline radioactive waste in storage tanks; the reduction, reoxidation, and diffusion of uranium forms in sediments; the hydraulic properties of unsaturated gravels; and the naturally occurring production of transport-enhancing mobile nanoparticles in the subsurface are some of the developing countries' most pressing subsurface contamination issues. To develop new methods and tools for estimating reactive transport parameters at the field scale and characterizing contamination sites, inverse modeling of reactive transport and joint hydrologic and geophysical inversion are studied.

Direct migratory of contaminants into groundwater from subsurface sources (e.g., pipelines, storage tanks) that are located in the saturated zone. Furthermore, pollutants may seep into the groundwater system from the surface through vertical leakage through well casing seals, wells that have been abandoned without following the correct procedures, or the improper disposal of contaminated well materials.

8.1.1. Governing Processes of Contaminant Transport

Advection, dispersion, and retardation are the three main mechanisms that control the movement of pollutants in groundwater. Retardation processes can slow the rate of movement of contaminants, but dispersion and density/viscosity differences may speed up the movement. Two or more fluids may be involved in some contamination issues. Examples include organic liquids and water in an aquifer, or air, water, and organic liquids in the unsaturated zone. Water flow in the saturated and unsaturated zones can be described with the help of tracer technology.

8.1.1.1. Advection

Advection is the term used to describe the movement brought about by groundwater flow. Darcy's law serves as the basis for calculating groundwater flow, or advection. Advective transport paths can be computed via particle tracking. By inserting a particle into the flow field and numerically integrating the flow path, particle tracking is a numerical technique.

8.1.1.2. Dispersion

The contaminant plume progressively becomes diluted due to dispersive spreading both within and across the direction of the main flow.

Aquifer heterogeneities cause a contaminant plume to disperse and spread. On a macroscopic level, dispersion results from changes in porosity and hydraulic conductivity. At the decimeter scale, preferential flow-paths resulting from variations in hydraulic conductivity can impact solute transport.

8.1.1.3. Retardation

Soil-growth and biodegradation are the two main processes that slow the movement of contaminants. An equilibrium sorption isotherm can be used to describe the solute's sorptive process if it proceeds more quickly than the flow velocity and reaches equilibrium with the sorbed phase.

The linear sorption isotherm can be described by the equation:

$$C^* = K_d C \qquad (1)$$

Where C^*= mass of solute sorbed per dry unit weight of solid (mg/kg), C= concentration of solute in solution in equilibrium with the mass of solute sorbed onto the solid (mg/l) and K_d = distribution coefficient (L/kg).

8.1.1.4. Non Aqueous Phase Liquids (NAPL)

Dense nonaqueous phase liquids, or DNAPLs, are organic liquids with densities higher than water. LNAPLs, or light nonaqueous phase liquids, are nonaqueous phase liquids with densities lower than that of water. Fuel spillages, such as those involving gasoline or jet fuel, are a common cause of LNAPL contamination.

8.1.2. Groundwater Flow and Contaminant Transport Modeling

Creating a conceptual model, choosing a computer code, and designing the model are the first stages in simulating groundwater flow and pollutant transport. A numerical groundwater flow model's definition is dependent on factors such as the field system's water sources and sinks, the geohydrologic system's data availability, and the system geometry i.e., the kinds and quantity of model layers;

(d) the hydraulic properties' temporal and spatial organization; and (e) boundary conditions. Finite differences schemes are used in the popular MODFLOW and MT3D solute transport numerical codes, which are regarded as being extremely dependable. A three-dimensional modular finite-difference model of U is called MODFLOW. S. The behavior of the groundwater system is described and predicted using the Geological Survey, which is widely utilized. Variable grid spacing is used by the program in both the x and y directions. Finding the set of parameter values that best fits the model's output to field observations can be done through parameter estimation. Using a finite difference grid system, the model first calculates drawdown, flow direction, and hydraulic heads on each nodal point. The MT3D model is run to simulate contaminant transport in a groundwater system, with the initial condition being the steady-state hydraulic heads determined by the model. A model can be used to forecast for management or other objectives once it has been calibrated.

8.2. SOURCES AND CONSEQUENCES OF GROUNDWATER CONTAMINATION

Groundwater is utilized for domestic, agricultural, and industrial purposes and is one of the world's main sources of fresh water. Roughly one-third of the world's population gets their drinking water from groundwater. In arid and semi-arid regions with limited surface water and precipitation, groundwater is an especially valuable resource (Li et al., 2017a). One of the main factors influencing a country's ability to develop sustainably is ensuring a steady and renewable supply of groundwater for consumption.

However, the quality of groundwater is seriously threatened by industrialization, urbanization, climate change, and agricultural practices. Human health, ecological services, and sustainable socioeconomic development are all at risk from contaminants like toxic metals, hydrocarbons, trace organic contaminants, pesticides, nanoparticles, microplastics, and other emerging contaminants (Li, 2020; Li and Wu, 2019). The theme of chemical contamination has been frequently reported in groundwater studies over the last thirty years. Although groundwater contamination poses a serious threat to human populations, it also offers researchers and decision-makers an excellent chance to learn more about the history of our subsurface aquifers and strategies for safeguarding their quantity and quality. The Critical Zone (CZ), which stretches from the top of the vegetation canopy to the bottom of the aquifer, includes fresh water aquifers as one of its most significant sections (Lin, 2010). A variety of studies have been conducted as part of the global effort to comprehend the structures, functions, and processes within the CZ, adding to our understanding of groundwater circulation and evolution (Sawyer et al., 2016; Goldhaber et al., 2014).

Due to the dissolution of natural mineral deposits in the Earth's crust, a large number of contaminants found in groundwater are geogenic in origin (Basu et al., 2014; Pandey et al., 2016; Rao Subba et al., 2020; He et al., 2020a). But now that the world's population is growing so quickly, along with urbanization, industrialization, agricultural output, and the economy, we have to deal with the detrimental effects of pollutants that have human origins. Rapidly developing economies, many of which are found in the eastern hemisphere, are the ones most impacted by these global shifts (Clement and Meunie, 2010; Hayashi et al., 2013; Lam et al., 2015). It is therefore fitting that the focus of this special issue, "The fate and consequences of groundwater contamination," is on research concerning the particular difficulties associated with contaminants in groundwater that have both geological and anthropogenic origins in a number of eastern hemisphere nations, such as China, India, Turkey, Bangladesh, Ethiopia, and Nigeria.

The articles in this special issue cover a wide range of topics, including the latest methods for detecting and tracking groundwater contaminants, innovative techniques for assessing risks to human populations from contaminated groundwater, the impact of groundwater contamination on the abiotic environment, and case studies and remedial actions for controlling contamination from natural and human sources. The co-editors hope that these articles will improve understanding of the causes and effects of groundwater contamination, offer strategies for remediation, and help protect groundwater quality.

8.2.1. Major Contaminants

Groundwater contamination is defined as the addition of undesirable substances to groundwater caused by human activities (Government of Canada, 2017). This can be caused by chemicals, road salt, bacteria, viruses, medications, fertilizers, and fuel. However, groundwater contamination differs from contamination of surface water in that it is invisible and recovery of the resource is difficult at the current level of technology (MacDonald and Kavanaugh 1994). Contaminants in groundwater are usually colorless and odorless. In addition, the negative impacts of contaminated groundwater on human health are chronic and are very difficult to detect (Chakraborti et al., 2015). Once contaminated, remediation is challenging and costly, because groundwater is located in subsurface geological strata and residence times are long (Wang et al., 2020; Su et al., 2020). The natural purification processes for contaminated groundwater can take decades or even hundreds of years, even if the source of contamination is cut off (Tatti et al., 2019).

Although there is a rapid increase in the classes of contaminants found in groundwater, there are three main categories into which they can be broadly classified: chemical, biological, and radioactive. Both manmade and natural sources may contribute to these pollutants (Elumalai et al., 2020). Mineral deposits, brackish water, low-quality surface waters, and saltwater are some of the natural sources of groundwater pollution. If human activities disturb the natural environmental balance, these natural sources could become serious sources of contamination. Examples of such activities include acid mine drainage from the extraction of minerals, leaching of hazardous chemicals from excessive irrigation, and depletion of aquifers that result in saltwater intrusion (Su et al., 2020; Wu et al., 2015; Li et al., In 2016 and 2018.

Ammonia nitrogen, nitrate, and nitrate are examples of common inorganic contaminants that contain nitrogen. The majority of nitrate comes from human sources, such as agriculture (i.e., manure, and fertilizers) as well as household wastewater (Hansen et al., 2017; Wu and He, 2019; He et al., 2019; Karunanidhi et al., 2019; Li et al., 2019a; Serio et al., 2018; Zhang et al., 2018). There have been numerous reports of groundwater nitrate contamination from all over the world. Major cations like Ca^{2+} and Mg^{2+} as well as anions and oxyanions like F^-, SO_4^{2-}, and Cl^- are common inorganic contaminants found in groundwater. Groundwater may also have higher levels of total dissolved solids (TDS), which is the total amount of inorganic and organic ligands in water. Although most of these pollutants have a natural origin, groundwater levels can also be elevated by human activity (Adimalla and Wu, 2019).

The natural environment and human populations are at risk for health problems due to toxic metals and metalloids. Metals like zinc (Zn), lead (Pb), mercury (Hg), chromium (Cr), and cadmium (Cd), as well as metalloids like selenium (Se) and arsenic (As), are among the chemical elements commonly found in groundwater. While some of these elements are necessary micronutrients at lower doses, exposures at high concentrations can cause severe poisoning (Hashim et al., 2011). For instance, He and Li (2020) found that exposure to hexavalent chromium (Cr^{6+}) can raise the risk of cancer. The US Environmental Protection Agency (EPA) and the International Agency for Research on Cancer (IARC) classify arsenic as a Group 1 human carcinogen. As3+ can react with proteins' and enzymes' sulfhydryl (–SH) groups to disrupt cellular processes and ultimately result in cell death (Abbas et al., 2018; Rebelo and Caldas, 2016). When hazardous metals from the environment get into the food chain, they are likely to be persistent and undergo moderate bioaccumulation (He and Li, 2020; Hashim et al., 2011).

The natural environment and human populations are at risk for health problems due to toxic metals and metalloids. Metals like zinc (Zn), lead (Pb), mercury (Hg), chromium (Cr), and cadmium (Cd), as well as metalloids like selenium (Se) and arsenic (As), are among the chemical elements commonly found in groundwater. While some of these elements are necessary micronutrients at lower doses, exposures at high concentrations can cause severe poisoning (Hashim et al., 2011). For instance, He and Li (2020) found that exposure to hexavalent chromium (Cr6+) can raise the risk of cancer. The US Environmental Protection Agency (EPA) and the International Agency for Research on Cancer (IARC) classify arsenic as a Group 1 human carcinogen. As3+ can react with proteins' and enzymes' sulfhydryl (–SH) groups to disrupt cellular processes and ultimately result in cell death (Abbas et al., 2018; Rebelo and Caldas, 2016). When hazardous metals from the environment get into the food chain, they are likely to be persistent and undergo moderate bioaccumulation (He and Li, 2020; Hashim et al., 2011).

In addition to coming from natural radionuclide deposits, man-made sources of radioactive contamination in groundwater include nuclear power plant waste, nuclear weapons testing, and incorrect disposal of medical radioisotopes (Dahlgaard et al., 2004; The Lytle Group, 2014; Huang et al., 2012). The human body can be exposed to radioactive materials in a number of ways, including through drinking water. Nonetheless, it is uncommon to find radioactive pollutants in groundwater at concentrations that pose a risk to public health.

Algae and microorganisms, such as bacteria, viruses, and protozoa, are examples of biological contaminants. More than 400 different species of bacteria and more than 100 different types of viruses have been identified as microbial contaminants in human and animal feces (Shen and Gao, 1995). While some of these microbial organisms are naturally occurring, others are microscopic species that coexist with natural algal species and engage in resource competition (Flemming and Wuertz, 2019; Lam et al., 2018). Drinking water tainted by microorganisms can cause a variety of illnesses in people, including severe diarrheal illnesses like cholera and typhoid. The COVID-19 virus is currently causing a pandemic that is affecting people everywhere in the world. According to the

Centers for Disease Control and Prevention (2020), respiratory droplets are the main way that this coronavirus spreads between individuals. But this virus can also pose a health risk to people through contaminated water (Bhowmick et al., 2020; Gautam and Lokhandwala, 2020). Because of eutrophication, algae are frequently found in surface waters like lakes and reservoirs, but they are rarely found in groundwater at high biomass levels.

8.2.2. Consequences of Groundwater Contamination

Contamination of groundwater can have an effect on socioeconomic development, environmental quality, and human health. For instance, numerous studies have demonstrated that populations of humans are at risk for health problems due to high concentrations of metals, nitrate, fluoride, and persistent organic pollutants (Wu et al., 2020). Because they are more vulnerable to the effects of these contaminants than adults, infants and children should pay particular attention to this (He et al., 2020b; Karunanidhi and associates, 2016; Wu and Sun. 2020; Mthembu et al., 2020; Ji et al., 2020; Rao Subba et al., 2020; Zhou et al., 2020). For example, high nitrate concentrations in the drinking water used to make baby formulas cause blue baby syndrome, also known as infant methemoglobinemia. Because of its effects on the food production system, groundwater contamination may also have an impact on human health. Human health risks can arise from the accumulation of toxic elements in cereals and vegetables due to irrigation with groundwater contaminated by heavy metals and wastewater containing persistent contaminants (Jenifer and Jha, 2018; Yuan et al., 2019; Njuguna et al., 2019). The quality of lands and forests can also be adversely affected by groundwater contamination. Degradation of land quality and soil contamination can result from contaminated groundwater. For instance, high groundwater salinity is one of the main factors influencing soil salinization in many arid agricultural areas (Wu et al., 2014). In the root zone, soluble salts and other impurities, including heavy metals, can build up and hinder the growth of plants. Surface water-groundwater interactions can also carry contaminants from groundwater, which can worsen the quality of surface water (Teng et al., 2018).

A balance between the rate of natural resource renewal and human demand is necessary for sustainable economic development (Li et al., 2017b). Perhaps the most precious natural resource is freshwater. Chronic groundwater contamination, on the other hand, has the potential to decrease freshwater availability, upsetting the equilibrium between water supply and demand and resulting in socioeconomic problems and even wars. Contamination-induced water shortages could eventually contribute to conflicts among residents (Schillinger et al., 2020), potentially postponing a country's socioeconomic advancement. Collaboration between social scientists and natural scientists is necessary because groundwater contamination is a social as well as environmental issue.

8.2.3. Case Study: Major Groundwater

In many urban areas across the world, groundwater is a vital supply for both industrial and residential applications. The primary causes of natural groundwater contamination are accident spills and leaks from surface and subterranean storage tanks, as well as industrial area effluents. Estimating the spatial extent of contamination is crucial when such contamination is found. Traditionally, the process of determining the level of contamination involves collecting numerous samples from various points within predetermined time and financial constraints. This typically necessitates the installation of multiple observation wells. As a result, these procedures can be highly expensive, particularly if higher resolution measurements are needed.

In an attempt to forecast what would happen to the jet fuel that spilled from above-surface storage tanks in Karachi, Pakistan, a three-dimensional model of contaminant transport was created. The product that had dissolved sawp into the groundwater system and began to spread beyond the site because the tanks were located in a sandy layer. The modeling process included the groundwater system's steady-state simulation, a transient simulation covering the period from January 1986 to December 2015, and jet fuel calibration carried out in relation to various groundwater system parameters. A modular three-dimensional finite-difference groundwater model (PMWIN) Mod Flow and solute transport model (MT3D) were used to simulate the fuel in a hypothetical 1986–2001 period. Calibration was accomplished and matched to known conditions following a realistic distribution of piezometric heads within the aquifer system; as a result, the solute transport component was coupled to the flow. Contour maps of jet fuel concentration illustrate the growing plume over time, which was nearly obvious in the years before.

Transverse dispersion of the contaminant plume, or spreading, can be predicted using two-dimensional (2-D) solute transport models. Furthermore, 2-D models are appropriate in situations where the source of contamination may be located within or close to a continuously pumping well's radius of influence. Three-dimensional (3-D) numerical models ought to be employed solely in cases where copious data pertaining to vertical and horizontal heterogeneity, as well as spatial variability in contaminant concentrations, are accessible. For the purpose of understanding the dynamics of the jet fuel leak from above-ground storage tanks in the Karachi, Pakistan, metropolitan area, a localized contaminant transport model for groundwater is developed. Refined hydrocarbons of the kerosene type—mixtures of benzene, toluene, ethyl benzene, and isomers of xylene—make up jet fuel. As a result of ongoing seepage from the storage depot, hypothetical monitoring wells were constructed to estimate the concentrations of jet fuel over a predetermined time period. Despite the lack of specific information on the seepage history, it is anticipated that seepage from the storage tanks occurred for more than ten years based on the results of an electrical resistivity sounding survey (ERSS) about the nature of the subsurface lithologies and the conclusions of earlier researchers from Mott MacDonald Pakistan (MMP). The subsurface resistivity values that are assigned to various geological materials are obtained using ERSS. MMP carried out an assessment of the environmental harm caused by oil pollution to the soil and groundwater, and recommended corrective actions for Shell Pakistan's airfields, installations, and depots across the nation. The composition of jet fuel is given in Appendix A (Table 8.1).

8.2.4. Site Description

The project site is located between longitudes 67° 07′ 20" and 67° 10′ 30" and latitudes 24° 52′ 20" and 24° 54′ 20". The land surface elevation ranges from approximately 15 to 33 meters above mean sea level (masl). In the far south, the Malir River drains into the Arabian Sea (Figure

8.1). The typical lithology of the site is silty to sandy clay from 0 to 15 ft (4.6 m) bls, gravelly sand from 15 to 43 ft (4.6 to 13 m) bls, and clayey to silty sand from 43 to 80 ft (13 to 24 m) bls. The region is arid with an average annual rainfall of about 200 mm (7.9 in). Out of this, only 10% is considered to recharge the aquifer system (6.34 x 10-9 m/sec).

Within the storage site, the vadose zone is contaminated with up to 1300 parts per million (ppm) of total petroleum hydrocarbons (TPH). To determine the concentration of hydrocarbon compound, soil samples were taken at various points three feet (1 m) below the base of the land (bls) and analyzed. The soil samples that are connected to the storage area have shown higher concentrations of TPH. The plume of contaminants moves southwestward along the hydraulic gradient.

The two-dimensional model that was created to explain the regional migration and spreading of light nonaqueous-phase liquids (LNAPLs) that were leaked or spilled from underground storage tanks. The two-dimensional contaminant transport models were combined with the nonaqueous-phase liquids (NAPL) transport model to forecast soil gas and groundwater contamination from an LNAPL migrating on the water table. The finite-difference method was used to simulate the migration of LNAPL and the contamination of groundwater. The model was used to simulate subsurface jet fuel contamination. The findings showed that volatilization had the greatest impact on LNAPL migration. Furthermore, the free-product plume and the local geology had an impact on the dissolved plume's movement and spread. Twenty years after the spill, the majority of the released mass was still in a free LNAPL phase. A significant amount of soil and groundwater were contaminated as a result of the LNAPL's migration over such an extended period of time.

The facility consisted of 20 buried steel tanks with a capacity of about 12.5 million gallons each. The tanks contain jet-fuel and diesel fuel marine. The bottoms of the tanks are situated about 80 feet above the basal water table. The geology of the area is primarily basaltic lava flows. Investigations found evidence of releases from several tanks. Two borings were drilled to identify and monitor potential migration of contamination to the potable water source. A numerical model of the regional hydrogeology at the Red Hill Fuel Storage Facility (RHFSF) was developed to simulate the fate and transport of potential contamination from the jet-fuel tanks and the effect on the saltwater/freshwater transition zone of various pumping scenarios.

To characterize the release of contaminants at the slurry-soil interface following slurry application on the surface and irrigation or rainfall, column experiments were carried out. A single-parameter release function appears to be able to represent the release of substances from slurry based on the shape of the release curves. They contrasted the solute transport predictions made under two different assumptions: (a) rectangular pulse input; and (b) input defined by the release function.

The Method of Characteristics (USGS-MOC) model of the United States Geological Survey was used to implement the parameter identification (PI) process that was developed.

The test results demonstrated that, in ideal circumstances, the suggested algorithm could correctly identify transmissivity and dispersivity. The increased effectiveness of the model calibration process allowed for the successful extension of application to field conditions.

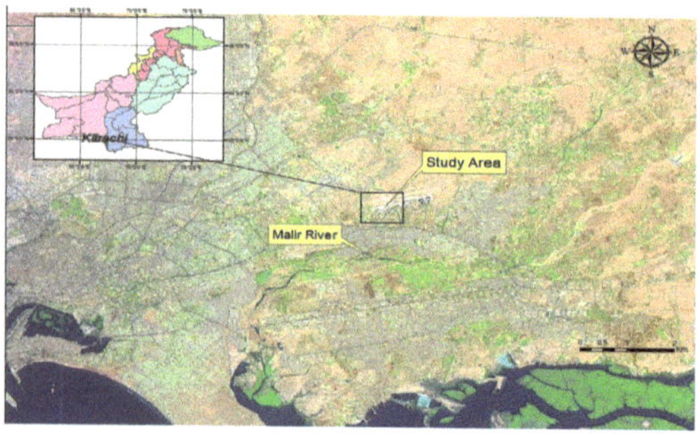

Figure 8.1. *Location of the study area.*

Source: https://cdnintech.com/media/chapter/42669/1512345123/media/image2.jpeg

The groundwater's hydrocarbon plumes by installing a large number of monitoring wells. An alternate method used to map petroleum contaminants underground was electromagnetic induction surveying. EM34-XL ground conductivity meter surveys were carried out at a coal mining site near Gillette, Wyoming. The survey's data were cross-checked against the known concentrations of diesel compounds found in groundwater. The electromagnetic survey data and groundwater data had a perfect correlation that allowed for the creation of a site model that identified subsurface diesel plumes. The study's findings suggested that this geophysical method was a useful instrument for determining the sources and plumes of subsurface petroleum hydrocarbons at contaminated sites.

8.2.5. Model Conceptualization and Simulation

The groundwater flow system is divided into two layers. The upper layer consists mainly of silty sand and is 15 feet (4.6 meters) thick, with the water table as its upper boundary. The lower layer is predominantly clayey sand and 65 feet (20 meters) thick. Both layers are unconfined and receive recharge from infiltrating rain, but only over permeable surfaces. A small stream on the east side of the model area acts as a drain for groundwater, which flows from northeast to southwest. All boundaries except for the stream are artificial and are not constant head or constant flow boundaries. The processes that control the groundwater flow are: (i) recharges from infiltrated rainfall; (ii) flow entering the model across the eastern boundary (also across the northern boundary); (iii) flow reaching the stream; (iv) flow leaving the model across the southern boundary; and (v) pumping from one well near tank no. 9. Values of hydraulic conductivities (K) for the layers are taken from the literature. The K value for depth range 16–31 ft (4.9–9.4m) is taken as 1.7×10^{-6} ft/sec (5.2×10^{-7} m/sec) and for depth range 31–97 ft (9.4–30 m) as 1.5×10^{-5} ft/sec (4.6×10^{-6} m/sec).

Many physical and chemical processes, such as advection, dispersion, volatilization, dissolution, biodegradation, and sorption, are involved in the transport and destiny of hydrocarbons. A solute's rate of movement during chemical reactions may be significantly slower than the average groundwater flow rate. This study simulates the sorption process, which involves both adsorption and partitioning into soil organic matter or organic solvents, to slow down the movement of dissolved hydrocarbons. For the simulation, the MT3D program was utilized.

It uses a linear isotherm to simulate partitioning of a contaminant species between the porous media and the fluid phase due to sorption. This sorption process is approximated by the following equilibrium relationship between the dissolved and adsorbed phases:

$$S = K_d C \qquad (2)$$

where *S* is the concentration of the adsorbed phase (M/M), *C* is the concentration of the dissolved phase (M/L3) and K_d is the sorption or distribution coefficient (L^3/M). K_d values for organic materials are commonly calculated as the product of the fraction of organic carbon in the soil, *foc*, and the organic carbon partitioning coefficient, K_{oc}, or K_d = *foc Koc*. *Koc* values are contaminant specific and reported in various sources. The *foc* in the uncontaminated soil was estimated to range from 0.001 to 0.02 based on guidelines by. Assuming the linear isotherm, the retardation factor (*R*) is expressed as follows:

$$R = 1 + (r_b/n_{eff})\ Kd \quad (3)$$

where r_b is the bulk density of the porous material (M/L3) and n_{eff} is the effective porosity. For jet fuel, the distribution coefficient K_d is taken as 0.004415 ft^3/kg. With these values,

$$R = 1+ [(48\ kg/ft^3)/0.25] \times 0.004415 = 1.848$$

8.2.6. Numerical Ground Water Flow Modeling

Processing ModFlow for Windows (PMWIN5), a modular 3-D finite-difference groundwater model, is used to configure the flow field. The model consists of 41 columns and 39 rows in each layer (Figure 8.2).

The size of cells is 410 ft x410 ft (125 m x 125 m) outside the fuel storage domain and 205 ft x 205 ft (62.5 m x 62.5 m) within the storage domain. Automatic calibration of the water table was made with algorithm - UCODE and a perfect match obtained with the known condition prior to developing the transport model. Using the steady-state hydraulic heads calculated by PMWIN5 as the initial condition, the solute transport model MT3D was run to simulate the dispersion of the dissolved jet-fuel plume. The dispersion coefficient and the retardation factor R for every cell in the finite-difference grid were the modified parameters. Ten monitoring wells have had concentration-time curves computed. The groundwater flow model and simulation result are retrieved from PMWIN5 using PMPATH. The groundwater flow paths, travel times, and time-related capture zones resulting from pumping a neighboring well at the storage facility are calculated using a semi-analytical particle-tracking scheme. The PMWIN5 utility package was used as a preprocessor to modeling and creating input data files. A groundwater information system was established with all data in binary and/or ASCII files that could be exported to other software programs before the modeling work began.

8.2.7. Locations of Hypothetical Wells

The dissolved phase jet-fuel plume was traced using a combination of ten hypothetical monitoring wells (Figure 8.2) known as MW-1 through MW-10. The wells served to identify lithology, observe water levels, and monitor concentrations of organic compounds. The wells extend to a depth of 80 ft (24 m).

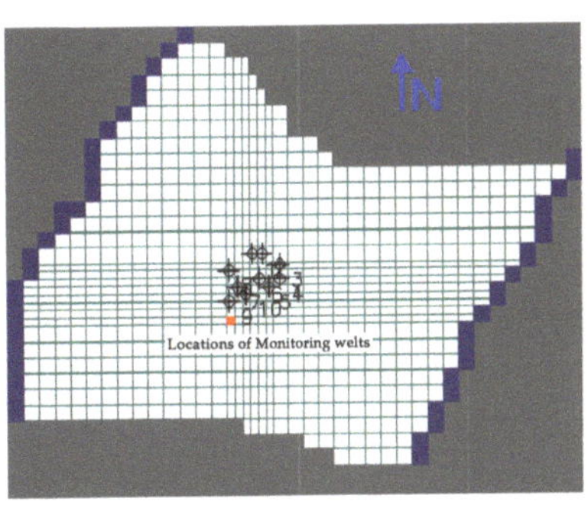

Figure 8.2. *Model design indicating finite difference grid and locations of hypothetical monitoring wells.*

Source: https://cdnintech.com/media/chapter/42669/1512345123/media/image6.png.

In addition, actual well was completed to a depth of 100 ft (30.5 m) near storage tank no.9 in case of emergency need. In the modeling study, this well was used to track the time-related capture zone. The general layout of the storage tanks over the finite-difference grid is shown in Figure 8.3. The location of the pumping well is marked as a small red square in Figures 8.2 and 8.3.

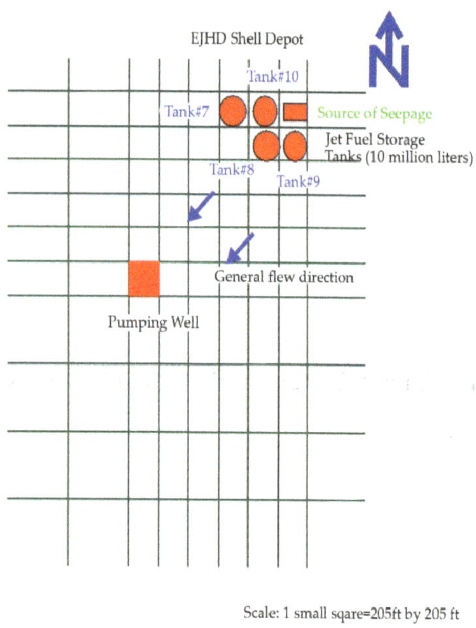

Figure 8.3. *Layout of the storage tanks marked on a finite-difference grid.*

Source: https://cdnintech.com/media/chapter/42669/1512345123/media/image7_w.jpg.

8.2.8. Model Calibration

The steady-state conditions were used to calibrate the model. Given the possibility that the fuel seepage may have begun as early as 1986, groundwater flow modeling was initiated in January 1971, coinciding with the opening of the storage facility. The only sources of input during the steady-state phase are infiltrated rainfall and constant head boundaries, which run east and west along the model. Every output is directed toward the stream and the model's southwest boundary, which are constant head boundaries. By slightly modifying the K values of both layers, the differences between the known and simulated heads were calibrated to be less than 0.30 feet (0.091 meters). Groundwater flow was transiently calibrated using the time-varying hydraulic head values. During the calibration, variables like the recharge rates for each stress period, the hydraulic heads in the stream and along the model boundaries, the storage properties of the aquifer, the pumping rates, and the time-dependent capture zone were changed. The recharge from infiltration was set at 10% of monthly rainfall in order to maintain objectivity and consistency. Effective porosity of the aquifer was varied between 15% and 25% until the value of 25% was determined to be the best predictor for the model. Constant pumping rates of 1500 US gallons/hr (1.58 x 10^{-3} m^3/sec) and 1000 US gallons/hr (1.05 x 10^{-3} m^3/sec) were used in layer 1 and layer 2 respectively.

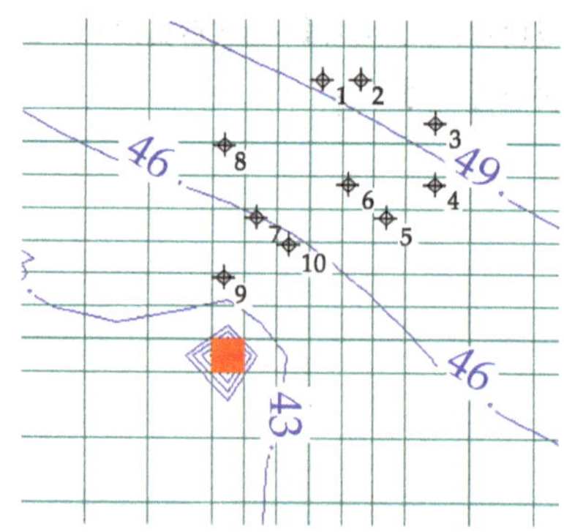

Figure 8.4. *Cone of Depression visible around pumping well developed in layer 1.*

Source: https://cdnintech.com/media/chapter/42669/1512345123/media/image8.png.

The period from 1986 through 2000 was divided into seven stress periods, each of 2 years in duration. From 2000 to 2001, one stress period was assigned. The length of a stress period was made equal to the number of days in that month. The effect of the pumping well is clearly visible as a cone of depression (Figure 8.4). The drawdown was determined to be 14.0 ft (4.3 m) near the storage facility. Every recharging cell receives a uniform recharge from permeating rainfall, according to the model. Effective porosity, hydraulic conductivity, and recharge may differ over the course of the simulation, but the model should have generated a reasonable configuration of the groundwater flow pattern overall. Figures 8.5 and 8.6 depict the time-related capture zones that are created as a result of continuous pumping. The input and output components of the water balances, which were computed for every year of the simulated period, perfectly matched.

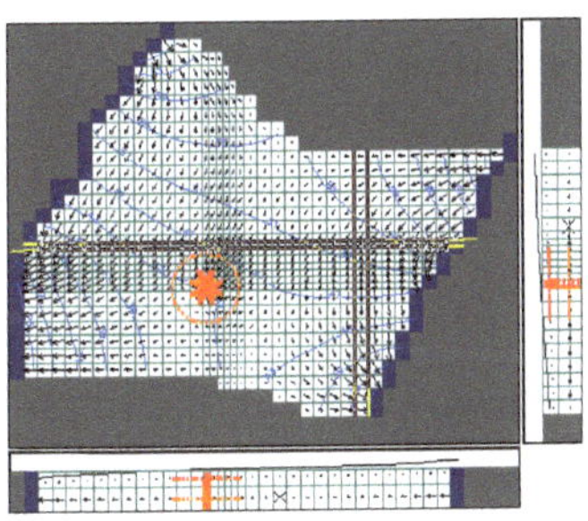

Figure 8.5. *Capture zone of the pumping well with arrows indicating flow directions.*

Source: https://cdnintech.com/media/chapter/42669/1512345123/media/image9.png.

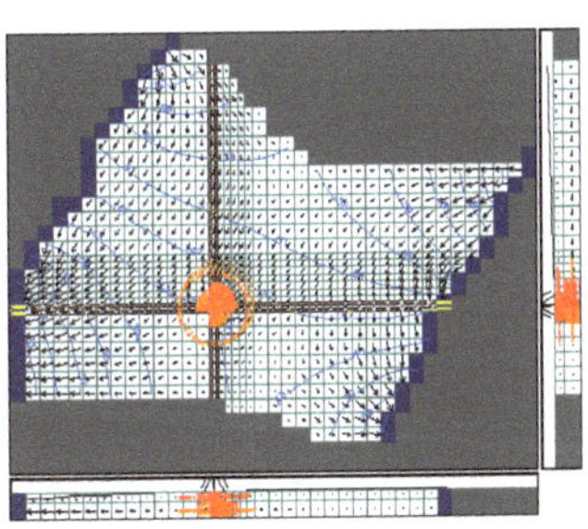

Figure 8.6. *Days-capture zone calculated by PMPATH.*

Source: https://cdnintech.com/media/chapter/42669/1512345123/media/image10.png.

8.2.9. Calibration of Plume Dispersion

For simulation of the movement of dissolved jet fuel, lateral hydraulic conductivity values equal to 0.864 ft/day (0.263 m/day) for layer 1 and 1.29 ft/day (0.393 m/day) for layer 2 were accepted, while the vertical hydraulic conductivity were taken as 0.0864 ft/day (0.0263 m/day) and 0.129 ft/day (0.0393 m/day) for each layer, respectively . The hydraulic gradient and flow-net were obtained by running the flow component of the model derived from water level information in the previous study.

The value for longitudinal dispersion, according to the United States Environmental Protection Agency (USEPA) and the Georgia Environmental Protection Division (GAEPD), should be one-tenth of the distance from the point of contamination entry into the groundwater system to the receptor downstream (a well, stream, or other point of compliance). The separation between the storage location to the pumping well is roughly 30 point 48 meters (100 feet). As advised, the longitudinal dispersion value in each calibration run was set to 10 feet (3 meters). Additionally, the transverse dispersion value for a solute transport model should equal one-third of the longitudinal dispersion, according to USEPA and GAEPD recommendations.

For this model, transverse dispersion would equal 3.3 ft (1.0 m). In the simulation of the fate of jet fuel, the transverse dispersion coefficient was varied within a range of 2.0 ft to 3.3 ft (0.61 m to 1.0 m). In the model a value of 0.001 ft^2/day (1 x 10^{-5} cm^2/sec) was used for molecular diffusion. With a retardation factor of 1.80, dissolved jet fuel takes 1.33 years to travel a lateral distance of about 70 to 80 ft (21 to 24 m) in groundwater beneath tank no. 8. The best value of the microbial decay coefficient for jet fuel is estimated to be 1–10/day with a microbial yield coefficient for oxygen of 0.52.

8.2.10. Strategy Development for Release of Jet Fuel

Ten million liters of jet fuel were stored in storage tanks, and the results of the prior integrity test revealed no loss. It is unclear when the leak started, but inventory records suggested that it didn't exist prior to tank integrity testing. The product has been found in numerous soil samples collected within a few tens of feet of the tank as well as in a number of hypothetical monitoring wells, most notably MW-2 and MW-4. The modeling does not place a high priority on the initial concentration of jet fuel entering the system. The total mass of jet fuel entering the system in a specific time interval is given by multiplying the influx (in L^3/T) by the concentration (in M/L^3). For the purpose of calibrating the jet fuel input, the initial concentration used, based upon field data , varied from 0.095 to 0.19 g/ft^3 (0.0027 to 0.0054 g/m^3). The initial mass of jet fuel, as simulated by the model, was equal to each of four cells "injecting" at a mass rate of 95 to 190 g/ft^3 (2.7 to 5.4 g/m^3) following the initial period of 15 years during which no groundwater contamination was assumed (Table 8.1).

Table 8.1. *Strategy Developed for the Plume Modeling Scenarios*

Phase	Stress period	Condition
Safe Period	15 years (1971 to, 1986)	No leakage
Hazardous Period	10 years (1986 to, 1996)	Low to moderate leakage
Risk Assessment	5 years (1996–2001)	Moderate leakage
Future Prediction	14 years (2001–2015)	Accretion in leakage

Using steady-state hydraulic heads as initial conditions, the evolution of the plume was modeled over nine stress periods as a result of continuous seepage from cells (18,16,1; tank7), (19,16,1; tank 10), (19,17,1; tank 8), and (20,17,1; tank9) as shown in Table 8.2.

Table 8.2. *Stress Period Used in Time-Dependent Solute Transport Modeling of Jet Fuel*

Stress Period	Time interval(years)	Elapsed Time(sec)	Period
1	2	6.30×10^7	1986–88
2	2	12.60×10^7	1988–90
3	2	18.92×10^7	1990–92
4	2	25.23×10^7	1990–94
5	2	3.15×10^8	1994–96
6	2	3.78×10^8	1996–98
7	2	4.41×10^8	1998–00
8	1	4.73×10^8	2000–01
9	14	9.14×10^8	2001–15

8.2.10.1. Calibration Scenario

Parameters describing various processes are used after calibration with different combination of parameters (Table 8.3).

Table 8.3. *Preliminary and Final Values of Parameters Used in Modeling*

Parameters	Value
Longitudinal Dispersion	10 ft
Transverse Dispersion	3.3 ft
Molecular Diffusion	0.001 ft^2/day
Distribution Coefficient	0.004415 ft^3/kg
Retardation Factor (*R*)	1.80 to 1.848
Decay Coefficient	1 x10^{-9} day^{-1}
Hydraulic Conductivity *K* (Layer-1)	1x10^{-5} ft/sec (0.864 ft/day)
Hydraulic Conductivity *K* (Layer-2)	1.49x10^{-5} ft/sec (1.29 ft/day)
Effective Porosity (Layer-1)	0.25
Effective Porosity (Layer-2)	0.30

The release of the jet fuel is simulated in four cells, all along columns 18 to 20 from row 16 to row 17.

The area of injection is equal to 42025 ft^2 (3,904 m^2). The concentration of jet fuel at the source (95 to 190 mg/ft^3 [2.7 to 5.4 g/m^3]) maintained constant throughout the designated "hazardous period" simulation period (1986–1996).

The concentration was increased slightly (about 0.01%) from 1996 through 2011 and further up to longer time duration of 4 years, i.e., up to 2015. The plume simulations are shown in Figure 8.7.

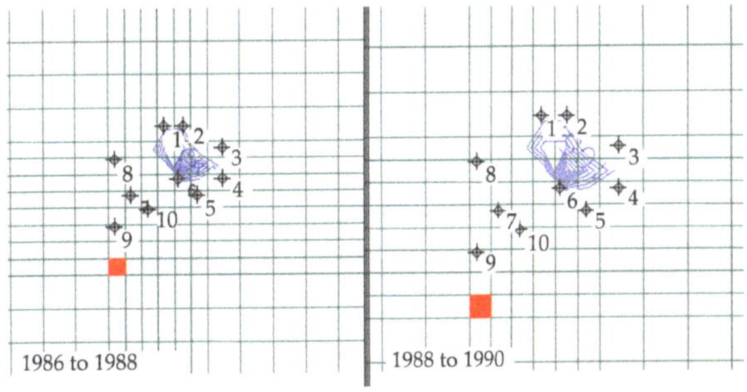

Figure 8.7. *Simulated Jet-fuel plumes (1986 to 1988 and 1988 to 1990).*

Source: https://cdnintech.com/media/chapter/42669/1512345123/media/image11.png.

The jet-fuel break-through curves for the hypothetical monitoring wells are shown in Figure 8.8. Conventionally, determination of the extent and level of contamination is undertaken by taking multiple measurements in wells. However, higher spatial resolution generally requires installation of monitoring wells, which is costly.

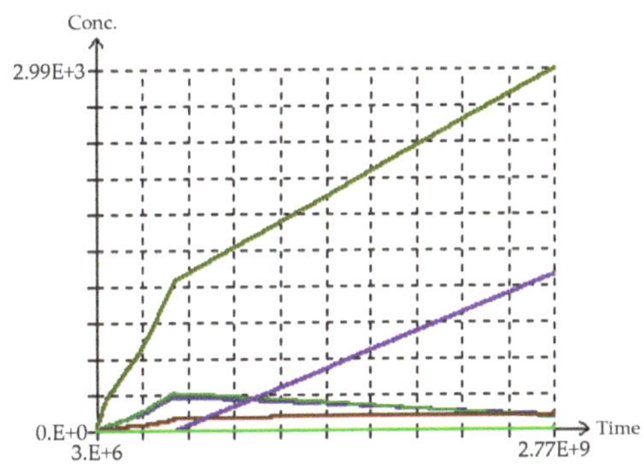

Figure 8.8. *Concentration versus time based on data from 10 monitoring wells.*

Source: https://cdnintech.com/media/chapter/42669/1512345123/media/image12.png.

The concentrations in the remaining wells were substantially lower than the modeled concentrations at MW-1, MW-2, and MW-6.

In MW-6, the highest concentration level measured was 2990 µg/ft^3. Figure 8.9 depicts the 2001 plume spreading.

The groundwater flow direction forms the major axis of the elliptical plume shape. Advection and longitudinal dispersion produce this shape. Transverse dispersion and molecular diffusion cause the plume to spread laterally. Molecular diffusion causes the plume to spread upward. The plume moves in the direction of the stream, which is still far to the west. By the end of 2015, the plume's effects were becoming noticeable, and the monitoring wells MW-4, MW-5, and MW-6 showed higher jet fuel concentrations (Figure 8.10). The ensuing plume seems to be expanding more in the direction of groundwater flow but along the elliptical route.

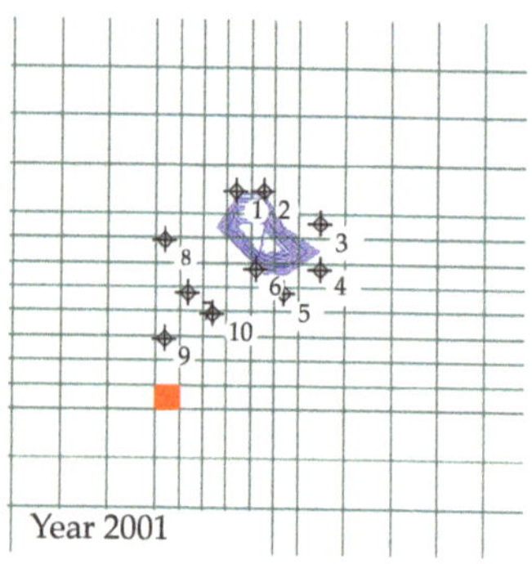

Figure 8.9. *Extent of simulated plume in 2001.*

Source: https://cdnintech.com/media/chapter/42669/1512345123/media/image13.png.

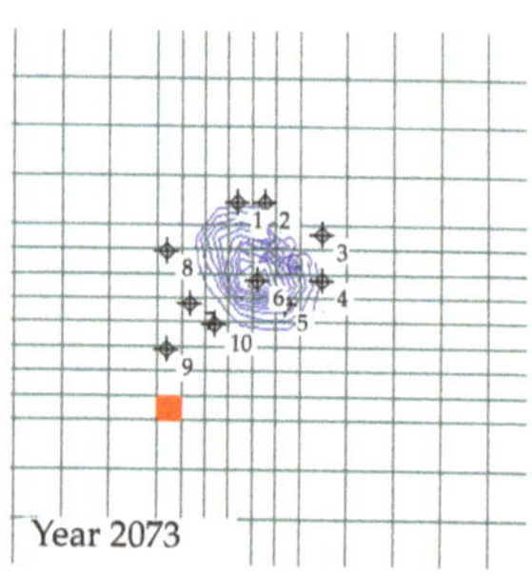

Figure 8.10. *Extent of plume spreading in year 2015.*

Source: https://cdnintech.com/media/chapter/42669/1512345123/media/image14.png.

8.2.10.2. Conclusions and Recommendation

The modeling indicates that there hasn't been any significant expansion or movement of the jet-fuel plume. It faces southwest to northeast and is situated less than 250 feet (76.22 meters) beyond the storage tanks.

Although the concentration found in the simulated monitoring wells is high, no negative effects are anticipated because the groundwater is brackish and therefore unlikely to be used. However, it is anticipated that over time, the level of concentration will rise due to the ongoing process of jet fuel leaking from the storage depot.

Regionally, the expansion of jet fuel will only affect a small localized area and have a noticeable impact over a longer geographic coverage.

It is necessary to conduct an interdisciplinary study of the mechanisms governing the movement and fate of hydrocarbons underground.

Concentrations ought to be carried out in wells both inside and below the plume, since field data would strengthen the case for what happens to jet fuel in groundwater. Wells require routine monitoring in order to maintain effective control over variations in groundwater chemistry. In order to prevent the future release of jet fuel, defective storage depots must be repaired.

CASE STUDY

GROUNDWATER QUALITY: THE CASE OF FLINT, MICHIGAN

Groundwater is a crucial resource for drinking water, irrigation, and industrial processes. The quality of groundwater directly affects public health, the environment, and the economy. This case study focuses on the groundwater quality issues in Flint, Michigan, a city that faced a significant water crisis due to contamination.

Background

Flint, Michigan, is an industrial city with a population of approximately 100,000. Historically, the city sourced its drinking water from Lake Huron via the Detroit Water and Sewerage Department (DWSD). In April 2014, to save money, the city switched its water source to the Flint River, treating the water at the Flint Water Treatment Plant.

Problem Identification

The switch to Flint River water, combined with inadequate treatment and aging infrastructure, led to severe groundwater contamination. The primary issues were:

- **Lead Contamination:** Corrosive Flint River water leached lead from old pipes into the drinking water.
- **Bacterial Contamination:** There were outbreaks of Legionnaires' disease linked to the contaminated water supply.
- **Chemical Contaminants:** Various industrial pollutants, including trihalomethanes (THMs), were found in the water.

Causes of Contamination

The main factors contributing to the groundwater quality issues in Flint included:

- **Corrosion Control Failure:** The water treatment plant did not apply corrosion inhibitors, which are essential to prevent lead leaching from pipes.
- **Aging Infrastructure:** Flint's water distribution system contained many old lead pipes and fixtures.
- **Industrial Pollution:** The Flint River had a history of industrial pollution, making it a challenging water source to treat adequately.

Impact on Public Health and Environment

The groundwater contamination had severe implications:

- **Health Issues:** Elevated blood lead levels were found in children, leading to developmental delays, learning difficulties, and other health problems. The Legionnaires' disease outbreak caused multiple deaths.
- **Environmental Impact:** The polluted groundwater affected local ecosystems, including aquatic life in the Flint River.
- **Economic Impact:** The crisis led to significant financial costs, including infrastructure repairs, medical expenses, and loss of trust in public institutions.

Mitigation and Response

The response to the Flint water crisis involved several steps:

- **Switching Back to DWSD:** Flint returned to using Lake Huron water via DWSD in October 2015.
- **Corrosion Control Measures:** Enhanced corrosion control treatments were implemented.
- **Pipe Replacement Programs:** Efforts to replace lead service lines throughout the city began.
- **Public Health Interventions:** Blood lead level monitoring and health services were expanded for affected residents.

Lessons Learned

The Flint water crisis highlighted key lessons for groundwater quality management:

- **Importance of Corrosion Control:** Proper treatment to prevent pipe corrosion is critical in maintaining water quality.
- **Infrastructure Investment:** Regular maintenance and upgrading of water infrastructure are essential to prevent contamination.
- **Transparent Governance:** Clear communication and transparency in public health decisions are vital for maintaining public trust.
- **Regulatory Oversight:** Strong regulatory frameworks and enforcement are necessary to ensure safe drinking water.

Conclusion

The Flint water crisis serves as a stark reminder of the importance of maintaining high standards in water quality management. Ensuring safe groundwater requires vigilant monitoring, robust infrastructure, and proactive governance to protect public health and the environment.

CLASS ACTIVITY

For a class activity on groundwater quality, students can conduct a simulated groundwater contamination experiment. Divide the class into small groups and provide each group with clear containers filled with sand and gravel to represent an aquifer. Have them introduce different colored dyes to represent various contaminants and pour water to simulate rainfall. Students will observe how contaminants spread through the aquifer over time. Afterward, each group can present their findings, discussing the factors that influenced contaminant movement and potential real-world implications. This hands-on activity will illustrate the complexity of groundwater contamination and the importance of monitoring and remediation.

SUMMARY

- Groundwater quality and contaminant hydrogeology are critical fields within environmental science, focusing on the assessment and management of water quality in subterranean aquifers.
- Groundwater and Contaminant Hydrology has a range of research relating to the transport and fate of contaminants in soils and groundwater.
- Contaminant hydrology is the study of processes that affect both ground and surface water pollution. It draws on the principles of hydrology and chemistry.
- Groundwater is a major source of fresh water for the global population and is used for domestic, agricultural, and industrial uses.
- Biological contaminants include algae and microbial organisms, such as bacteria, viruses, and protozoa.

REVIEW QUESTIONS

1. What are the primary physical, chemical, and biological parameters used to assess groundwater quality?
2. How do contaminants such as heavy metals and nitrates typically enter groundwater systems?
3. Explain the role of hydrogeology in understanding contaminant movement in groundwater.
4. What are the main differences between point source and non-point source pollution in the context of groundwater contamination?

5. Describe at least three common remediation techniques used to treat contaminated groundwater.
6. What are some advanced analytical techniques used to detect and quantify contaminants in groundwater?
7. How do agricultural practices contribute to groundwater contamination, and what mitigation strategies can be implemented?
8. What are the potential health risks associated with drinking contaminated groundwater?
9. Explain the concept of natural attenuation and its role in groundwater contamination management.
10. Discuss the importance of regulatory frameworks and public education in the sustainable management of groundwater resources.

REFERENCES

1. Abbas, G., Murtaza, B., Bibi, I., Shahid, M., Niazi, N. K., Khan, M. I., Amjad, M., Hussain, M., & Natasha. (2018). Arsenic uptake, toxicity, detoxification, and speciation in plants: Physiological, biochemical, and molecular aspects. *International Journal of Environmental Research and Public Health, 15*(59). https://doi.org/10.3390/ijerph15010059.
2. Adimalla, N. (2021). Application of entropy weighted water quality index (EWQI) and pollution index of groundwater (PIG) to assess the groundwater quality for drinking purposes: A case study. *Archives of Environmental Contamination and Toxicology, 80*(6), 1–10.
3. Adimalla, N., & Li, P. (2019). Occurrence, health risks and geochemical mechanisms of fluoride and nitrate in groundwater of the rock-dominant semi-arid region, Telangana State, India. *Human and Ecological Risk Assessment: An International Journal, 25*, 81–103.
4. Adimalla, N., & Qian, H. (2021). Geospatial distribution and potential noncarcinogenic health risk assessment of nitrate contaminated groundwater in Southern India: A case study. *Archives of Environmental Contamination and Toxicology, 80*, 107–119. https://doi.org/10.1007/s00244-020-00762-7.
5. Adimalla, N., & Wu, J. (2019). Groundwater quality and associated health risks in a semi-arid region of south India: Implication to sustainable groundwater management. *Human and Ecological Risk Assessment: An International Journal, 25*, 191–216.
6. Adithya, V. S., Chidambaram, S., Prasanna, M. V., Venkatramanan, S., Tirumalesh, K., Thivya, C., & Thilagavathi, R. (2021). Health risk implication and spatial distribution of radon in groundwater along the lithological contact in south India. *Archives of Environmental Contamination and Toxicology, 80*(10), 1–11.
7. Ahmed, A., Nawaz, R., Woulds, C., & Drake, F. (2020). Influence of hydro-climatic factors on future coastal land susceptibility to erosion in Bangladesh: A geospatial

modelling approach. *Journal of Geovisualization and Spatial Analysis, 4*(6). https://doi.org/10.1007/s41651-020-00050-x.

8. Al-Abadi, A. M., Ghalib, H. B., & Al-Mohammdawi, J. A. (2020). Delineation of groundwater recharge zones in Ali Al-Gharbi District, Southern Iraq using multi-criteria decision-making model and GIS. *Journal of Geovisualization and Spatial Analysis, 4*(9). https://doi.org/10.1007/s41651-020-00054-7.

9. Alshayef, M. S., Javed, A., & Mohammed, A. M. B. (2019). Appraisal of potential hydrocarbon zones in Masila Oil Field, Yemen. *Journal of Geovisualization and Spatial Analysis, 3*(17). https://doi.org/10.1007/s41651-019-0043-0.

10. Ambade, B., Sethi, S. S., Kumar, A., & Sankar, T. K. (2021). Health risk assessment, composition and distribution of polycyclic aromatic hydrocarbons (PAHs) in drinking water of southern Jharkhand, East India. *Archives of Environmental Contamination and Toxicology, 80*(1), 120–133.

11. Basu, A., Saha, D., Saha, R., Ghosh, T., & Saha, B. (2014). A review on sources, toxicity, and remediation technologies for removing arsenic from drinking water. *Research on Chemical Intermediates, 40*, 447–485.

12. Bhowmick, G. D., Dhar, D., Nath, D., Ghangrekar, M. M., Banerjee, R., Das, S., & Chatterjee, J. (2020). Coronavirus disease 2019 (COVID-19) outbreak: Some serious consequences with urban and rural water cycle. *NPJ Clean Water, 3*(32). https://doi.org/10.1038/s41545-020-0079-1.

13. Centers for Disease Control and Prevention. (2020). Global COVID-19. https://www.cdc.gov/coronavirus/2019-ncov/global-covid-19/index.html (accessed on 18 July 2024).

14. Chakraborti, D., Rahman, M. M., Mukherjee, A., Alauddin, M., Hassan, M., Dutta, R. N., Pati, S., Mukherjee, S. C., Roy, S., Quamruzzman, Q., Rahman, M., Islam, T., Sorif, S., Selim, Md., Islam, M. R., & Hossain, M. M. (2015). Groundwater arsenic contamination in Bangladesh—21 years of research. *Journal of Trace Elements in Medicine and Biology, 31*, 237–248.

15. Chandrasekar, T., Keesari, T., Gopalakrishnan, G., Karuppannan, S., Senapathi, V., Sabarathinam, C., & Viswanathan, P. M. (2021). Occurrence of heavy metals in groundwater along the lithological interface of K/T boundary, peninsular India: A special focus on source, geochemical mobility and health risk. *Archives of Environmental Contamination and Toxicology, 80*(1), 183–207.

16. Çiner, F., Sunkari, E. D., & ☒enba☒, B. A. (2021). Geochemical and multivariate statistical evaluation of trace elements in groundwater of Ni☒de Municipality, South-Central Turkey: Implications for arsenic contamination and human health risks assessment. *Archives of Environmental Contamination and Toxicology, 80*, 164–182. https://doi.org/10.1007/s00244-020-00759-2.

17. Clement, M., & Meunie, A. (2010). Is inequality harmful for the environment? An empirical analysis applied to developing and transition countries. *Review of Social Economy, 68*, 223–232.

18. Dahlgaard, H., Eriksson, M., Nielsen, S. P., & Joensen, H. P. (2004). Levels and trends of radioactive contaminants in the Greenland environment. *Science of The Total Environment, 331*, 53–67.

19. Dutta, N., Haldar, A., & Gupta, A. (2021). Electrocoagulation for arsenic removal: Field trials in West Bengal, India. *Archives of Environmental Contamination and Toxicology, 80*(9), 1–11.

20. Elumalai, V., Nethononda, V. G., Manivannan, V., Rajmohan, N., Li, P., & Elango, L. (2020). Groundwater quality assessment and application of multivariate statistical analysis in Luvuvhu catchment, Limpopo, South Africa. *Journal of African Earth Sciences, 171*, 103967. https://doi.org/10.1016/j.jafrearsci.2020.103967.

21. Flemming, H., & Wuertz, S. (2019). Bacteria and archaea on Earth and their abundance in biofilms. *Nature Reviews Microbiology, 17*, 247–260.

22. Goldhaber, M. B., Mills, C. T., Morrison, J. M., Stricker, C. A., Mushet, D. M., & LaBaugh, J. W. (2014). Hydrogeochemistry of prairie pothole region wetlands: Role of long-term critical zone processes. *Chemical Geology, 387*, 170–183. https://doi.org/10.1016/j.chemgeo.2014.08.023.

23. Government of Canada. (2017). Groundwater contamination. Retrieved from: https://www.canada.ca/en/environment-climate-change/services/water-overview/pollution-causes-effects/groundwater-contamination.html (accessed on 18 July 2024).

24. Gwenzi, W., & Chaukura, N. (2018). Organic contaminants in African aquatic systems: Current knowledge, health risks, and future research directions. *Science of the Total Environment, 619–620*, 1493–1514.

25. Haji, M., Karuppannan, S., Qin, D., Shube, H., & Kawo, N. S. (2021). Potential human health risks due to fluoride contaminants of groundwater using multi-techniques approaches in Bilate River Basin of Southern Main Ethiopian Rift, Ethiopia. *Archives of Environmental Contamination and Toxicology, 80(1–2).*

26. Hansen, B., Thorling, L., Schullehner, J., Termansen, M., & Dalgaard, T. (2017). Groundwater nitrate response to sustainable nitrogen management. *Scientific Reports,* 7, 8566. https://doi.org/10.1038/s41598-017-07147-2.

27. Hashim, M. A., Mukhopadhyay, S., Sahu, J. N., & Sengupta, B. (2011). Remediation technologies for heavy metal contaminated groundwater. *Journal of Environmental Management, 92*, 2355–2388.

28. Hayashi, A., Akimoto, K., Tomoda, T., & Kii, M. (2013). Global evaluation of the effects of agriculture and water management adaptations on the water-stressed population. *Mitigation and Adaptation Strategies for Global Change, 18*, 591–618.

29. He, S., & Wu, J. (2019). Hydrogeochemical characteristics, groundwater quality and health risks from hexavalent chromium and nitrate in groundwater of Huanhe Formation in Wuqi County, northwest China. *Exposure and Health, 11*, 125–137.

30. He, X., & Li, P. (2020). Surface water pollution in the middle Chinese Loess Plateau with special focus on hexavalent chromium (Cr6+): Occurrence, sources, and health risks. *Exposure and Health, 12*, 385–401.

31. He, X., Li, P., Ji, Y., Wang, Y., Su, Z., & Elumalai, V. (2020). Groundwater arsenic and fluoride and associated arsenicosis and fluorosis in China: Occurrence, distribution, and management. *Exposure and Health, 12*, 355–368.

32. He, X., Li, P., Wu, J., Wei, M., Ren, X., & Wang, D. (2020a). Poor groundwater quality

and high potential health risks in the Datong Basin, northern China: Research from published data. *Environmental Geochemistry and Health, 43, 791–812*. https://doi.org/10.1007/s10653-020-00520-7.

33. He, X., Wu, J., & He, S. (2019). Hydrochemical characteristics and quality evaluation of groundwater in terms of health risks in Luohe aquifer in Wuqi County of the Chinese Loess Plateau, northwest China. *Human and Ecological Risk Assessment: An International Journal, 25*, 32–51.
34. Huang, C., Lin, T., Chiao, L., & Chen, H. (2012). Characterization of radioactive contaminants and water treatment trials for the Taiwan Research Reactor's spent fuel pool. *Journal of Hazardous Materials, 233–234*, 140–147.
35. International Association of Hydrogeologists. (2020). Groundwater—More about the hidden resource. Retrieved from: https://iah.org/education/general-public/groundwater-hidden-resource (accessed on 18 July 2024).
36. Jenifer, M. A., & Jha, M. K. (2018). Comprehensive risk assessment of groundwater contamination in a weathered hard-rock aquifer system of India. *Journal of Cleaner Production, 201*, 853–868.
37. Ji, Y., Wu, J., Wang, Y., Elumalai, V., & Subramani, T. (2020). Seasonal variation of drinking water quality and human health risk assessment in Hancheng City of Guanzhong Plain, China. *Exposure and Health, 12*, 469–485.
38. Jurado, A., Vàzquez-Suñé, E., Carrera, J., de Alda, M. L., Pujades, E., & Barceló, D. (2012). Emerging organic contaminants in groundwater in Spain: A review of sources, recent occurrence and fate in a European context. *Science of the Total Environment, 440*, 82–94.
39. Kannan, R., Venkateswaran, S., Vijay Prabhu, M., Suresh, R., & Kalaiyarasi, A. (2019). Demarcation of groundwater possible zones in a hard rock terrain, Nagavathi watershed of Dharmapuri District, Tamil Nadu, India—Geophysical and geoinformatics approach. *Journal of Geovisualization and Spatial Analysis, 3*(2). https://doi.org/10.1007/s41651-018-0025-7
40. Karunanidhi, D., Aravinthasamy, P., Deepali, M., Subramani, T., Bellows, B. C., & Li, P. (2020). Groundwater quality evolution based on geochemical modeling and aptness testing for ingestion using entropy water quality and total hazard indexes in an urban-industrial area (Tiruppur) of Southern India. *Environmental Science and Pollution Research, 28*, 18523–18538. https://doi.org/10.1007/s11356-020-10724-0.
41. Karunanidhi, D., Aravinthasamy, P., Deepali, M., Subramani, T., Shankar, K., & Kumar, M. (2021). Groundwater pollution and human health risks in an industrialized region of southern India: Impacts of the COVID-19 lockdown and the monsoon seasonal cycles. *Archives of Environmental Contamination and Toxicology, 80*(1), 259–276.
42. Karunanidhi, D., Aravinthasamy, P., Subramani, T., Wu, J., & Srinivasamoorthy, K. (2019). Potential health risk assessment for fluoride and nitrate contamination in hard rock aquifers of Shanmuganadhi River basin, South India. *Human and Ecological Risk Assessment: An International Journal, 25*, 250–270.
43. Keesari, T., Pant, D., Roy, A., Sinha, U. K., Jaryal, A., Singh, M., & Jain, S. K. (2021). Fluoride geochemistry and exposure risk through groundwater sources in

northeastern parts of Rajasthan, India. *Archives of Environmental Contamination and Toxicology, 80*(2), 1–14.

44. Khan, M. H., Nafees, M., Muhammad, N., Ullah, U., Hussain, R., & Bilal, M. (2021). Assessment of drinking water sources for its quality, human health risks and pollution source(s) apportionment-a case study. *Archives of Environmental Contamination and Toxicology, 80*(1), 41–54.
45. Lam, S., Nguyen-Viet, H., Tuyet-Hanh, T. T., Nguyen-Mai, H., & Harper, S. (2015). Evidence for public health risks of wastewater and excreta management practices in Southeast Asia: A scoping review. *International Journal of Environmental Research and Public Health, 12*, 12855–12863.
46. Lam, T. P., Lee, T. M., Chen, C. Y., & Chang, J. S. (2018). Strategies to control biological contaminants during microalgal cultivation in open ponds. *Bioresource Technology, 252*, 180–187.
47. Lapworth, D. J., Baran, N., Stuart, M. E., & Ward, R. S. (2012). Emerging organic contaminants in groundwater: A review of sources, fate and occurrence. *Environmental Pollution, 163*, 287–303.
48. Lapworth, D. J., Baran, N., Stuart, M. E., Manamsa, K., & Talbot, J. (2015). Persistent and emerging micro-organic contaminants in chalk groundwater of England and France. *Environmental Pollution, 203*, 214–225.
49. Lesser, L. E., Mora, A., Moreau, C., Mahlknecht, J., Hernández-Antonio, A., Ramírez, A. I., & Barrios-Piña, H. (2018). Survey of 218 organic contaminants in groundwater derived from the world's largest untreated wastewater irrigation system: Mezquital Valley, Mexico. *Chemosphere, 198*, 510–521.
50. Li, P. (2020). To make the water safer. *Exposure and Health, 12*, 337–342.
51. Li, P., & Wu, J. (2019). Sustainable living with risks: Meeting the challenges. *Human and Ecological Risk Assessment: An International Journal, 25*, 1–10.
52. Li, P., He, X., & Guo, W. (2019). Spatial groundwater quality and potential health risks due to nitrate ingestion through drinking water: A case study in Yan'an City on the Loess Plateau of northwest China. *Human and Ecological Risk Assessment: An International Journal, 25*, 11–31.
53. Li, P., He, X., Li, Y., & Xiang, G. (2019). Occurrence and health implication of fluoride in groundwater of Loess aquifer in the Chinese Loess Plateau: A case study of Tongchuan, northwest China. *Exposure and Health, 11*(2), 95–107. https://doi.org/10.1007/s12403-018-0278-x.
54. Li, P., Qian, H., & Zhou, W. (2017). Finding harmony between the environment and humanity: An introduction to the thematic issue of the Silk Road. *Environmental Earth Sciences, 76*, 105. https://doi.org/10.1007/s12665-017-6428-9.
55. Li, P., Qian, H., Wu, J., Chen, J., Zhang, Y., & Zhang, H. (2014). Occurrence and hydrogeochemistry of fluoride in shallow alluvial aquifer of Weihe River, China. *Environmental Earth Sciences, 71*, 3133–3145.
56. Li, P., Tian, R., Xue, C., & Wu, J. (2017). Progress, opportunities and key fields for groundwater quality research under the impacts of human activities in China with

a special focus on western China. *Environmental Science and Pollution Research, 24*, 13224–13234.

57. Li, P., Wu, J., & Qian, H. (2016). Preliminary assessment of hydraulic connectivity between river water and shallow groundwater and estimation of their transfer rate during dry season in the Shidi River, China. *Environmental Earth Sciences, 75*, 99. https://doi.org/10.1007/s12665-015-4949-7.
58. Li, P., Wu, J., Tian, R., He, S., He, X., Xue, C., & Zhang, K. (2018). Geochemistry, hydraulic connectivity and quality appraisal of multilayered groundwater in the Hongdunzi coal mine, northwest China. *Mine Water and the Environment, 37*, 222–237.
59. Lin, H. (2010). Earth's critical zone and hydropedology: Concepts, characteristics, and advances. *Hydrology and Earth System Sciences, 14*, 25–45.
60. Lokhandwala, S., & Gautam, P. (2020). Indirect impact of COVID-19 on environment: A brief study in Indian context. *Environmental Research, 188*, 109807. https://doi.org/10.1016/j.envres.2020.109807.
61. Lytle, D. A., Sorg, T., Wang, L., & Chen, A. (2014). The accumulation of radioactive contaminants in drinking water distribution systems. *Water Research, 50*, 396–407.
62. MacDonald, J. A., & Kavanaugh, M. C. (1994). Restoring contaminated groundwater: An achievable goal? *Environmental Science & Technology, 28*(8), 362A–368A.
63. Marghade, D., Malpe, D. B., Duraisamy, K., Patil, P. D., & Li, P. (2020). Hydrogeochemical evaluation, suitability, and health risk assessment of groundwater in the watershed of Godavari basin, Maharashtra, Central India. *Environmental Science and Pollution Research, 28*, 18471–18494. https://doi.org/10.1007/s11356-020-10032-7.
64. Meffe, R., & Bustamante, I. (2014). Emerging organic contaminants in surface water and groundwater: A first overview of the situation in Italy. *Science of the Total Environment, 481*, 280–295.
65. Mthembu, P. P., Elumalai, V., Brindha, K., & Li, P. (2020). Hydrogeochemical processes and trace metal contamination in groundwater: Impact on human health in the Maputaland coastal aquifer, South Africa. *Exposure and Health, 12*, 403–426.
66. Njuguna, S. M., Makokha, V. A., Yan, X., Gituru, R. W., Wang, Q., & Wang, J. (2019). Health risk assessment by consumption of vegetables irrigated with reclaimed wastewater: A case study in Thika (Kenya). *Journal of Environmental Management, 231*, 576–581.
67. Ololade, I. A., Arogunrerin, I. A., Oladoja, N. A., Ololade, O. O., & Alabi, A. B. (2021). Concentrations and toxic equivalency of polycyclic aromatic hydrocarbons (PAHs) and polychlorinated biphenyl (PCB) congeners in groundwater around waste dumpsites in South-West Nigeria. *Archives of Environmental Contamination and Toxicology, 80*(1), 134–143.
68. Pandey, H. K., Duggal, S. K., & Jamatia, A. (2016). Fluoride contamination of groundwater and its hydrological evolution in District Sonbhadra (U.P.) India. *Proceedings of the National Academy of Sciences, India Section A: Physical Sciences, 86*, 81–93.

69. Prakash, R., Srinivasamoorthy, K., Sundarapandian, S. M., Nanthakumar, C., Gopinath, S., Saravanan, K., & Vinnarasi, F. (2021). Submarine groundwater discharge from an urban estuary to southeastern Bay of Bengal, India: Revealed by trace element fluxes. *Archives of Environmental Contamination and Toxicology, 80*, 208–233. https://doi.org/10.1007/s00244-020-00774-3.

70. Rebelo, F. M., & Caldas, E. D. (2016). Arsenic, lead, mercury and cadmium: Toxicity, levels in breast milk and the risks for breastfed infants. *Environmental Research, 51*, 671–688.

71. Ren, X., Li, P., He, X., Su, F., & Elumalai, V. (2021). Hydrogeochemical processes affecting groundwater chemistry in the central part of the Guanzhong Basin, China. *Archives of Environmental Contamination and Toxicology, 80*, 74–91. https://doi.org/10.1007/s00244-020-00772-5.

72. Sawyer, A. H., Michael, H. A., & Schroth, A. W. (2016). From soil to sea: The role of groundwater in coastal critical zone processes. *WIREs Water, 3*, 706–726.

73. Schillinger, J., Özerol, G., Güven-Griemert, ⊠., & Heldeweg, M. (2020). Water in war: Understanding the impacts of armed conflict on water resources and their management. *WIREs Water, 7*, e1480. https://doi.org/10.1002/wat2.1480.

74. Schulze, S., Zahn, D., Montes, R., Rodil, R., Quintana, J. B., Knepper, T. P., Reemtsma, T., & Berger, U. (2019). Occurrence of emerging persistent and mobile organic contaminants in European water samples. *Water Research, 153*, 80–90.

75. Serio, F., Miglietta, P. P., Lamastra, L., Ficocelli, S., Intini, F., De Leo, F., & De Donno, A. (2018). Groundwater nitrate contamination and agricultural land use: A grey water footprint perspective in Southern Apulia Region (Italy). *Science of the Total Environment, 645*, 1425–1431.

76. Shen, J., & Gao, Y. (1995). *Groundwater and the Environment*. Wuhan: China University of Geoscience Press.

77. Shukla, S., & Saxena, A. (2021). Appraisal of groundwater quality with human health risk assessment in parts of Indo-Gangetic alluvial plain, North India. *Archives of Environmental Contamination and Toxicology, 80*, 55–73. https://doi.org/10.1007/s00244-020-00771-6.

78. Sorensen, J. P. R., Lapworth, D. J., Nkhuwa, D. C. W., Stuart, M. E., Gooddy, D. C., Bell, R. A., Chirwa, M., Kabika, J., Liemisa, M., Chibesa, M., & Pedley, S. (2015). Emerging contaminants in urban groundwater sources in Africa. *Water Research, 72*, 51–63.

79. Su, Z., Wu, J., He, X., & Elumalai, V. (2020). Temporal changes of groundwater quality within the groundwater depression cone and prediction of confined groundwater salinity using Grey Markov model in Yinchuan area of northwest China. *Exposure and Health, 12*, 447–468.

80. Subba Rao, N., Dinakar, A., Kumari, B. K., Karunanidhi, D., & Kamalesh, T. (2021). Seasonal and spatial variation of groundwater quality vulnerable zones of Yellareddygudem watershed, Nalgonda district, Telangana State, India. *Archives of Environmental Contamination and Toxicology, 80*, 11–30. https://doi.org/10.1007/s00244-020-00783-2.

81. Subba Rao, N., Ravindra, B., & Wu, J. (2020). Geochemical and health risk evaluation of fluoride rich groundwater in Sattenapalle Region, Guntur district, Andhra Pradesh, India. *Human and Ecological Risk Assessment: An International Journal, 26*, 2316–2348.

82. Tatti, F., Papini, M. P., Torretta, V., Mancini, G., Boni, M. R., & Viotti, P. (2019). Experimental and numerical evaluation of groundwater circulation wells as a remediation technology for persistent, low permeability contaminant source zones. *Journal of Contaminant Hydrology, 222*, 89–100.

83. Teng, Y., Hu, B., Zheng, J., Zhai, Y., & Zhu, C. (2018). Water quality responses to the interaction between surface water and groundwater along the Songhua River, NE China. *Hydrogeology Journal, 26*, 1591–1607.

84. Wang, D., Wang, L., Yang, Q., Yu, K., & Ma, H. (2021). Hydrogeochemistry assessment of shallow groundwater and its health threat on human in northwestern Ordos basin, China. *Archives of Environmental Contamination and Toxicology, 80*(1), 1–15.

85. Wang, D., Wu, J., Wang, Y., & Ji, Y. (2020). Finding high-quality groundwater resources to reduce the hydatidosis incidence in the Shiqu County of Sichuan Province, China: Analysis, assessment, and management. *Exposure and Health, 12*, 307–322.

86. Wu, J., & Sun, Z. (2016). Evaluation of shallow groundwater contamination and associated human health risk in an alluvial plain impacted by agricultural and industrial activities, mid-west China. *Exposure and Health, 8*, 311–329.

87. Wu, J., Li, P., & Qian, H. (2015). Hydrochemical characterization of drinking groundwater with special reference to fluoride in an arid area of China and the control of aquifer leakage on its concentrations. *Environmental Earth Sciences, 73*, 8575–8588.

88. Wu, J., Li, P., Qian, H., & Fang, Y. (2014). Assessment of soil salinization based on a low-cost method and its influencing factors in a semi-arid agricultural area, northwest China. *Environmental Earth Sciences, 71*(8), 3465–3475. https://doi.org/10.1007/s12665-013-2736-x.

89. Wu, J., Zhang, Y., & Zhou, H. (2020). Groundwater chemistry and groundwater quality index incorporating health risk weighting in Dingbian County, Ordos basin of northwest China. *Geochemistry, 80*(4), 125607. https://doi.org/10.1016/j.chemer.2020.125607.

90. Yuan, Y., Xiang, M., Liu, C., & Theng, B. K. G. (2019). Chronic impact of an accidental wastewater spill from a smelter, China: A study of health risk of heavy metal(loid)s via vegetable intake. *Ecotoxicology and Environmental Safety, 182*, 109401. https://doi.org/10.1016/j.ecoenv.2019.109401.

91. Zhang, Y., Wu, J., & Xu, B. (2018). Human health risk assessment of groundwater nitrogen pollution in Jinghui canal irrigation area of the loess region, northwest China. *Environmental Earth Sciences, 77*, 273.

92. Zhao, H., Song, F., Su, F., & Shen, Y., & Li, P. (2021). Removal of cadmium from contaminated groundwater using a novel silicon/aluminum nanomaterial: An experimental study. *Archives of Environmental Contamination and Toxicology, 80*, 234–247. https://doi.org/10.1007/s00244-020-00784-1.

CHAPTER

9

Groundwater Pollution Remediation and Protection

LEARNING OBJECTIVES

After studying this chapter, you will be able to:

- Identify the groundwater pollution remediation techniques
- Define groundwater pollution protection strategies in industrialized countries
- Explain groundwater protection strategies in developing countries

KEY TERMS FROM THIS CHAPTER

Air stripping
Bioremediation
Contaminants
In-situ chemical oxidation
Monitored natural attenuation (MNA)
Phytoremediation
Pump-and-treat
Aquifer
Contaminant plume
Groundwater
Ion exchange
Permeable reactive barriers
Pollution
Remediation

INTRODUCTION

Groundwater pollution remediation and protection are critical aspects of environmental management, focusing on maintaining the quality and safety of groundwater resources, which are vital for drinking water, agriculture, and industrial use. Groundwater, stored in aquifers, can become contaminated by various pollutants, including industrial chemicals, agricultural runoff, leaking underground storage tanks, and improper waste disposal. These contaminants can degrade water quality, posing significant risks to human health and ecosystems. Therefore, effective remediation and protection strategies are essential to safeguard these vital resources.

Remediation of groundwater pollution involves identifying the sources and extent of contamination and then implementing appropriate techniques to remove or neutralize pollutants. Common remediation methods include pump-and-treat, where contaminated water is extracted and treated above ground; in-situ treatments, which involve injecting chemicals or biological agents directly into the groundwater to degrade pollutants; and containment strategies like physical barriers or permeable reactive barriers that prevent the spread of contaminants. Each method has its advantages and limitations, and the choice depends on the specific contaminants, geological conditions, and the extent of pollution.

Protection of groundwater resources involves proactive measures to prevent contamination from occurring. This includes regulatory approaches like establishing protective zones around water wells, implementing land use controls, and enforcing strict regulations on the disposal and handling of hazardous materials. Public education and community engagement are also crucial, as they raise awareness about the sources of pollution and encourage practices that protect groundwater. Additionally, regular monitoring of groundwater quality helps in early detection of potential contamination, allowing for timely interventions to prevent widespread pollution.

Overall, the combined efforts of remediation and protection are essential to ensure the sustainability and safety of groundwater resources. As the demand for clean water continues to grow with increasing population and industrial activities, the importance of effective groundwater management cannot be overstated. This involves not only technical solutions but also policy frameworks, public participation, and continuous research to adapt to emerging challenges in groundwater pollution and protection.

9.1. GROUNDWATER POLLUTION REMEDIATION TECHNIQUES

Groundwater pollution remediation and protection involve a range of strategies and techniques aimed at cleaning up contaminated groundwater and preventing future contamination. Groundwater, a crucial source of drinking water and irrigation, can be contaminated by various pollutants, including chemicals, heavy metals, and pathogens. Effective remediation and protection require a combination of scientific, engineering, regulatory, and community-based approaches.

9.1.1. Pump and Treat Remediation

Pump and treat remediation is a widely used technology for groundwater cleanup, especially for the removal of contaminants such as organic solvents, petroleum hydrocarbons, and heavy metals. This technique involves the extraction of contaminated groundwater, its treatment above ground, and the subsequent discharge or reinjection of the treated water. Despite its widespread use, pump and treat systems can be complex, requiring careful design, operation, and monitoring to achieve desired cleanup levels. (Figure 9.1).

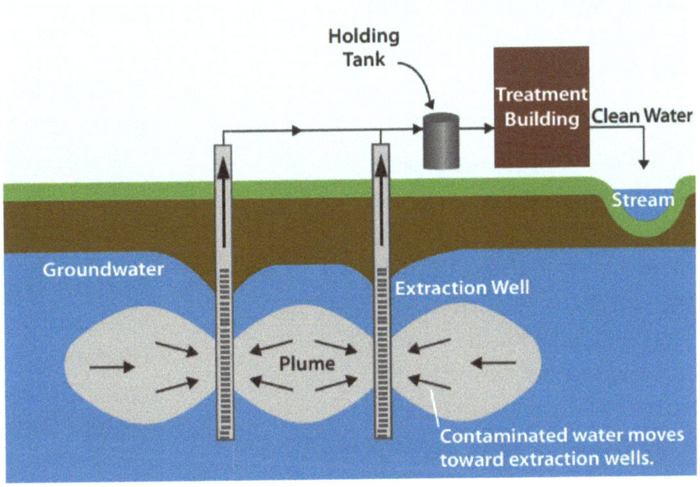

Figure 9.1. *Example of a Pump and Treat System with Two Extraction Wells.*

Source: https://19january2021snapshot.epa.gov/sites/static/files/2015-04/documents/a_citizens_guide_to_pump_and_treat.pdf.

9.1.1.1. Principles of Pump and Treat Remediation

1. *Basic Concept*

   The pump and treat method operates on the principle of hydraulic control and contaminant mass removal. Contaminated groundwater is extracted from the subsurface using wells or trenches. The extracted water is then conveyed to an above-ground treatment system where contaminants are removed. The cleaned water is either discharged to surface water bodies, used for irrigation, or reinjected into the aquifer.

2. *Hydraulic Control*

   Hydraulic control is crucial in pump and treat systems to prevent the spread of contamination. By strategically placing extraction wells, a hydraulic gradient is created, directing contaminated groundwater towards the extraction points. This helps to contain the plume of contamination and prevents it from spreading to uncontaminated areas.

3. *Contaminant Removal*

   Above-ground treatment technologies vary depending on the type of contaminants present. Common treatment methods include air stripping, carbon adsorption, chemical precipitation, ion exchange, and biological treatment. The selection of the treatment technology depends on the nature and concentration of contaminants, regulatory requirements, and site-specific conditions.

9.1.1.2. Design Considerations

1. *Site Characterization*

A thorough site characterization is essential to design an effective pump and treat system. Key aspects of site characterization include:

- **Hydrogeological Conditions:** Understanding the aquifer properties, groundwater flow directions, and hydraulic conductivity.
- **Contaminant Distribution:** Determining the extent, concentration, and type of contaminants present in the subsurface.
- **Geochemical Conditions:** Assessing the chemical composition of groundwater which can affect treatment processes and contaminant behavior.

2. *Well Design and Placement*

The design and placement of extraction wells are critical for effective hydraulic control and contaminant capture. Factors to consider include:

- **Well Spacing:** Optimizing well spacing to balance between maximizing contaminant capture and minimizing operational costs.
- **Screen Interval:** Placing the well screen in zones with the highest contaminant concentrations and hydraulic conductivity.
- **Pumping Rates:** Determining optimal pumping rates to achieve hydraulic control without causing excessive drawdown or disturbing the aquifer structure.

9.1.1.3. Treatment System Design

The above-ground treatment system must be tailored to the specific contaminants and site conditions. Common components of a treatment system include:

- **Pre-treatment:** Removal of particulates and adjusting pH to enhance the efficiency of subsequent treatment processes.
- **Primary Treatment:** The main process for contaminant removal, such as air stripping for volatile organic compounds or activated carbon adsorption for dissolved organics.
- **Post-treatment:** Polishing steps to remove any residual contaminants and ensure compliance with discharge standards.

9.1.1.4. Operational Strategies

1. *System Start-Up*

The initial phase of operation involves system start-up and stabilization. Key activities include:

- **Baseline Monitoring:** Establishing pre-operation contaminant levels and groundwater conditions.
- **System Testing:** Verifying the performance of extraction wells and treatment equipment.
- **Operational Adjustments:** Making necessary adjustments to pumping rates and treatment processes based on initial performance data.

2. *Long-Term Operation*

Effective long-term operation requires continuous monitoring and optimization. Strategies include:

- **Performance Monitoring:** Regular sampling and analysis of groundwater and treated water to ensure system effectiveness.
- **Operational Adjustments:** Modifying pumping rates, well configurations, or treatment processes in response to changing conditions or performance issues.
- **Maintenance:** Routine maintenance of pumps, wells, and treatment equipment to prevent failures and maintain efficiency.

3. *System Optimization*

Optimizing pump and treat systems can improve performance and reduce costs. Techniques include:

- **Dynamic Pumping:** Adjusting pumping rates and schedules based on contaminant levels and hydraulic conditions.
- **Pulse Pumping:** Intermittent pumping to enhance contaminant recovery and reduce operational costs.
- **Treatment Upgrades:** Incorporating advanced treatment technologies or additional processes to enhance contaminant removal.

9.1.1.5. Advantages and Limitations

1. *Advantages*

- **Proven Technology:** Extensive track record of successful applications for a wide range of contaminants.
- **Flexibility:** Can be tailored to site-specific conditions and contaminants.

- **Immediate Impact:** Provides immediate hydraulic control and contaminant removal.
- **Scalability:** Systems can be scaled up or down based on the size and complexity of the contamination.

2. Limitations

- **High Costs:** Installation, operation, and maintenance can be expensive, particularly for long-term projects.
- **Limited Efficiency for Low-Permeability Zones:** Difficulty in extracting contaminants from low-permeability soils and bedrock.
- **Extended Time Frames:** Remediation can take years or decades to achieve regulatory cleanup levels.
- **Potential for Contaminant Rebound:** Residual contamination can cause rebound effects after system shutdown.

9.1.2. Permeable Reactive Barriers

Since the late 1980s, research has concentrated on alternative, in situ strategies like permeable reactive barriers (PRBs) in response to the realization that the pump-and-treat approach can prove costly and frequently ineffective. In summary, PRBs are created by extracting material from a section of the aquifer and then filling it in with a permeable mixture that will react with the contaminant. While trench construction is the most common method for installing PRBs, barriers have also been built by injecting reactive materials into the ground or creating aquifer fractures and then filling them with reactive materials (Hocking et al., 2000; Richardson & Nicklow, 2002). The contaminant is kept inside the barrier by the reactive material that is present in it. Hexavalent chromium, uranium, and technetium have all been treated with PRBs that contain zero-valent iron, or iron filings (Blowes et al., 2000) as well as PCE and TCE, chlorinated ethers. Sulfate, iron, nickel, cobalt, and zinc are among the dissolved elements connected to acid mine drainage that have been eliminated using solid phase organic carbon, which takes the form of municipal compost. In this way, dissolved nutrients, such as phosphate and nitrate, have also been eliminated from agricultural drainage and home septic system effluent (Blowes et al., 2000).

A variety of techniques have been employed to treat inorganic and organic contaminants, including biodegradation, precipitation, sorption to encourage organic matter partitioning and ion exchange, chemical (abiotic) degradation, and manipulation of the redox potential to improve biological reductive dechlorination and alter the chemical speciation of metals.

Reactive barriers with "continuous wall" and "funnel and gate" designs, as shown in Figure9.2, are the most widely used types of PRBs. According to Starr and Cherry (1994), funnel and gate PRBs are made up of low hydraulic conductivity cut-off walls, such as sheet piles and slurry walls, with gaps that house in situ reactors for the removal of contaminants. In order to stop plumes from growing further, funnel and gate systems can be put in place in front of them or right downgradient of contaminant source areas to stop contaminants from starting plumes. Groundwater flows primarily through the high conductivity gaps of the gates due to the modification of the groundwater flow pattern by cut-off walls, also known as the funnel. An uninterrupted wall of permeable material mixed with reactive material, such as a trench filled with reagent and pea gravel, allows continuous PRBs to cross the pollutant plume. While some PRBs have been positioned as deep as 40 meters, most have been positioned at comparatively shallow depths of 10 to 20 meters.

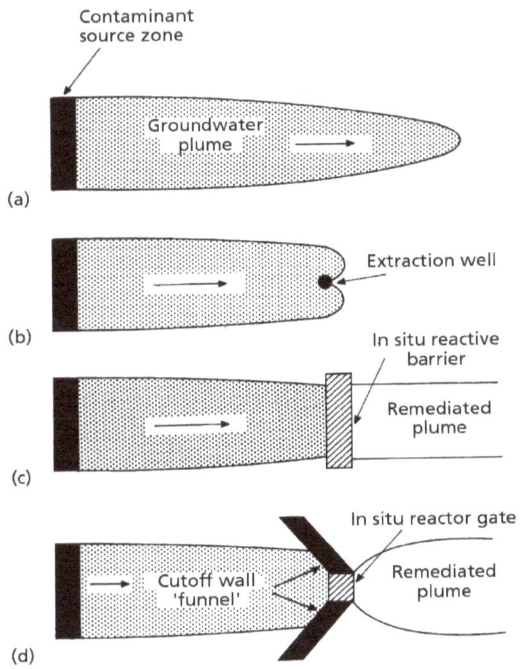

Figure 9.2. *Three options for remediation of contaminated groundwater. (a) Unremediated contaminant plume. (b) Pumpand-treat system. (c) In situ 'continuous wall' reactive barrier. (d) In situ 'funnel and gate' reactive barrier. In the case of the funnel and gate system, a balance must be achieved between maximizing the size of the capture zone for a gate and maximizing the retention time of contaminated groundwater in the gate. In general, capture zone size and retention time are inversely related.*

Source: Neven Kresic; Hydrogeology 101: Introduction to Groundwater Science and Engineering; ISBN: 979-8-218-06984-1.

For PRB remediation to be successful, long operating and maintenance times—possibly decades—will probably be necessary. However, in contrast to the pump-and-treat remediation method, PRBs have low operating and maintenance requirements, which makes such long-term clean-up feasible. A shift from oxidizing to reducing conditions, for example, could alter the in situ biological and geochemical environment inside the PRB and down its gradient. This could lead to secondary reactions and the precipitation of mineral phases like hydroxides and carbonates. Long-term studies are necessary for a complete assessment of PRBs because biofouling, chemical precipitation, and gas production can all have an impact on the long-term hydraulic and chemical performance of PRBs.

9.1.3. Monitored Natural Attenuation

Natural attenuation is characterized by a range of physical, chemical, or biological processes that, in the presence of favorable circumstances and without the assistance of humans, act to lower the mass, toxicity, mobility, volume, or concentration of contaminants in soil or groundwater (US EPA, 1997).

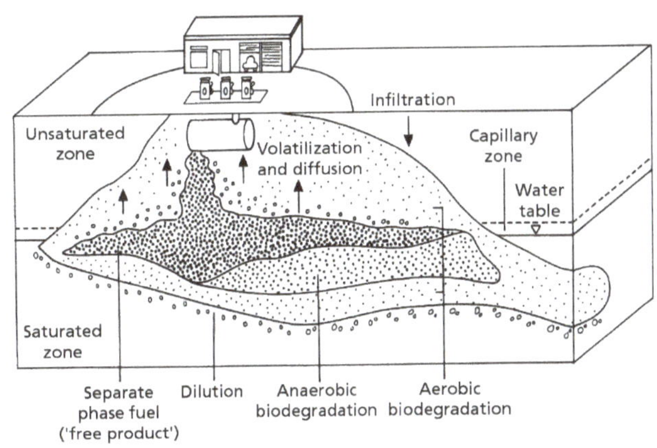

Figure 9.3. *Conceptual diagram illustrating the important natural attenuation processes that affect the fate of petroleum hydrocarbons in aquifers.*

Source: Neven Kresic; Hydrogeology 101: Introduction to Groundwater Science and Engineering; ISBN: 979-8-218-06984-1.

Biodegradation, dispersion, dilution, sorption, volatilization, radioactive decay, and chemical or biological stabilization, transformation, or destruction of contaminants are some examples of these in situ processes (Figure 9.3).

Reliance on monitored natural attenuation (MNA) looks appealing as a substitute for more costly pump-and-treat and engineered solutions to groundwater contamination, but detractors argue that MNA conveniently avoids the high costs of installing clean-up systems. Whether the regulatory goal is to clean up a pollutant plume to drinking water standards or if a less strict, risk-based goal, like stopping a plume from spreading, applies determines whether MNA as a strategy is feasible. The application of MNA as a corrective measure for BTEX compounds (benzene, toluene, ethylbenzene, and xylene) has grown significantly since the mid-1990s. For contaminants such as radionuclides, heavy metals, nitroaromatics, and chlorinated solvents, natural attenuation has been suggested; however, more investigation and comprehension of these substances are needed before the method can be deemed reliable (Bekins et al., 2001a). A significant level of comprehension of the subsurface processes needs to be developed before applying MNA at a contaminated site. Therefore, the primary cost will probably move from designing and running an active pump-and-treat system or a passive PRB to a thorough analysis and modeling of the site to comprehend the groundwater flow naturally occurring there and the biogeochemical processes that are attenuating the pollution.

More investigation into the real-world problems surrounding the long-term natural attenuation performance is essential to the success of MNA in the future. The effects of active remediation efforts on the natural attenuation process, the creation of long-term monitoring networks to confirm the effectiveness of natural attenuation, and the demonstration of the aquifer's natural attenuation capacity over the course of the contaminant source's lifetime should all be covered by this research.

9.1.4. In-Situ Bioremediation

In-situ bioremediation is an environmentally friendly and cost-effective technique for treating contaminated groundwater. This approach leverages the natural metabolic processes of microorganisms to degrade contaminants directly at the contamination site, eliminating the need for excavation or water extraction.

9.1.4.1. Principles of In-Situ Bioremediation

1. *Biological Processes*

   Microorganisms, such as bacteria and fungi, are pivotal to in-situ bioremediation. These organisms metabolize contaminants, breaking them down into less harmful substances through various metabolic pathways. The key biological processes include:

- **Aerobic Respiration:** Uses oxygen as the terminal electron acceptor to degrade organic contaminants into carbon dioxide and water.
- **Anaerobic Respiration:** Utilizes alternative electron acceptors like nitrate, sulfate, or iron in the absence of oxygen to break down contaminants.
- **Cometabolism:** Involves the transformation of contaminants by microorganisms while they metabolize another compound.

2. Biostimulation and Bioaugmentation

- **Biostimulation**: Involves the addition of nutrients or electron acceptors to enhance the activity of native microorganisms.

- **Bioaugmentation:** Involves the introduction of specialized microbial strains that possess specific abilities to degrade contaminants that indigenous microorganisms cannot.

9.1.4.2. Types of In-Situ Bioremediation

1. *Aerobic Bioremediation*

- **Principle:** Relies on oxygen to degrade contaminants.
- **Applications:** Effective for petroleum hydrocarbons, BTEX compounds, and certain pesticides.
- **Implementation:** Methods include air sparging, injection of oxygen-releasing compounds, or biosparging.

2. *Anaerobic Bioremediation*

- **Principle:** Utilizes electron acceptors other than oxygen, such as nitrate or sulfate.
- **Applications:** Suitable for chlorinated solvents, nitroaromatic compounds, and some metals.
- **Implementation:** Includes injecting electron donors like lactate or molasses to stimulate anaerobic microbial activity.

3. *Intrinsic Bioremediation*

- **Principle:** Natural attenuation processes where indigenous microorganisms degrade contaminants without human intervention.
- **Applications:** Used where natural conditions support sufficient biodegradation.
- **Implementation:** Requires extensive monitoring to ensure contaminants are being degraded effectively.

4. *Phytoremediation*

- **Principle:** Uses plants to enhance bioremediation processes.
- **Applications:** Suitable for shallow groundwater contaminated with metals and organic pollutants.
- **Implementation:** Plants are selected for their ability to uptake, accumulate, or degrade specific contaminants.

9.1.4.3. Design Considerations

1. *Site Characterization*

   Detailed site characterization is critical for effective bioremediation design. Key components include:

- **Hydrogeological Assessment:** Evaluating groundwater flow patterns, hydraulic conductivity, and aquifer properties.
- **Contaminant Profiling:** Mapping the extent, concentration, and types of contaminants.
- **Microbial Community Analysis:** Identifying indigenous microorganisms capable of degrading contaminants.
- **Geochemical Analysis:** Assessing pH, temperature, redox potential, and nutrient availability.

2. *Bioremediation Strategy Selection*

   Choosing the appropriate bioremediation strategy depends on site-specific conditions:

- **Biostimulation vs. Bioaugmentation:** Deciding whether to stimulate native microbes or introduce specialized strains.
- **Aerobic vs. Anaerobic Conditions:** Determining the optimal conditions based on contaminant types and site characteristics.

- **Nutrient and Electron Acceptor Delivery:** Designing systems for the effective delivery of amendments.

3. *Treatment Design*

Designing an effective treatment system involves:

- **Injection Well Placement:** Strategic placement for optimal distribution of nutrients and electron acceptors.
- **Injection Methodology:** Techniques such as direct-push technology, permeable reactive barriers, or injection wells.
- **Monitoring Plan:** Developing a robust monitoring plan to track contaminant levels, microbial activity, and geochemical changes.

9.1.4.4. Implementation Strategies

1. *Pilot Testing*

Conducting a pilot test is essential before full-scale implementation:

- **Objective:** To evaluate the feasibility and optimize the design.
- **Activities:** Includes small-scale injection of amendments and monitoring for contaminant degradation.

2. *Full-Scale Implementation*

Steps involved in full-scale implementation include:

- **Installation of Injection Wells:** Ensuring proper placement and construction of wells.
- **Amendment Injection:** Controlled delivery of biostimulants or bioaugmentation cultures.
- **Monitoring and Optimization:** Continuous monitoring to assess performance and make necessary adjustments.

9.1.4.5. Monitoring and Optimization

1. *Performance Monitoring*

Regular monitoring is crucial to ensure the effectiveness of bioremediation:

- **Contaminant Levels:** Periodic sampling and analysis to track reductions.
- **Microbial Activity:** Monitoring microbial populations and activity using molecular techniques.
- **Geochemical Parameters:** Tracking changes in pH, redox potential, and nutrient levels.

2. Optimization Techniques

Optimizing bioremediation involves:

- **Adjusting Amendment Delivery:** Modifying injection rates and schedules based on monitoring data.
- **Dynamic Response to Conditions:** Adapting the treatment strategy to changing site conditions.
- **Integrating Advanced Technologies:** Utilizing real-time sensors and data analytics for enhanced control.

9.1.4.6. Advantages and Limitations

1. *Advantages*

- **Cost-Effective:** Lower costs compared to ex-situ methods due to elimination of excavation and transport.
- **Minimal Site Disruption:** In-situ methods cause less disturbance to the site and surrounding areas.
- **Sustainability:** Utilizes natural processes, reducing the need for chemical treatments.

- **Versatility:** Effective for a wide range of organic and inorganic contaminants.

2. *Limitations*

- **Site-Specific:** Effectiveness is highly dependent on site conditions, including microbial populations and hydrogeology.
- **Time-Consuming:** Bioremediation processes can take months to years to achieve desired cleanup levels.
- **Monitoring Intensive:** Requires extensive and ongoing monitoring to ensure effectiveness.
- **Potential for Contaminant Rebound:** Risk of contaminant levels rebounding if conditions change or treatment is not fully optimized.

9.2. GROUNDWATER POLLUTION PROTECTION STRATEGIES IN INDUSTRIALIZED COUNTRIES

9.2.1. Groundwater Vulnerability Mapping and Aquifer Resource Protection

As seen in Figure 9.4, the nature of the overlying soil cover, the existence and kind of overlying superficial deposits, the makeup of the aquifer's geological strata, the depth of the unsaturated zone, and the thickness of confining deposits all affect how vulnerable groundwater is to pollution from the surface. Using this method, the Environment Agency in England and Wales created 53 regional maps of groundwater vulnerability that illustrate vulnerability classes derived from the 1:100,000 scale overlay of soil and hydrogeological data (Figure 9.5).

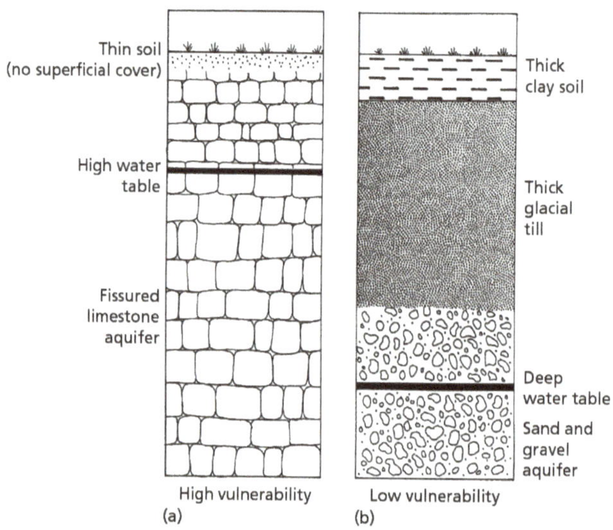

Figure 9.4. *Illustration of two situations of contrasting groundwater vulnerability to surface-derived pollution. In (a) the unconfined, fissured limestone aquifer with a permeable soil cover and high water table (thin unsaturated zone) has a high apparent vulnerability. In (b) the sand and gravel aquifer, overlain by a low permeability soil and glacial till cover, has a low apparent vulnerability.*

The maps are included in the groundwater resource protection strategy of the Environment Agency (Robins et al., 1994; Environment Agency, 1998) with the goal of promoting the growth of potentially polluting activities in those regions where they will cause the least amount of concern. The management of diffuse pollution can be easily linked to aquifer vulnerability zones using regional maps. A geographical information system (GIS) can be used to easily manipulate the overlay operation of soils and hydrogeological data to produce customized groundwater vulnerability maps, like the nitrate vulnerability map displayed in Figure 9.6.

Source: Neven Kresic; Hydrogeology 101: Introduction to Groundwater Science and Engineering; ISBN: 979-8-218-06984-1.

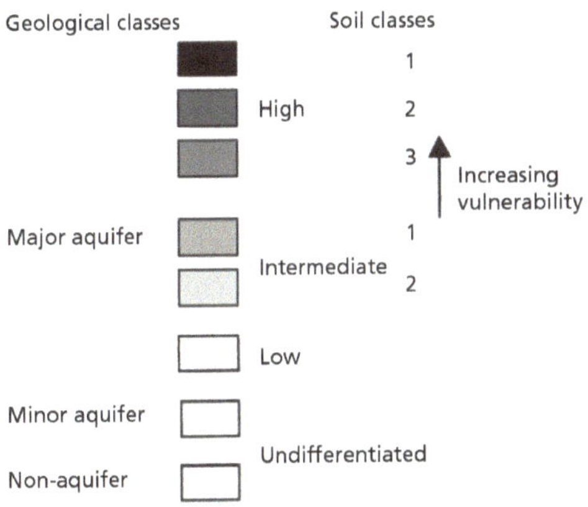

Figure 9.5. *Groundwater vulnerability classification scheme used by the Environment Agency in England and Wales derived from the overlay of information on geological strata and soil type. Major aquifers of regional importance include highly permeable formations usually with significant fracturing. Minor aquifers and non-aquifers are classed as having variable or negligible permeability that generally support local or very small abstractions. Soil classes are divided on the basis of leaching potential (high, intermediate and low) depending on the physicochemical properties of soil types with respect to diffuse source pollutants and liquid discharges.*

Source: Neven Kresic; Hydrogeology 101: Introduction to Groundwater Science and Engineering; ISBN: 979-8-218-06984-1.

Similar methods for assessing groundwater vulnerability have been created by the Environmental Protection Agency in the US. These methods allow for a methodical assessment of the potential for groundwater pollution at any given site (Aller et al., 1987). The system consists of two main parts: first, mappable units called hydrogeologic settings are designated; second, a relative rating system called DRASTIC is superposed. Each hydrogeologic setting has inherent physical properties that influence the potential for groundwater pollution. The impact of the vadose zone (I), depth to water (D), net recharge (R), aquifer media (A), soil media (S), topography (slope) (T), and hydraulic conductivity of the aquifer (C) are the most significant mappable factors taken into consideration to control the groundwater pollution potential. Three important components make up the numerical ranking system that is used for the DRASTIC factors: weights, ranges, and ratings. On a scale of 1 to 5, with 5 being the most important, weights correspond to the relative importance of each of the seven factors. Each factor's impact on the potential for pollution is broken down into ranges (or significant media types) for each factor. Subsequently, a rating system ranging from 1 to 10 is employed to distinguish between each range in terms of its potential for pollution. Every range is assigned a single value to the factor, with the exception of factors A and I's ranges, which have been assigned a variable and a typical rating respectively. The following equation is then used to provide each hydrogeologic setting with a relative numerical value:

$$\text{Pollution potential} = D_R D_W + R_R R_W + A_R A_W + S_R S_W + T_R T_W + I_R I_W + C_R C_W$$

where R is rating and W is weight. The greater the DRASTIC score the greater the pollution potential. The scores are applied to their respective hydrogeologic settings and are mapped.

Figure 9.6. *Specific groundwater nitrate vulnerability map for England and Wales. The vulnerability classes are derived from a GIS overlay operation of: (i) simulated mean nitrate concentrations in land drainage assuming a uniform nitrogen loading of 100 kg N ha−1; (ii) soil types; (iii) presence or absence of low permeability superficial deposits; and (iv) aquifer types. Regions of high groundwater vulnerability to nitrate pollution (classes 1–4) are in areas of major aquifers.*

Source: Neven Kresic; Hydrogeology 101: Introduction to Groundwater Science and Engineering; ISBN: 979-8-218-06984-1.

The aforementioned methods of mapping apparent and particular groundwater vulnerability have the drawback of only offering a regional overview, which leaves room for interpretation regarding the true extent of the threat to the groundwater resource at the local level. The only reliable way to determine the true vulnerability is to conduct site-specific field investigations that provide supporting data. However, by influencing land-use management, groundwater vulnerability maps can aid in the achievement of water quality goals by helping planners understand the potential for groundwater pollution.

9.2.2. Source Protection Zones

In the definition of groundwater source protection zones, the proximity of a hazardous activity to a point of groundwater abstraction (including springs, wells and boreholes) is one of the most important factors in assessing the pollution threat to an existing groundwater source. In principle, the entire recharge area in the vicinity of a groundwater source should be protected, but this is unrealistic on socioeconomic grounds. In this situation, a system of zoning of the recharge area, or protection area, is desirable and this approach has been adopted in Europe and the United States. For example, in the Netherlands, abstraction of drinking water supplies is concentrated in wellfields tapping mainly uniform, horizontally layered aquifers of unconsolidated sands and clays. As illustrated in Figure 9.7 and Table 9.1, the zoning system includes a first zone based on a delay time of 60 days from any point below the water table in order to protect against pathogenic bacteria and viruses and rapidly degrading chemicals. This zone typically extends some 30–150 m from an individual borehole. For the continuity of water supplies in the event of a severe pollution incident requiring remedial action, and in order to exclude public health risks, a delay time of at least 10 years is needed in the next zone. In many cases, even 10 years is not sufficient to guarantee the continuity of safe water supplies, and a protection zone of 25 years is necessary. The 10- and 25-year protection zones extend to about 800 m and 1200 m from the borehole, respectively, and constitute the source protection area.

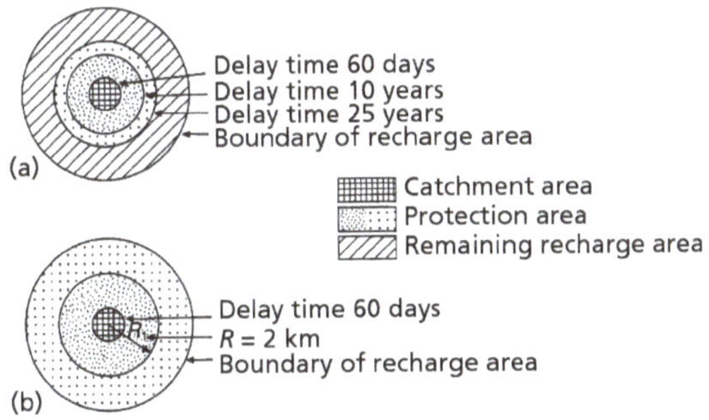

Figure 9.7. *Examples of protection zones for groundwater sources in (a) a porous, permeable aquifer and (b) a fissured, karstic aquifer.*

Source: Neven Kresic; Hydrogeology 101: Introduction to Groundwater Science and Engineering; ISBN: 979-8-218-06984-1.

Table 9.1. *Land-Use Restrictions for the Source Protection Zones Shown in Figure 9.7*

Catchment area	Protection area	Remaining recharge area
60 days and ≥30 m	10- and 25-year delay-time or 2 kilometres	
Protection against pathogenic bacteria and viruses and against chemical pollution sources	Protection against hardly degradable chemicals	Soil and groundwater protection rules
Only activities in relation to water supply are admissible	As a rule, the following are not admissible: Transport and storage of dangerous goods Industrial sites Waste disposal sites Building Military activities Intensive agriculture and cattle breeding Quarrying Waste water disposal	

Source: Neven Kresic; Hydrogeology 101: Introduction to Groundwater Science and Engineering; ISBN: 979-8-218-06984-1.

Delineating the area where an abstraction well draws its water and preventing potentially hazardous activities from occurring there are the goals of the United States' Wellhead Protection Program (United States Environmental Protection Agency, 1993). While the second area, known as the well capture zone or zone of contribution (ZOC), is defined as the area surrounding a pumping well that includes all areas or features that supply groundwater to the well, the first area, known as the zone of influence (ZOI), is almost synonymous with the cone of depression (Figure 9.8).

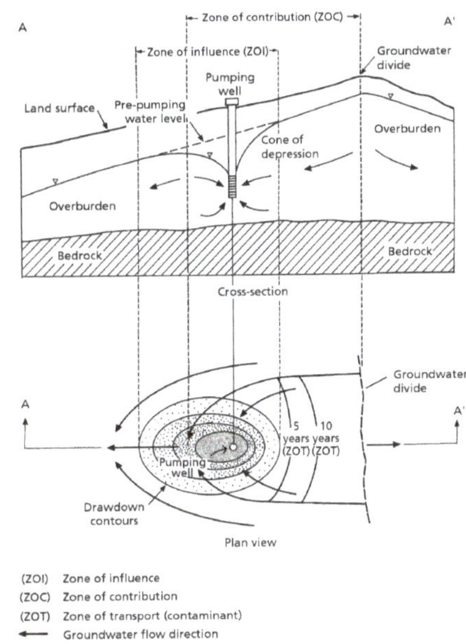

Figure 9.8. *Conceptual model of a wellhead protection area and associated terminology. As shown, the zone of contribution (ZOC) and the zone of influence (ZOI) do not coincide. For the ZOC, groundwater is removed from the pumping well from only a relatively small portion of the downstream area of the well, but it may extend as far as the groundwater divide on the upgradient side of the well. By contrast, the downgradient portion of the groundwater within the ZOI is not drawn towards the pumping well but continues downgradient, while the ZOI does not extend to the upgradient limit of the ZOC. However, experience shows that if the ZOC is small, then the ZOC and ZOI will generally overlap.*

Source: Neven Kresic; Hydrogeology 101: Introduction to Groundwater Science and Engineering; ISBN: 979-8-218-06984-1.

The location and hydraulic loading of the contaminant source, the characteristics and boundaries of the aquifer, and the well design all affect the ZOI and ZOC's size and shape. The zone of contaminant transport (ZOT) can be used to further define the ZOC.

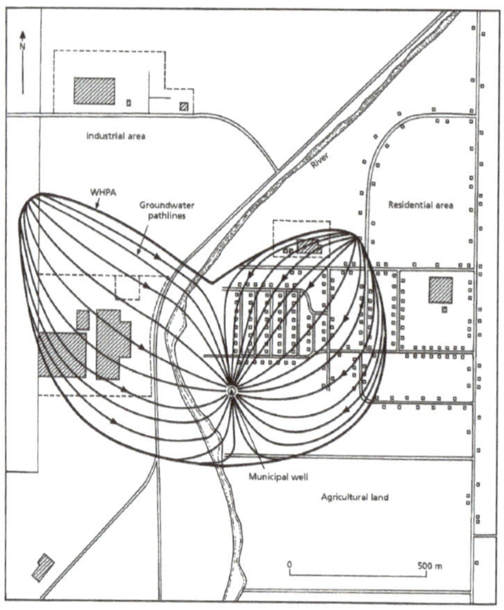

Figure 9.9. *Wellhead protection area (WHPA) defined using a fully three-dimensional numerical model (MODFLOW and MODPATH) to simulate groundwater flow and contaminant migration using particle tracking. In this hypothetical example, the municipal well pumps at a rate of 250 m3 day−1. The aquifer comprises sand and gravel with a transmissivity of 26 m2 day−1 and is 30 m thick. The well partially penetrates the aquifer with the well screen extending 20–30 m below ground surface. The well is located close to a river that is situated within an area of alluvial deposits with a transmissivity of 0.1 m2 day−1 and a thickness of 10 m. The groundwater model explicitly represents the alluvial unit overlying the sand and gravel aquifer near the river. The results of the simulation define a capture zone configuration that suggests that groundwater protection efforts should be concentrated in the areas to the west and east of the river.*

Source: Neven Kresic; Hydrogeology 101: Introduction to Groundwater Science and Engineering; ISBN: 979-8-218-06984-1.

It is typically depicted as isochrones, or contours with equal travel times, which show how long it takes a contaminant from a source inside the ZOC to reach a pumping well (Figure 9.8). The groundwater flow velocity, the characteristics of the contaminants, and the characteristics and composition of the aquifer material all affect the travel time. There are various cost and complexity levels associated with mapping wellhead protection area (WHPA) criteria; these range from arbitrary radii to numerical flow and transport models (Figure 9.9), such as the aquifer's ability to absorb pollutants (Livingstone et al., 1995). Creating a management zone, an attenuation zone, and a remedial action zone are the main goals of wellhead protection. In more complicated hydrogeologic environments, like confined aquifers and fractured rocks (US EPA, 1991a), specific guidance is available.

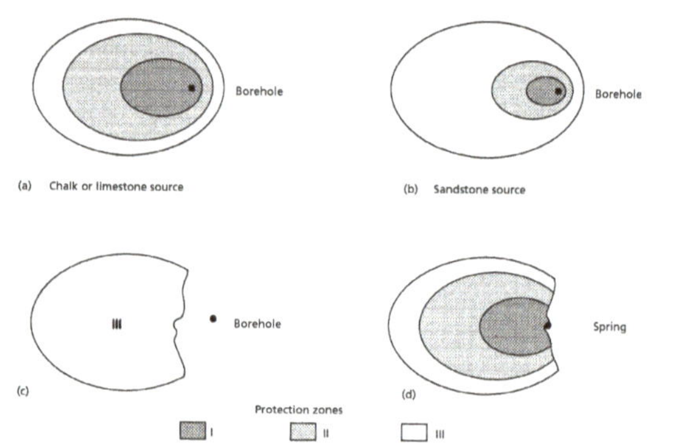

Figure 9.10. *Schematic illustration of source protection zones showing the relationship between Zones I, II and III and the groundwater source in four idealized hydrogeological situations representing: (a) a low effective porosity limestone aquifer; (b) a high effective porosity sandstone aquifer; (c) a confined aquifer; (d) a spring. In reality, the size, shape and relationship of the zones will vary significantly depending on the soil, geology, amount of recharge and volume of water abstracted.*

Source: Neven Kresic; Hydrogeology 101: Introduction to Groundwater Science and Engineering; ISBN: 979-8-218-06984-1.

In England and Wales, the Environment Agency (1998) has established source protection zones (SPZs) that are applied to public water supplies (there are nearly 2000 major sources) and private water supplies, including bottled water, and commercial food and drink production. As illustrated schematically in Figure 9.10, the orientation, shape and size of the SPZs are determined by the hydrogeological characteristics and the direction of groundwater flow around each source. Steady-state groundwater flow modelling is used to define three zones (Zones I, II and III) in each SPZ and a set of groundwater protection policy statements set out the acceptability of various polluting activities in each zone, for example landfill operations and the application of liquid effluents to land.

Zone I, or the inner source protection zone, is located immediately adjacent to the groundwater source, and is designed to protect against the impacts of human activity which might have an immediate effect upon the source. The area is defined by a 50-day travel time from any point below the water table to the source and as a minimum 50-metre radius from the source. This rule of thumb is used in other countries and is based on the presumed time taken for biological contaminants to decay in groundwater. The land immediately adjacent to the source and controlled by the operator of the source is included within this zone.

Zone II, or the outer protection zone, is the area around the source defined by a 400-day travel time, and is based on the requirement to provide delay and attenuation of slowly degrading pollutants. In high groundwater storage aquifers, such as sandstones, it is necessary, in order to provide adequate attenuation, to define further the outer protection zone to be the larger of either the 400-day travel time area or the recharge catchment area calculated using 25% of the long-term abstraction rate for the source.

Zone III, or the source catchment, is the remaining catchment area of a groundwater source, and is defined as the area needed to support an abstraction from long-term annual groundwater recharge. For wells and boreholes, the source catchment area is defined by the authorized abstraction rate while, for springs, it is defined by the best known value of average annual total discharge. In practice, the size of Zone III will vary from tens to a few thousands of hectares depending on the volume of groundwater abstraction and the amount of recharge. In areas where the aquifer is confined beneath impermeable cover, the source catchment may be some distance from the actual abstraction.

9.2.3. Risk Assessment Methods

Risk assessment techniques are becoming more and more important for managing aquifers. They are used in the decision-making process when deciding which aquifer remediation technology to use in situations where pollution has already occurred, like in contaminated land areas, as well as when siting new containment facilities, like municipal landfills. The Royal Society (1992) and the United States National Research Council (1983) both published important works that define risk assessment. Petts et al., (1997) stated that risk assessment is a process comprising hazard identification, hazard assessment, risk estimation and risk evaluation and, in general, is the study of decisions subject to uncertain consequences.

A basic risk assessment calculation can be performed by the determination of a toxicological index, I_{tox}, for a given site using the following equation:

$$I_{tox} = \sum_{i=1}^{n} c_i / \mathrm{LAC}_i$$

where i = 1, n represents the contaminant constituent, c_i is the measured concentration of constituent i and LAC is the limit of the admissible concentration of constituent i. An example calculation of I_{tox} values for a large pulp and paper mill complex in north-west Russia is given in Table 9.2.

Table 9.2. *Calculation of the Toxicological Index, I_{tox}, For Three Sites Located at the Pulp and Paper Mill Complex at Sjasstroj, North-West Russia*

Sample number	Sample date	Constituent, *i*	BOD	COD	SO_4^{2-}	Cl^-	NH_4^+	NO_2^-	NO_3^-	Fe	Al	Phenols	Surfactants	I_{tox}
		LAC_i	3	30	500	350	2	3	45	0.3	0.5	0.001	0.1	1
I	09/99			3044	160	15	5.6	0.04	1.0	5	90			301
II	11/99		5.4	61	13	42.7	3.5	0.005	0.45	53	0.2	0.001	0.1	185
	02/00		0.9	44	6.9	83	5	0.002	0.15	2.3	4.6	0.003	0.13	25
III	02/00		0.6	7.7	5.8	4	1.45	0.005	0.0	0.6	4.5	0.002	0	14

Source: Neven Kresic; Hydrogeology 101: Introduction to Groundwater Science and Engineering; ISBN: 979-8-218-06984-1.

Analytical results are given in mg L^{-1}.

I, excess sludge; II, groundwater close to active sludge basin in upper sand aquifer; III, groundwater in lower aquifer; LACi, Russian limit of the admissible concentration of constituent, I; BOD, biochemical oxygen demand; COD, chemical oxygen demand.

In calculating I_{tox}, constituents are chosen arbitrarily, mainly as a function of laboratory and financial capabilities, such that the importance of different compounds in terms of their hazard potential is not evaluated. The results of the risk assessment allow a comparative, quantitative assessment of analytical results for different measurement points but are neither source nor target related. For the calculations shown in Table 9.2, it is clear that all three samples are predicted to be at a high potential risk given the values of I_{tox} in excess of 1 and would therefore suggest that remedial action is necessary. However, shortcomings of the data presented in Table 9.2 are that substances with no toxicological potential, for example chemical oxygen demand (COD) for sample 1, can determine the outcome of the toxicological index calculation, and that concentrations of phenols, surfactants and the biological oxygen demand (BOD) were not always available.

More advanced methods of assessing the risk of groundwater pollution use a cost-effective, tiered approach to risk assessment and acknowledge the source-pathway-target paradigm. The pathway term in contaminated land studies refers to the path taken by the contaminant from the source of pollution to a receiving well or borehole receptor. This is determined by risk assessment. Separating the pathway into an exposure pathway connecting the receptor to an ecological or human target and an environmental pathway connecting the source and groundwater receptor is a refinement. While the exposure pathway evaluates the effects of exposure to contaminated water, the environmental pathway's goal is to ascertain the contaminant concentration at an abstraction site. Tiered approaches to risk assessment are commonly used in order to maximize the use of limited resources, including time and money. In terms of site-specific risk assessment, higher tiers correspond to higher levels of sophistication. In general, four levels of assessment can be identified:

- **Tier 1:** Initial inquiries include a qualitative desk research, identification of the source-pathway-target chains, a small amount of invasive research, and sampling.
- **Tier 2:** Site characterization: semiquantitative; some intrusive site investigation and sampling; prioritization and screening methods.
- **Tier 3:** Generic risk assessment methods include computer simulations and modeling, stochastic approaches, intrusive site investigations, environmental pathway assessment, qualitative and semiquantitative methods, and comparison of estimated contaminant concentrations with generic guidelines and standards.
- **Tier 4:** Quantitative risk assessment: use of derived contaminant concentrations for exposure pathway assessment; exposure assessment models; integrated approaches to risk assessment.

Although the procedure can be used for any contaminant and release scenario, the American Society for Testing and Materials standard guide to RBCA (risk-based corrective action) for site investigation and remediation at petroleum-contaminated sites is based on this kind of tiered approach. The framework consists of a stage for site characterization, which is followed by a three-tier approach that takes equal consideration of environmental and human health risks. Every tier is evaluated as it is finished to see if any additional information is needed.

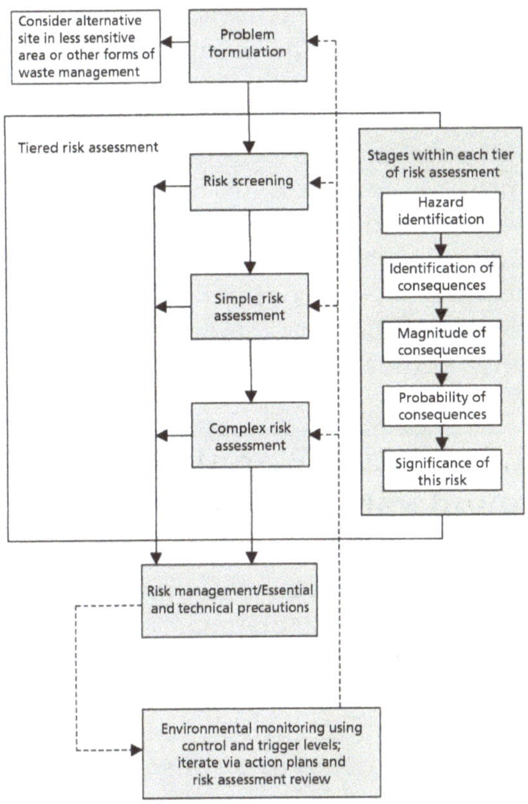

Figure 9.11. *Framework for a tiered approach to hydrogeological risk assessment for application in waste management. Two levels of risk assessment are recognized in the scheme. Simple risk assessment consists of quantitative calculations, typically deterministic analytical solutions using conservative (worst-case) input parameters, assumptions and methods. Complex risk assessment consists of quantitative, stochastic (probabilistic) techniques applied to analytical solutions using site-specific characterization data.*

Source: Neven Kresic; Hydrogeology 101: Introduction to Groundwater Science and Engineering; ISBN: 979-8-218-06984-1.

If so, an evaluation of the corrective action is necessary; if not, the next level of assessment is carried out. The Environment Agency in England and Wales assesses the hydrogeological risk of landfills using a tiered method, as shown in Figure 9.11.

A practical example of a Tier 3 risk assessment is provided by Davison et al., (2002), who presented a management tool to identify the best use for urban groundwater pumped from a user-defined location. A probabilistic catchment zone model and land-use model are combined to provide contaminant source data that are then used in a pollution risk model to calculate a probability distribution for the concentration of a contaminant at a selected pumped borehole for comparison with water quality standards.

9.2.4. Spatial Planning and Groundwater Protection

It has become evident that stricter regulations on land-use activities, integrated into local and regional spatial planning, are necessary if groundwater quality is to improve. The problem of nitrate and pesticides leaching from regions of intensively managed arable farming and cattle grazing affects wide areas of western Europe and other regions of the world. This kind of strategy has been used in Denmark and the United Kingdom, and it offers insights for future development of approaches to lessen the effects of diffuse groundwater pollution.

Beginning in 1990, the Pilot Nitrate Sensitive Area (NSA) program in England covered 80% of the land area, or 35,000 ha in 32 catchments, by the time the Main NSA program was implemented in 1998. This voluntary, compensated agri-environment program was in place. Farmers who entered the program were given the option to choose between two payment levels. The Premium scheme offered higher compensation for converting arable land to grass with a total nitrogen input of less than 150 kg ha^{-1}. In the alternative Basic scheme, farmers received a lower payment for planting winter cover crops, regulating the amount and timing of inputs of organic manure, and supplying nitrogen fertilizer at levels below the suggested maximum. In terms of measured soil nitrate losses both before and during the scheme, Table 9.3 presents the findings of long-term monitoring of the scheme's effectiveness in the first ten catchments included in the Pilot NSA scheme.

Table 9.3. *Summary of Measured Soil Nitrate Losses Prior to (1990/91) and During the Pilot Nitrate Sensitive Area (NSA) Scheme*

Crop type	N loss winter 1990/91			Mean N loss in winters of 1992/3, 1993/4, 1994/5		
	kg N ha^{-1} (adjusted)	mg L^{-1} as NO_3^-	*n*	kg N ha^{-1} (adjusted)	mg L^{-1} as NO_3^-	*n*
Potatoes, sugar beet	92 ± 23	228 ± 52	10	38 ± 4	87 ± 9	46
Cereals	40 ± 4	102 ± 11	61	53 ± 3	85 ± 5	142
Grass	44 ± 7	102 ± 16	27	42 ± 8	69 ± 14	79
Premium scheme	–	–	0	6 ± 4	9 ± 6	70
All sites	65 ± 12	163 ± 29	108	47 ± 3	79 ± 5	380

Fluxes have been adjusted to mean rainfall conditions.

Source: Neven Kresic; Hydrogeology 101: Introduction to Groundwater Science and Engineering; ISBN: 979-8-218-06984-1.

Comparing the results to baseline values, there was an overall decrease of roughly 50% in nitrate concentrations and 28% in nitrogen fluxes out of the root zone. Between 1994 and 1996 and 1998 and 2000, the corresponding values for the additional catchments included in the Main NSA scheme are 34% and 16%, respectively. Although the Premium scheme option's adoption and conversion of arable land was restricted, it had the greatest impact on lowering nitrate leaching losses by at least 80% and significantly reduced overall nitrate losses in a catchment.

England has now established a Nitrate Vulnerable Zone (NVZ) program in response to the EU Directive on Diffuse Pollution by Nitrates (91/676/EEC; Council of European Communities, 1991) (Figure 9.12) with obligatory, unpaid measures founded on "good agricultural practice" currently covering roughly 55% of England's land area. The lands that drain into polluted waters, such as surface waters and ground waters that contain or may contain nitrate concentrations exceeding 50 mg L^{-1}, as well as naturally occurring freshwater lakes and other freshwater bodies, estuaries, coastal waters, and marine waters that are eutrophic or may soon become so if preventative action is not taken, have been identified as the NVZ areas.

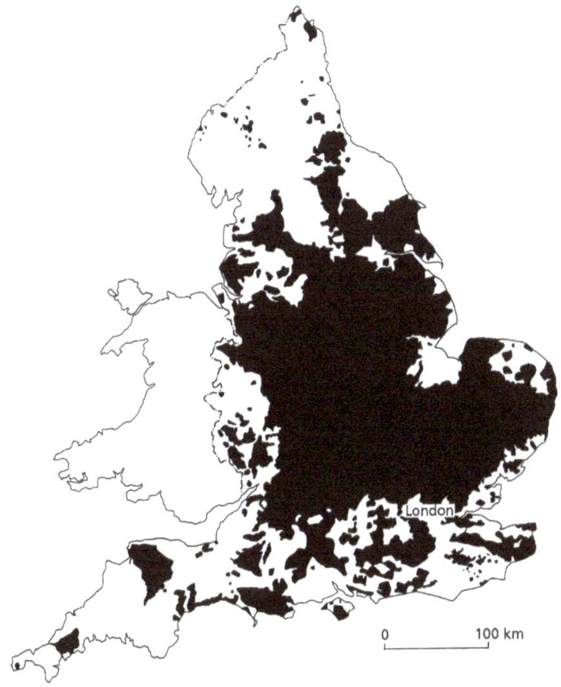

Figure 9.12. *Areas of Nitrate Vulnerable Zones (NVZs) in England. The following rules apply under this mandatory, uncompensated scheme based on 'good agricultural practice': (i) closed periods for inorganic nitrogen (fertilizer) applications during the autumn and winter and for organic nitrogen (manure) applications during the autumn for both arable and grass lands; (ii) nitrogen limits applied for inorganic and organic nitrogen applied to arable and grass lands that do not exceed crop requirements; (iii) spreading controls to restrict fertilizer and manure applications on steep slopes or close to water courses; (iv) slurry storage for manure during the autumn closed period; and (v) record keeping of agricultural practices for at least 5 years.*

Source: Neven Kresic; Hydrogeology 101: Introduction to Groundwater Science and Engineering; ISBN: 979-8-218-06984-1.

9.3. GROUNDWATER PROTECTION STRATEGIES IN DEVELOPING COUNTRIES

Groundwater constitutes a vital component of the water supply system in many developing countries. It supports various sectors, including agriculture, which is a backbone of these economies. Protecting groundwater is therefore essential for sustaining livelihoods, ensuring food security, and promoting economic development. However, the protection of this resource is fraught with challenges such as over-extraction, contamination from industrial and agricultural activities, and the impacts of climate change. Effective groundwater management requires a multifaceted approach that includes robust policy frameworks, technological innovations, and active community participation.

9.3.1. Challenges to Groundwater Protection

9.3.1.1. Over-Extraction

One of the primary threats to groundwater resources in developing countries is over-extraction, driven largely by agricultural demands. For instance, in India, the Green Revolution led to a significant increase in groundwater use for irrigation, resulting in declining water tables in many regions. The lack of regulation and enforcement exacerbates this problem, leading to unsustainable extraction rates.

9.3.1.2. Contamination

Groundwater contamination poses a severe threat to public health and the environment. Sources of contamination include industrial discharge, agricultural runoff containing pesticides and fertilizers, and inadequate waste management systems. In Bangladesh, for example, arsenic contamination in groundwater has affected millions of people, highlighting the critical need for effective contamination control measures.

9.3.1.3. Climate Change

Climate change impacts groundwater both directly and indirectly. Changes in precipitation patterns can affect recharge rates, while increased temperatures may lead to higher evaporation rates, reducing surface water availability and increasing reliance on groundwater. Moreover, extreme weather events such as floods and droughts can disrupt groundwater recharge and quality.

9.3.1.4. Governance Issues

Effective groundwater management is often hampered by governance challenges, including weak regulatory frameworks, lack of institutional capacity, and limited public awareness. In many developing countries, groundwater resources are managed by multiple agencies with overlapping responsibilities, leading to inefficiencies and lack of coordination.

9.3.2. Policy and Regulatory Frameworks

9.3.2.1. Legislation and Regulation

Establishing robust legal frameworks is essential for sustainable groundwater management.

Legislation should define groundwater rights, regulate extraction, and set standards for water quality. For example, South Africa's National Water Act (1998) provides a comprehensive framework for water resource management, emphasizing the sustainable use of water resources, including groundwater.

9.3.2.2. Institutional Strengthening

Building the capacity of institutions responsible for groundwater management is crucial. This includes training personnel, improving data management systems, and enhancing inter-agency coordination. Strong institutions can enforce regulations more effectively and respond to emerging challenges.

9.3.2.3. Community Involvement

Engaging local communities in groundwater management can lead to more sustainable practices. Community-based management approaches empower local users to monitor and manage their resources, ensuring that extraction rates are sustainable and contamination risks are minimized. Programs such as India's Participatory Groundwater Management (PGWM) have demonstrated the effectiveness of community involvement.

9.3.3. Technological and Management Approaches

9.3.3.1. Sustainable Extraction Techniques

Implementing techniques that promote sustainable groundwater extraction is essential. These include:

- **Drip Irrigation:** Reduces water use in agriculture by delivering water directly to the plant roots, minimizing wastage.
- **Rainwater Harvesting:** Captures and stores rainwater for use during dry periods, reducing reliance on groundwater.
- **Managed Aquifer Recharge (MAR):** Involves artificially recharging aquifers using surface water, treated wastewater, or stormwater.

9.3.3.2. Contamination Prevention and Mitigation

Preventing and mitigating contamination requires a combination of regulatory measures and technological solutions:

- **Pollution Control Regulations:** Enforcing strict regulations on industrial discharges and agricultural runoff.

- **Bioremediation:** Using microorganisms to degrade pollutants in groundwater.
- **Constructed Wetlands:** Natural systems that filter contaminants from runoff before it infiltrates groundwater.

9.3.3.3. Groundwater Recharge

Enhancing groundwater recharge is vital for maintaining water levels. Techniques include:

- **Recharge Pits and Trenches:** Structures that facilitate the infiltration of surface water into the ground.
- **Permeable Pavements:** Surfaces that allow rainwater to percolate into the soil, enhancing recharge.
- **Afforestation:** Planting trees to increase soil infiltration and reduce runoff.

9.3.3.4. Data Collection and Monitoring

Accurate data on groundwater levels and quality is crucial for effective management. Modern technologies such as remote sensing, Geographic Information Systems (GIS), and Internet of Things (IoT) sensors can provide real-time data, aiding in decision-making and policy implementation.

9.3.4. Example

9.3.4.1. India

India faces severe groundwater challenges due to over-extraction and contamination. The government has implemented several initiatives to address these issues, such as:

- **Atal Bhujal Yojana (ABY):** A program aimed at sustainable groundwater management through community participation and technology-driven solutions.
- **National Aquifer Mapping and Management Program (NAQUIM):** An initiative to map aquifers and provide data for better management practices.

9.3.4.2. Kenya

In Kenya, groundwater is a crucial resource for arid and semi-arid regions. Strategies for groundwater protection include:

- **Water Resource Users Associations (WRUAs):** Community-based organizations that manage water resources locally.
- **Borehole Rehabilitation:** Programs to repair and maintain boreholes, ensuring sustainable extraction.

9.3.4.3. Brazil

Brazil has implemented various measures to protect its groundwater resources, including:

- **National Water Agency (ANA):** An organization responsible for the management and protection of water resources.
- **Guarani Aquifer System:** An international initiative involving multiple countries to manage one of the largest aquifer systems in the world.

CASE STUDY

JET FUEL CLEAN-UP AT HEATHROW INTERNATIONAL AIRPORT

This case study concerns hydrocarbon contamination of groundwater adjacent to Technical Block L at Heathrow Airport. Heathrow Airport is built on the Taplow Terrace adjacent to the River Thames floodplain. The geology is formed by 4.5 m of coarse clean gravels overlying low permeability London Clay. The water table is shallow, about 2.5 m from the surface, with groundwater flow southwards beneath the airport towards the River Thames at Shepperton. A leak of jet fuel (kerosene, a light non-aqueous phase liquid, or LNAPL) occurred from a cracked fuel pipe leading to an engine maintenance facility, the leak having occurred over a number of years. The leak was discovered when fuel was observed floating on drainage water in a manhole north of Technical Block M (Figure 1). In response, a large concrete-lined well (Well 1), about 1.5 m in diameter, was installed close to the manhole and revealed about 10 cm of kerosene floating on the water table. As a first step in remediating the contaminated site, the leak was traced to the cracked pipe and the fracture repaired.

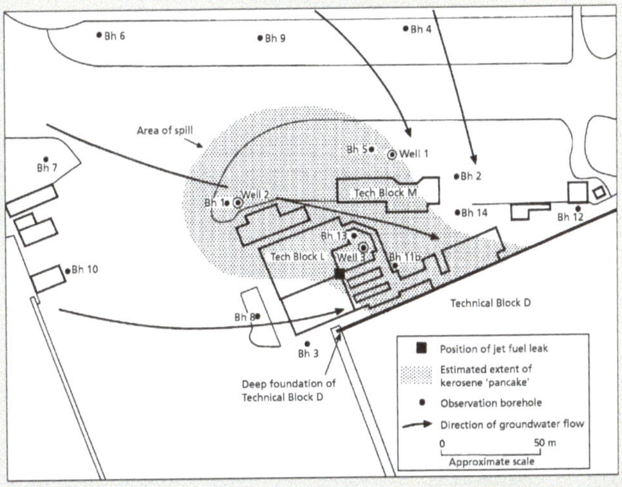

Figure 1. *Site of a jet fuel leak adjacent to Technical Block L at Heathrow Airport showing the estimated extent of the kerosene 'pancake' resting on the gravel aquifer water table.*

Source: Neven Kresic; Hydrogeology 101: Introduction to Groundwater Science and Engineering; ISBN: 979-8-218-06984-1.

A detailed site investigation, including the installation of 14 monitoring boreholes, showed that the 'pancake' of floating kerosene was about 100 m in diameter and at its thickest point measured a depth of 0.95 m in borehole 5, with further 'free product' measured in boreholes 1, 11 and 13 and in Well 1 (Figure 1). Odour was reported during the drilling of boreholes 3, 9 and 10, indicative of kerosene. No kerosene was detected in the outlying observation boreholes, including borehole 12.

The basic remediation structures used included large diameter wells lined with perforated concrete rings about 1.5 m in diameter (Figure 2). Wells 1 and 2, installed close to borehole 1 where a considerable thickness of fuel was shown to be floating on the water table, were used to begin recovery of the floating kerosene.

The kerosene was removed by floating oil-skimmer pumps. Surface-mounted centrifugal pumps, installed with their intakes in the two wells, were also used to lower the water table and encourage the kerosene to move towards the recovery wells. The waste water pumped from the wells was discharged to the drainage system of the airport, which leads to a balancing reservoir before flowing into the River Thames. The balancing reservoir provided settlement and dilution of the remediation waste water.

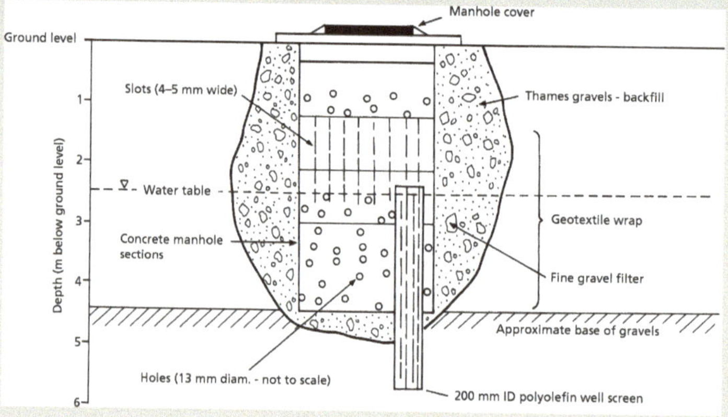

Figure 2. Construction details of recovery Well 2. The well was installed by excavating a pit by back-hoe as deeply as possible, about 2 m below the water table, lowering the perforated concrete rings into position and then backfilling around the rings using the gravel excavated from the pit.

Source: Neven Kresic; Hydrogeology 101: Introduction to Groundwater Science and Engineering; ISBN: 979-8-218-06984-1.

Groundwater levels were monitored regularly, and within 2 months the cone of depression in the water table produced by the pumping of Wells 1 and 2 encompassed the estimated area of the kerosene 'pancake'. Initially, the recovery rate was such that 19,200 L of kerosene were removed (Figure 3) and sold to be blended into commercial heating oil. The recovery rate then dropped substantially, yet the kerosene layer in borehole 11 still remained unaltered, suggesting that Wells 1 and 2 were not affecting the southern part of the kerosene 'pancake'. Later, Well 3 was installed. It is believed that Well 3 tapped a 'pool' of kerosene isolated from the effects of Wells 1 and 2 by the foundations of the Technical Blocks. A further 10,100 L of kerosene were removed from Well 3 (Figure 3), making a total recovery of 29,300 L of kerosene in 4 years at which point the removal of the original kerosene 'pancake' was considered to be complete. Although active remediation by pump-and-treat ended at this time, kerosene recovery using passive collectors, for example absorbent mops, in the three wells continued for about another year. The clean-up project officially ended in 1994.

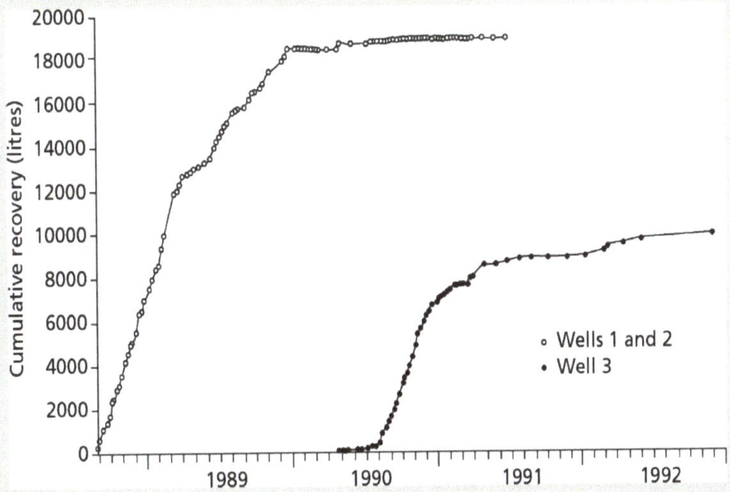

Figure 3. *Cumulative recovery of hydrocarbon from Wells 1, 2 and 3 at Heathrow Airport during clean-up of the gravel aquifer.*

Source: Neven Kresic; Hydrogeology 101: Introduction to Groundwater Science and Engineering; ISBN: 979-8-218-06984-1.

CLASS ACTIVITY

GROUNDWATER POLLUTION REMEDIATION

"Students will explore groundwater pollution remediation through a series of experiments and case studies. They'll investigate various methods such as bioremediation, pump-and-treat systems, and natural attenuation. Through hands-on activities, they'll analyze the effectiveness of different techniques and propose solutions for real-world contamination scenarios."

SUMMARY

- The remediation of groundwater can involve an attempt at the total clean-up of a contaminated aquifer or the containment of a groundwater pollution source. In certain circumstances, for example in areas of low risk of human or environmental exposure to contaminants, a further option may be to leave the aquifer to recover through natural attenuation processes
- The conventional pump-and-treat method of aquifer clean-up is to extract the contaminated groundwater and, following treatment to remove and possibly recover the contaminant source, for the treated water to be either injected into the aquifer or, if a discharge consent is obtained, released to a surface water course.
- Permeable reactive barriers are installed in trenches, but barriers have also been constructed by jetting reactive materials into the ground, or by generating fractures within an aquifer and filling the fractures with reactive materials. The reactive material contained in the barrier is selected to retain the contaminant within the barrier. PRBs containing zero-valent iron (iron filings) have been used to treat hexavalent chromium, uranium and technetium and chlorinated ethenes (PCE and TCE).
- The Environmental Protection Agency has developed a similar methodology to evaluate groundwater vulnerability designed to permit the systematic evaluation of the groundwater pollution potential at any given location.
- In the definition of groundwater source protection zones, the proximity of a hazardous activity to a point of groundwater abstraction (including springs, wells and boreholes) is one of the most important factors in assessing the pollution threat to an existing groundwater source. In principle, the entire recharge area in the vicinity of a groundwater source should be protected, but this is unrealistic on socioeconomic grounds.
- The problem of nitrate and pesticides leaching from regions of intensively managed arable farming and cattle grazing affects wide areas of western Europe and other regions of the world and it has become clear that more stringent controls on land-use activities, integrated into local and regional spatial planning, are required if groundwater quality is to improve.

REVIEW QUESTIONS

1. Explain the principle of pump and treat technology in groundwater remediation. What are its advantages and limitations?
2. How does in-situ bioremediation work, and what types of contaminants is it most effective against?
3. Describe the process of monitored natural attenuation (MNA) and the conditions under which it is considered a viable remediation strategy.

4. What are the primary regulatory frameworks and policies in place in industrialized countries to protect groundwater resources from contamination?
5. How do industrialized countries typically manage the risks associated with agricultural runoff to protect groundwater quality?
6. What are the main challenges faced by developing countries in implementing effective groundwater protection strategies?
7. Identify and explain at least three strategies that developing countries can adopt to improve groundwater protection despite limited resources.

REFERENCES

1. Aller, L., Bennett, T., Lehr, J. H., Petty, R. J., & Hackett, G. (1987). *DRASTIC: A Standardized System for Evaluating Ground Water Pollution Potential Using Hydrogeologic Settings*. NWWA/EPA Series. EPA/600/2-87/035. Washington, DC.
2. Bekins, B., Rittmann, B., & MacDonald, J. (2001a). Natural attenuation strategy for groundwater cleanup focuses on demonstrating cause and effect. *Eos, 82*(53), 57–58.
3. Blowes, D. W., Ptacek, C. J., Benner, S. G., McRae, C. W. T., Bennett, T. A., & Puls, R. W. (2000). Treatment of inorganic contaminants using permeable reactive barriers. *Journal of Contaminant Hydrology, 45*, 123–137.
4. Hocking, G., Wells, S. L., & Ospina, R. I. (2000). Deep reactive barriers for remediation of VOCs and heavy metals. In G. B. Wickramanayake, A. R. Gavaskar, & A. S. C. Chen (Eds.), *Chemical Oxidation and Reactive Barriers: Remediation of Chlorinated and Recalcitrant Compounds* (pp. 307–314). Battelle Press.
5. Livingstone, S., Franz, T., & Guiguer, N. (1995). Managing groundwater resources using wellhead protection programs. *Geoscience Canada, 22*, 121–128.
6. Mawdsley, J., Petts, G., & Walker, S. (1994). *Assessment of Drought Severity*. British Hydrological Society Occasional Paper No. 3. Institute of Hydrology.
7. Richardson, J. P., & Nicklow, J. W. (2002). In situ permeable reactive barriers for groundwater contamination. *Soil and Sediment Contamination, 11*, 241–268.
8. Robins, N. S., Adams, B., Foster, S. S. D., & Palmer, R. C. (1994). Groundwater vulnerability mapping: The British perspective. *Hydrogèologie, 3*, 35–42.
9. Wickramanayake, G. B., Gavaskar, A. R., & Chen, A. S. C. (Eds.). (2000). *Chemical Oxidation and Reactive Barriers: Remediation of Chlorinated and Recalcitrant Compounds*. Battelle Press.

CHAPTER

10

Groundwater Resources, Governance, and Management

LEARNING OBJECTIVES

After studying this chapter, you will be able to:

- Describe groundwater resources schemes
- Discuss about groundwater abstraction and river flows
- Explain wetland hydrogeology
- Define climate change and groundwater resources
- Elaborate groundwater governance

KEY TERMS FROM THIS CHAPTER

Aquifer storage
Catchment water resources management
Colmation layer
Groundwater
Groundwater development
Hydrogeological assessment
Reservoir
Riverbank filtration
Aquifer transmissivity
Climate change
Governance
Groundwater abstractions
Groundwater discharge
Hydrological cycle
Riparian environment
Water governance

INTRODUCTION

Groundwater resources, essential for global water supply, are increasingly vital in sustaining agriculture, industry, and domestic use. These subsurface water reserves, stored in aquifers, are crucial for providing a reliable source of fresh water, particularly in arid and semi-arid regions where surface water is scarce. Effective governance and management of groundwater are imperative to ensure its sustainable use, prevent over-extraction, and mitigate contamination.

Groundwater governance involves the frameworks, policies, and institutions that regulate the use and protection of groundwater resources. It encompasses the legal and regulatory mechanisms that govern access rights, extraction limits, and quality standards. Effective governance ensures equitable distribution, resolves conflicts among users, and protects against over-exploitation. Internationally, governance varies significantly, with some regions having well-established legal frameworks and others lacking comprehensive policies, leading to challenges in sustainable management.

Groundwater management, on the other hand, entails the practical implementation of policies and strategies to monitor, regulate, and preserve groundwater resources. This includes techniques such as artificial recharge, sustainable extraction practices, and contamination control measures. Advanced technologies like remote sensing, Geographic Information Systems (GIS), and hydrogeological modeling play a crucial role in enhancing groundwater management by providing accurate data for informed decision-making.

10.1. GROUNDWATER RESOURCES SCHEMES

Environmental conflicts may arise from the development of groundwater resources for industrial, agricultural, and public uses. Groundwater abstractions reduce the amount of freshwater habitats that depend on groundwater discharge by capturing recharge water that would otherwise flow to springs and rivers. Freshwater (or saline) ecosystems need a sufficient amount of water to be maintained in the modern era of integrated river basin management. This way, in order to achieve sustainable groundwater development, the fraction of available recharge required for environmental benefits as well as the fraction required for human and economic benefits are taken into account.

The evaluation and utilization of groundwater resources are essential components of hydrogeology. A large storage volume that can be developed gradually as needed; (ii) resilience to drought conditions due to the large storage volume; (iii) relatively low environmental impact of well-field developments; and (iv) no loss of storage volume due to evaporation are some of the benefits of groundwater resources over surface reservoir developments. Given the current understanding that groundwater and surface water resources should be managed in tandem, it is helpful to conceptualize this strategy within the framework of a water balance equation that compares water demand to abstraction needs and environmental requirements.

10.1.1. Water Balance Equation

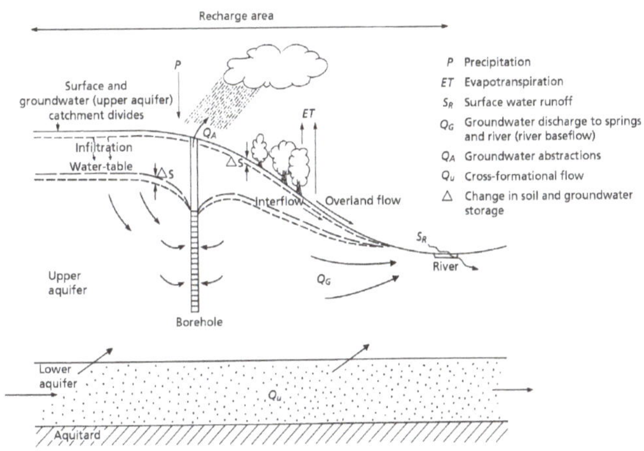

Figure 10.1. *Diagram showing the inputs, flowpaths and outputs within a river basin that comprise a catchment water balance. The total groundwater discharge, GR, is the sum of QG + QA + QU.*

Source. https://www.geokniga.org/bookfiles/geokniga-hydrogeologyprinciplesandpractice.pdf.

Precipitation (P) is the primary raw resource in a catchment. Water is either lost to evaporation and transpiration (commonly combined and called evapotranspiration, E_T) or flows through interflow and overland flow hydrological pathways to produce groundwater discharge (G_R) and surface water runoff (S_R).

Natural discharge to springs and rivers (the river baseflow, or Q_G) and artificial abstractions (Q_A) are included in the groundwater discharge component, which is sourced by groundwater recharge. It may be necessary to take into account additional elements of cross-formational groundwater flow (Q_U) depending on the catchment geology, as the groundwater catchment may or may not align with the surface catchment area. The catchment water balance equation for surface water and groundwater can now be expressed as follows, assuming that their catchments coincide (Figure 10.1).

$$P = ET + S_R + G_R \pm \Delta S \qquad (1)$$

On a short timescale of weeks to months, Equation 1 is balanced by changes in the water held in soil and groundwater storage (ΔS) as represented by changes in soil moisture content and groundwater levels. Over longer timescales of several years, changes in storage balance out to zero and, expressing $G_R = Q_G + Q_A$, Equation 1 becomes:

$$P - ET = S_R + Q_G + Q_A \qquad (2)$$

Table 10.1. *Groundwater Budget Information for the High Plains Aquifer Based on the United States Geological Survey's Regional Aquifer System Analysis Groundwater Model*

Budget parameter	Northern High Plains	Central High Plains	Southern High Plains
Primary inflows (in cubic metres per year)			
Recharge from precipitation on rangeland and streams	5.98×10^8†	4.66×10^8†	1.97×10^8
Recharge from precipitation on agricultural land‡	2.89×10^9	–	1.43×10^9
Groundwater irrigation return (pumpage minus crop demand)	2.31×10^9	2.07×10^9	3.61×10^9
Recharge from other human activities (e.g. seepage from reservoirs and canals)	2.31×10^9	–	–
Recharge from other aquifers across subunit boundary	–	1.88×10^7§	–
Totals	8.11×10^9	2.55×10^9	5.24×10^9
Primary outflows (in cubic metres per year)			
Total pumpage	6.48×10^9	6.89×10^9‖	8.59×10^9
Discharge to streams and shallow water-table areas	2.87×10^9	4.15×10^8	–
Discharge along eastern boundary	–	6.97×10^7	1.05×10^8
Totals	9.35×10^9	7.37×10^9	8.70×10^9
Net residual	-1.24×10^9	-4.82×10^9	-3.46×10^9

Source. https://www.geokniga.org/bookfiles/geokniga-hydrogeologyprinciplesandpractice.pdf.

* *Assumptions: Inflow/Outflow values determined using 1960–1980 estimates; base of aquifer modelled as no-flow boundary; vertical flow in aquifer considered negligible on regional scale. † Recharge distributed unevenly based on soil type. ‡ Additional recharge from precipitation on agricultural land because of changes in soil character due to tillage. § Flow only from northern and southern subunits to central subunit. ‖ Municipal and industrial pumpage is 3.2% of this amount.*

Stated differently, the distinction between evapotranspiration and precipitation, or effective precipitation, facilitates groundwater discharge as river baseflow, surface runoff, and borehole abstractions. The sum of S_R and Q_G determines a river's overall flow. The amount of effective precipitation (hydrological excess) and methods for calculating it are covered. It is obvious that any increase in groundwater abstraction (Q_A) will come at the expense of river baseflow (Q_G), potentially having an adverse effect on the environment, if Equation 2 is to be balanced. Generally speaking, the Q95 low flow statistic is a helpful indicator of the amount of water needed to maintain minimum acceptable river flows. This means that it is not desirable for the river flow to fall below the long-term average of flow, which is equal to or greater than for 95% of the time, during the baseflow recession period, when river flows are dominated by the groundwater input (Q_G). It is ideal for conflicting environmental (Q_G) and socioeconomic (Q_A) demands to not outpace the rate of groundwater recharge (G_R) in order to support sustainable groundwater development. There are three categories in which to define them: (i) $Q_{A}. < G_R - Q_G$ and further groundwater resources are available for exploitation; (ii) $Q_A = G_R - Q_G$, in which case the safe yield has been achieved; and (iii) When groundwater resources are over mined or exploited, $Q_A > G_R - Q_G$. In (iii), groundwater is extracted from aquifer storage to support abstractions, which may have long-term effects on groundwater levels (Luckey et al., 1986).

10.1.2. Large-Scale Groundwater Development Schemes

Compiling a water balance and comprehending aquifer conditions are essential components of water resources management. Table 10.1 for the United States High Plains Aquifer provides an example of a groundwater budget for a large-scale aquifer system, where the main inflows and outflows to the aquifer are tabulated. The Great Plains physiographic province includes the High Plains Aquifer, which covers 450,660 km^2 and is primarily composed of near-surface deposits from the late Tertiary or Quaternary era. The main hydrogeological unit under 347,060 km^2 is the Miocene Ogallala Formation, which is made up of a heterogeneous sequence of clay, silt, sand, and gravel. By using the High Plains Aquifer as a source of irrigation water, the mid-section of the United States has become one of the world's most important agricultural regions. The main crops are grains, particularly wheat, sorghum, and maize, as well as cotton and alfalfa. The 4.25 million swine (1997) and 15 million cattle that are raised over the aquifer are fed grains. Furthermore, 82% of the population residing within the aquifer's boundaries receives their drinking water from the aquifer (Dennehy et al., 2002).

As stated by Dennehy et al., According to (2002), the High Plains Aquifer's groundwater typically flows from west to east, naturally emptying into springs and streams. In locations where the water table is near the surface of the earth, evapotranspiration may occur. The main method of groundwater discharge, however, is pumping from multiple irrigation wells. Water levels have significantly decreased in many places due to abstractions that significantly outweigh recharge; for instance, in portions of Kansas, New Mexico, Oklahoma, and Texas, there have been declines of 30-43 m between the 1940s and 1980s.

The table displaying the groundwater budget reveals that all regions of the High Plains Aquifer are experiencing a deficit, as outflows exceed inflows. In the southern and central regions, outflows are double the inflows. Before irrigation was introduced, the aquifer was recharged by an average of 15 mm per year from precipitation, with small amounts of water being discharged to springs and rivers. Irrigation uses a significant amount of water, and since its development, 7% of the original water volume has been extracted from the aquifer. Without 30-40% of the pumped water infiltrating back into the aquifer as return flows, the decrease in storage would have been even greater. The declining groundwater levels pose a direct threat to the region's way of life, prompting efforts to implement water-efficient irrigation and best-management farming practices. These efforts may ultimately lead to a transition from irrigated agriculture to dryland farming, with far-reaching implications for the local economy detailed by Dennehy et al., (2002).

The High Plains Aquifer serves as an example of unsustainable groundwater development, but as long as the groundwater system is sufficiently understood to assess impacts, there is no fundamental reason why temporary overexploitation of aquifer storage for a specific benefit should not be permitted as part of a rational water resources management strategy. An instance of the intentional extraction of groundwater resources for the advantage of a country's economy is the Libyan Great Man-made River Project (GMRP). Under the GMRP, groundwater recharged during the last ice age's pluvial period is extracted from beneath the Libyan desert and transported to coastal towns and cities. Although the groundwater supply is obviously not a steady, renewable resource, it can provide the majority of Libya's population with a necessary resource for at least a few decades. According to Price (2002), the GMRP seems to satisfy the ethical standards for groundwater mining, which are as follows: (i) there is proof that pumping can continue for a considerable amount of time; (ii) the drawbacks of development are outweighed by the advantages; and (iii) decision-makers and users are aware that the resource will eventually run out. Price (2002), however, worries that it is not a prudent long-term strategy to use non-renewable groundwater for inefficient agriculture in an area where there is practically no recharge.

10.1.3. Regional-Scale Groundwater Development Schemes

The two large-scale groundwater developments mentioned above that assist irrigated agriculture tip the sustainability equation in favor of financial gain. Together with surface resources, efficient river basin management at the regional level can develop the significant storage volume in aquifers. This idea is not new; in fact, several schemes were developed in England and Wales after the Water Resources Act of 1963, primarily the Thames, Great Ouse, Severn, Itchen, and Waveney Schemes. In order to determine whether it would be feasible to regulate rivers by pumping groundwater into them for abstraction for water supply in the lower reaches while still ensuring an appropriate flow to satisfy all other environmental requirements, pilot studies were conducted (Downing, 1993).

An instance of a river regulation project, the Great Ouse Groundwater Scheme in eastern England, involved two sets of wells in a pilot area measuring 71.5 km^2 within the River Thet catchment: one set of wells located along the riverbank and another set situated further inland on higher terrain. A control area was designated in the nearby River Wissey catchment. A total of eighteen boreholes were drilled into the Chalk aquifer, producing 70×10^3 m^3 day^{-1}. When all the wells were in operation for 250 days, the pumping rate was three times higher than the average infiltration rate across the pilot area. (Figure 10.2).

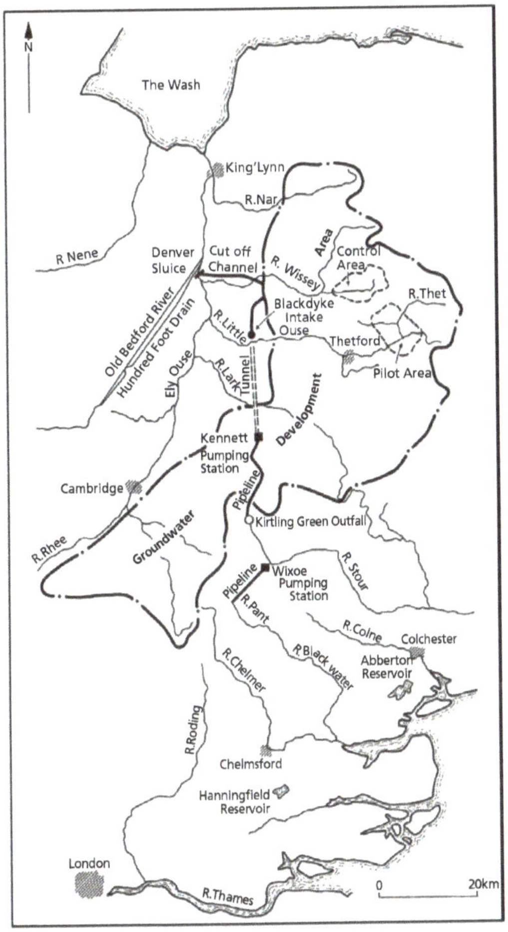

Figure 10.2. *Map of the location of the original development area of the Great Ouse Groundwater Scheme showing the various linked components of the current Ely Ouse Essex Water Transfer Scheme.*

Source. https://www.geokniga.org/bookfiles/geokniga-hydrogeologyprinciplesandpractice.pdf.

There were worries that the pilot are Over 80% of the River Thet's length had its groundwater levels drop below the riverbed in just six months as all of the river's baseflow was intercepted. Groundwater levels decreased by an average of 2.5 m, and after 200 days, stable conditions were attained as a result of induced groundwater flow across the pilot area's eastern boundary. Because the riparian zone is covered in relatively impermeable peats, silts, and clays, 10% of the pumped groundwater seeps through the riverbed.

a's Breckland Meres might be impacted by pumping the Chalk aquifer. These lakes, which are fed by groundwater, are significant from an ecological and scientific standpoint. Nonetheless, the pilot area's groundwater level drawdown of more than 1 m as a result of heavy pumping in 1970 was comparatively small (Backshall et al., 1972) (Figure 10.3).

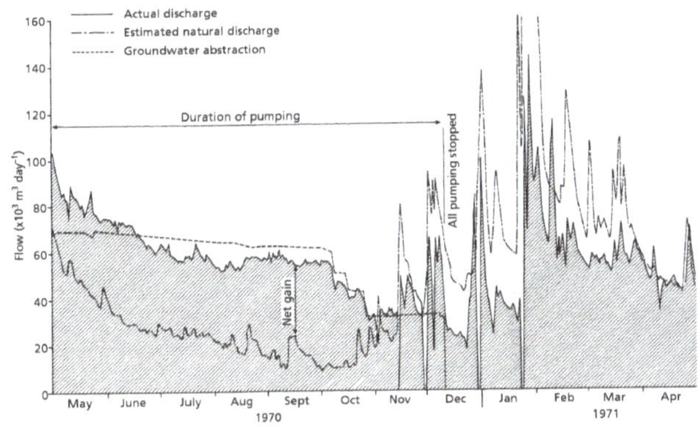

Figure 10.3. *Hydrograph records for 1970–1971 showing the net gain in flow of the River Thet from groundwater abstracted from Chalk boreholes in the pilot area of the Great Ouse Groundwater Scheme. As shown, the net gain is the difference between the actual discharge with groundwater abstraction and the estimated natural discharge without abstraction. As a result of measurements taken for identical 24-hour periods without allowance for travel times (or temporary storage effects), occasional zero flows (or even apparent losses) can arise.*

Source. https://www.geokniga.org/bookfiles/geokniga-hydrogeologyprinciplesandpractice.pdf.

The net benefit to the river can be used to measure the effectiveness of a river regulation plan. The characteristics of the aquifer and the separation between boreholes and natural spring discharges determine the actual net gain at any given time. The Great Ouse Groundwater Scheme's net gain fluctuated between 1970 and 1971, but in the summer months, values were roughly 70% of the amount pumped. As a result, the plan was successful in keeping river flows close to their mean values. The following winter saw a steady return of groundwater levels and river flows to their natural levels following the cessation of pumping.

In time for the start of the next baseflow recession period, natural river flows were restored. One characteristic of river regulation schemes is the recovery delay, which results from the groundwater storage being plowed during the summer regulation period being replenished first by winter groundwater recharge. If environmental effects are not to be irreversible, there must, by definition, be no net gain to the river over an extended period of time.

When the Great Ouse Groundwater Scheme was fully developed, nearly 350 new boreholes would be spread over a 2500 km^2 area. The Chalk groundwater resource could be developed in conjunction with surface runoff, which would produce 360×10^3 m^3 day^{-1} of water that could be exported to demand areas outside of the basin.

For the fully developed scheme, the Chalk aquifer drawdown was estimated to be 4 m during a drought with a 1 in 50-year return period.

Conflicts with other users of Chalk groundwater, such as irrigation water for valuable vegetable crops, and a shift in perspective toward protecting the environment, particularly the Breckland Meres, from the effects of groundwater drawdown were two reasons why such a large engineering program of works was not pursued.

The Ely Ouse Essex Water Transfer Scheme, a much larger project, now includes the boreholes dug for the pilot study in order to supply groundwater to meet the significant demand for public water supply in the southeast of England. (Figure 10.4).

Figure 10.4. *Transfer of water from the Ely Ouse River to the Cut-Off Channel at Denver. Water transferred from here travels to Abberton and Hanningfield reservoirs in support of water supply and river flows in the drier south-east of England.*

Source. https://www.geokniga.org/bookfiles/geokniga-hydrogeologyprinciplesandpractice.pdf.

The Ely Ouse Essex Water Transfer Scheme makes additional water available for the Essex rivers by allowing surplus water from the Ely Ouse (the River Great Ouse) to be transferred to the heads of the Essex rivers in the southeast of the Anglian region. Despite having low effective precipitation of less than 125 mm a^{-1}, Essex County, located north of London, has a sizable and growing population. The plan's ability to increase the capacity of current reservoirs without requiring the removal of agricultural land to build new ones is one of its many advantages. The plan transfers excess water from the eastern portion of the catchment to the Cut-Off Channel flood protection scheme at Denver. This includes Chalk groundwater resources that can be used to control the westward flow of the Chalk rivers, such as the Rivers Lark, Little Ouse, and Thet. Since the needs of the upstream rivers have now been satisfied, any excess water that would have otherwise been lost to tidal waters and ultimately the Wash could potentially be transferred. The Denver Impounding Sluice gate is intended to allow for a reversal of flow by raising the river channel's water level by about 0.6 meters. The water is sent to the Blackdyke Intake, which is about 25 km southeast, in the opposite direction. From here, the water is diverted into a 20-kilometer tunnel that ends at Kennett. From there, it is pumped via a 14-km pipeline to the River Stour at Kirtling Green. A portion of this discharge is pumped 10.3 km to the River Pant after being drawn off at Wixoe, 13.7 km downstream. Water moved from Denver is transported 148 km to Hanningfield Reservoir and 141 km to Abberton Reservoir. Watercourse exits are used for two thirds of these distances.

10.1.4. Artificial Storage and Recovery Schemes

The last phase of surface water and groundwater schemes development is artificial recharge, which involves replenishing treated wastewater-filled basins with fresh water before returning the water to the aquifer. Despite being used in other nations, frequently in an unregulated manner and possibly endangering the long-term quality of groundwater (for instance, in China and Mexico; Foster et al., 1999), the practice of artificial recharge—which involves returning treated water from sewage treatment plants to rivers—is uncommon in the United Kingdom. Aquifer storage and recovery, or ASR, is a relatively new approach to artificial recharge that aims to supply water during periods of high demand.

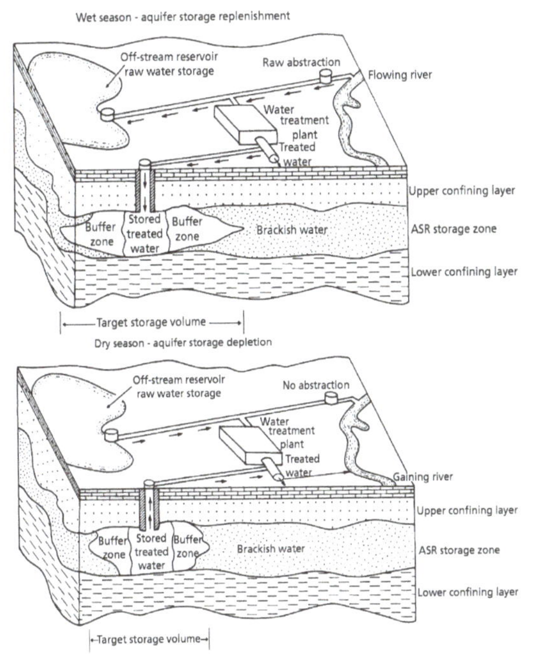

Figure 10.5. *Schematic diagram showing the operation of an aquifer storage and recovery (ASR) scheme. ASR involves storage of available, principally surface or drinking water-quality water during the wet season through boreholes completed in a brackish-water aquifer, with subsequent retrieval from these same boreholes during dry periods. The freshwater forms a 'bubble' within the aquifer around the ASR well and can be retrieved when needed. The hydrogeological characteristics of a successful ASR storage zone include: moderate aquifer hydraulic conductivity; confinement above and below by low permeability strata; and water quality as fresh as possible to limit mixing with the native brackish water. After the United States Geological Survey.*

Source. https://www.geokniga.org/bookfiles/geokniga-hydrogeologyprinciplesandpractice.pdf.

The ASR technique is based on the idea of replenishing aquifers with potable water through boreholes, then using those same boreholes to retrieve stored water during periods of drought or high demand. These kinds of projects work by uprooting the natural groundwater, which essentially forms an underground reservoir of water that is almost fit for human consumption. For supply purposes, the amount of water recovered from the aquifer is typically equal to or less than the amount injected (Eastwood and Stanfield, 2001). Thus, it is thought that ASR schemes provide a sustainable way to develop groundwater (Figure 10.5).Water quality parameters are typically the basis for determining the amount of water that can be recovered from an ASR scheme. The volume of water recovered after the water quality parameter is met divided by the volume of water injected is used to compute recovery efficiency. Recovery efficiencies approaching 100% are not unusual. Pyne (1995) provides comprehensive information on the development of the ASR program, system design, and both technical and non-technical issues.

More than 20 ASR schemes were in operation in the US by the middle of the 1990s, and many more were in the works. The reason ASR is so popular is that by injecting high-quality water into the native aquifer, improvements in water quality are achieved. This lower cost of using these aquifers for supply makes ASR more advantageous than other resource options. Using ASR technology as part of the Everglades Restoration Project in South Florida is one ambitious plan (United States Geological Survey, 2002). With a combined capacity exceeding 6×10^6 m^3 day^{-1}, between 300 and 330 ASR boreholes are planned. These boreholes will store freshwater in the deep brackish Upper Floridan Aquifer (the ASR zone) by using surface water that is currently discharged to the tide during the wet season. In order to continue providing the Everglades with a sufficient supply of high-quality water, the stored water will be retrieved during dry spells. The compatibility of the injected water with the aquifer water, the impact of large volumes of injected water on the confining unit, the efficiency of how much water will be recovered, and the environmental effects of the recovered water are the main uncertainties associated with this first-ever large-scale implementation of ASR. In response to these worries, a phased implementation strategy for ASR is being used to assess its viability and efficacy as a regional water storage alternative.

In the UK, attempts at ASR have yielded varying degrees of success. ASR was fully tested on a confined Chalk aquifer in the southern English county of Dorset, and the results showed that there were no negative environmental effects; in any case, the effects are absorbed by aquifer storage regardless of the injection and recovery scenario. However, disappointingly low recovery efficiencies (<15%) were obtained at a fluoride concentration of 2 mg L^{-1}, which is equal to half of the background groundwater fluoride concentration (Eastwood and Stanfield, 2001). Unlike the Dorset scheme, ASR has been effectively implemented in North London's confined Chalk and Tertiary Basal Sands aquifer.

10.1.5. Riverbank Filtration Schemes

Due to their relative ease of shallow groundwater exploitation, generally high production capacity, and close proximity to demand areas, alluvial aquifers hydraulically connected to a water course are preferred sites for producing drinking water in many countries worldwide (Doussan et al., 1997). While being close to a river can guarantee much higher pumping and recharge rates, riverbank well-field exploitation may result in water quality issues (Bertin and Bourg, 1994). Despite these issues, 50% of the potable supplies in the Slovak Republic, 45% in Hungary, 16% in Germany, and 5% in the Netherlands come from groundwater derived from seeping river water. In Germany, riverbank filtration is the main source of drinking water in Düsseldorf, which is located on the Rhine, and provides 75% of the water supply to the City of Berlin.The water supply industry in the United States has embraced the widely defined regulatory concept of "groundwater under the direct influence" (GWUDI) of surface water. Each State, Tribe, or other regulatory agent will define and implement GWUDI differently based on local conditions. This category of groundwater sources is thought to be susceptible to contamination by pathogens that are found in surface water, particularly pathogenic protozoa that are resistant to disinfection, like Cryptosporidium. The US Environmental Protection Agency's Proposed Long Term 2 Enhanced Surface Water Treatment Rule (LT2ESWTR) contains current efforts to regulate riverbank filtration, with a focus on the elimination of microbial pathogens (2001).

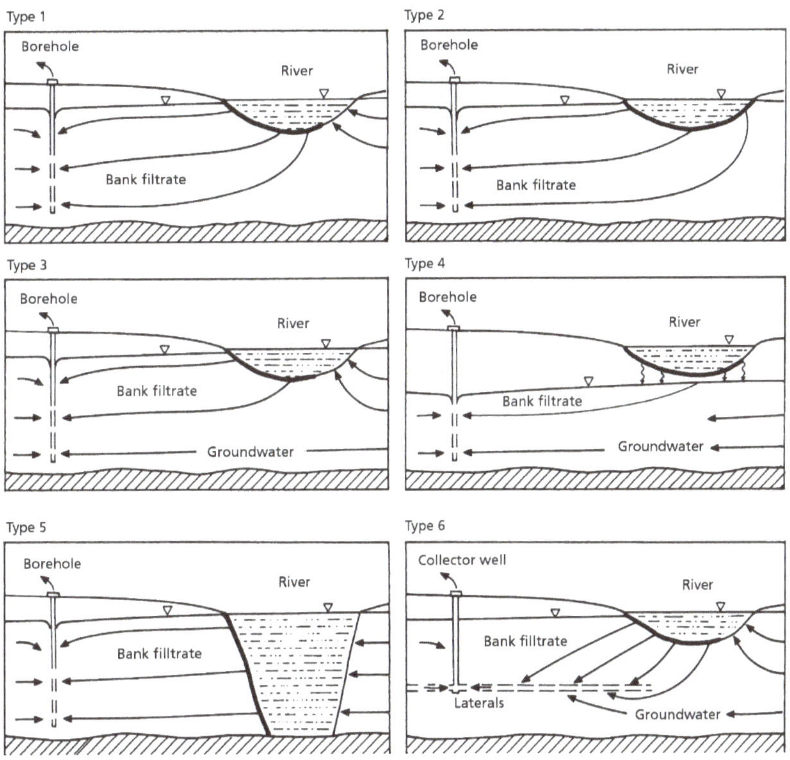

Figure 10.6. *Schematic representation of types of flow conditions at riverbank filtration sites. The majority of riverbank filtration schemes are of Type 1. Groundwater flow beneath the river (Types 3, 4 and 6) is typically neglected at most sites. The formation of unsaturated conditions beneath the river occurs if groundwater abstraction rates are not adapted to the hydraulic conductivity of the river bed or if the hydraulic conductivity of the river bed material becomes clogged due to surface water pollution inputs (Type 4). At some sites, the river bed cuts into the confining layer (Type 5). Collector wells are used with laterals at different depths, of different lengths and directions. Type 6 gives only one example with a lateral towards the river.*

Source. https://www.geokniga.org/bookfiles/geokniga-hydrogeologyprinciplesandpractice.pdf.

Riverbank filtration can happen naturally or be artificially induced by drawing water from nearby boreholes to lower the groundwater table below the surface water level. Figure 10.6 displays typical flow conditions linked to various riverbank filtration scheme types.

The knowledge of the catchment zones, infiltration zones, mixing proportions in the pumped raw water, flowpaths, and flow velocities of the bank filtrate is necessary for the quantitative and qualitative management of riverbank filtration systems.

The description of flow conditions during riverbank filtration typically relies on hydrogeological modeling and the interpretation of waterlevel measurements. The development of the colmation layer at the interface of surface and groundwater is a significant factor. Because of clogging from the input and precipitation of sediment particles, microorganisms, and colloids, as well as the precipitation of iron and manganese oxyhydroxides, calcium carbonates, and gas bubbles, this layer has a decreased hydraulic conductivity. According to Schubert (2002), permeability of clogged areas is not constant and varies with dynamic hydrology, especially after periods of flooding. Thus, a major determinant of the volume of bank filtrate is the hydraulic conductivity of the river bed.

The natural attenuation processes of riverbank filtration, as opposed to conventional surface water abstraction, can offer the following benefits: partial removal of adsorbable compounds; equilibration of temperature fluctuations and concentrations of dissolved constituents in the bank filtrate; and elimination of suspended solids, particles, biodegradable compounds, bacteria, viruses, and parasites. The formation of hydrogen sulfide and other sulfur compounds with an unpleasant odor due to shifting redox conditions is one of the undesirable consequences of riverbank filtration on water quality, along with increases in hardness, ammonium, and dissolved iron and manganese concentrations.

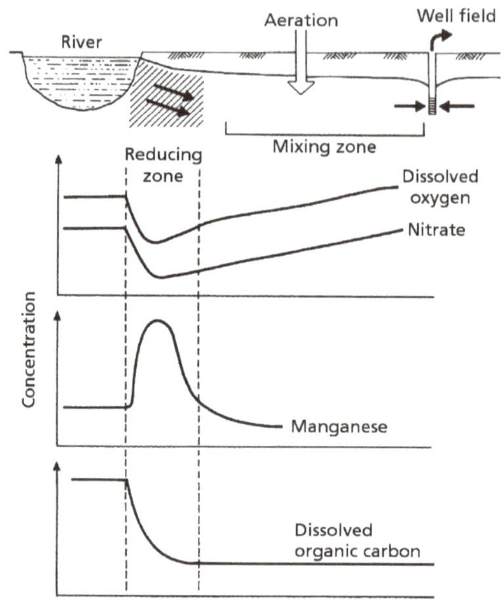

Figure 10.7. *Schematic diagram depicting the evolution of dissolved oxygen, nitrate, dissolved manganese and dissolved organic carbon along a flowpath during riverbank filtration. Dissolved oxygen becomes significantly depleted in the river bed sediments after a few metres of infiltration. Under these anoxic conditions, the microbial activity of denitrifying bacteria further decreases the groundwater redox potential leading to mobilization of the surface coatings of manganese and iron oxy-hydroxides causing a significant reduction in water quality. With further distance from the river bed, microbial activity decreases as a result of a decline in available electron donors. If the groundwater abstraction is non-continuous, or there are strong fluctuations in the river water level, a zone could potentially develop near the well where riverbank filtrate or groundwater flow temporarily results in fluctuations in microbial activity and redox conditions that can affect manganese and iron reduction and precipitation.*

Source. https://www.geokniga.org/bookfiles/geokniga-hydrogeologyprinciplesandpractice.pdf.

Within two main zones—the biologically active colmation layer, where intensive degradation and adsorption processes occur within a short residence time—and along the main flowpath between the river and abstraction borehole, where degradation rates and sorption capacities are lower and mixing processes greater—come together to produce the beneficial attenuation processes. Production wells are typically located more than 50 meters from the river or lake bank, and travel times typically range from 20 to 300 days (Grischek et al., 2002). (Figure 10.7).

At the location of a riverbank filtration system in southwest Ohio, travel times based on specific conductance measurements varied from roughly 20 hours to three months (Sheets et al., 2002), with shorter and more reliable travel times achieved when pumping continuously.

The quality of river water and consequently bank filtrate can be negatively impacted by diffuse sources of pollution in catchment runoff, particularly from agricultural activities. Grischek et al., (1998) study provided evidence of the possibility of denitrification at an aquifer site consisting of sand and gravel on the River Elbe in eastern Germany. The aquifer serves as an electron donor for both dissolved and solid organic carbon. Verstraeten et al., (2002) examined the concentrations of triazine and acetamide herbicides at a Nebraska well field on the River Platte. It found that the parent compounds were reduced by 76% of the river water value (with riverbank filtration accounting for 33% of this reduction), but that following ozonation treatment and riverbank filtration, there were increases in the concentrations of particular metabolite compounds.

A class of pollutants known as polar organic molecules, which include certain pharmaceutically active compounds (PhACs) that are virtually unaltered when released from municipal sewage treatment plants, are becoming a greater concern for riverbank filtration systems. Heberer (2002) detailed monitoring studies conducted in Berlin where PhACs, including ibuprofen, diclofenac, and clofibric acid (a blood lipid regulator used in human medical care), were found in groundwater samples from aquifers close to contaminated water courses at individual concentrations of up to several $\mu g\ L^{-1}$. Thus, it is evident that a number of drug residues are not completely removed during subsurface recharge (Heberer, 2002).

Further research is needed to quantify chemical reaction rates and microbial degradation, particularly in the river bed, as well as to examine the significance and behavior of biofilms, the fate of persistent compounds like PhACs and micropollutants, the mobility, adsorption, and inactivation of viruses, pathogens, and protozoa, and the effects of pH and redox controls. To guarantee the long-term viability of riverbank filtration systems, risk quantification techniques, such as the identification of pertinent compounds and metabolites, and suitable alarm systems, must be developed (Hiscock and Grischek, 2002).

10.1.6. Horizontal Well Schemes

It has long been known that drilling horizontal tunnels (adits) that radiate outward from a well or borehole shaft below the water table can boost well yields. The Chalk region of southeast England, Belgium, and the Netherlands is home to numerous sizable groundwater sources that are connected to adit systems, which are generally 1.2 m wide and 1.8 m high. There are 13.6 km of adits in Brighton, which is located on England's south coast, and 18 km in east London. In the Chalk aquifer, the adits were typically driven to cross the main fracture or fissure directions (Downing et al., 1993). An adit's groundwater flow can be either piped or open channeled. In south-east England's chalk, adits are often full of water under pressure. Under these circumstances, Darcy's law (see Eq. 5) is not appropriate, and flow in aquifer–adit systems must be modeled using different techniques (Zhang and Lerner, 2000).

Horizontal and slanted wells have been studied for a variety of hydrogeological scenarios more recently, thanks to advancements in drilling technology (Chen et al., 2003; Park and Zhan, 2003), as well as for use in environmental applications like contaminated aquifer vapor extraction (Plummer et al., 1997; Zhan and Park, 2002). Screened sections of horizontal wells can be arranged parallel to the direction of horizontal flow. The interception of vertical components of groundwater flow, improved control over the dynamics of the water table, improved contact between well screens and horizontal aquifer units, and ease of drilling operations near infrastructure (airport runways, roads, buildings, etc.) are just a few benefits of these wells; and the potential installation of lengthy screen segments in aquifers with restricted thickness.

10.2. GROUNDWATER ABSTRACTION AND RIVER FLOWS

The relationship between river flows and groundwater is essential to maintaining the riparian ecosystem, but it is also one of the more unpredictable hydrogeological scenarios. The intricate relationship between rivers and groundwater as well as the unpredictability of the hydraulic connection at a specific site are the causes of this challenge. Therefore, it is technically difficult to decide how groundwater abstraction will affect rivers.

The degree of connection between a river and an aquifer, which is determined by the material properties of the river bed and river bank sediments, as well as the extent to which the river's channel intersects the saturated portion of the aquifer, are significant determinants of the water flux between these bodies of water. There are three main categories of hydrogeological conditions that result in flow between a river and an aquifer.

The retention of water in the floodplain deposits around a river can have a significant impact on the rate of change in river flow as well as the attenuation of short-term fluctuations in river flow, such as a flood event. The additional water saturates the alluvial sediments and fills the available bank storage as the river stage and groundwater level rise. This bank storage is released as the water levels drop, which may help to temporarily lessen the negative effects of nearby abstractions on the river flow.

In the long term, and usually within one or two years for boreholes a few hundred metres from a river, groundwater abstraction will deplete the river flow at a rate equal to the pumping rate. The river flow depletion consists of two components:

1. interception of flow that would otherwise reach the river;
2. induced recharge from the river.

Pumping generally results in a gradual reduction of river flow, which will accelerate the closer the abstraction point is to the river. The transmissivity and storage coefficient of the aquifer are additional factors that affect the degree of depletion. The River Colne valley, located north of London, provides a striking example of how groundwater abstraction affects river flows. Direct groundwater abstractions from the underlying Chalk aquifer provided the substantial growth in residential areas here and elsewhere on the outskirts of London with water during the first half of the 20th century. Several rivers were significantly impacted by this quick development, most notably the Rivers Ver and Misbourne in the Chiltern Hills, which are tributaries of the River Colne. (Figure 10.8).

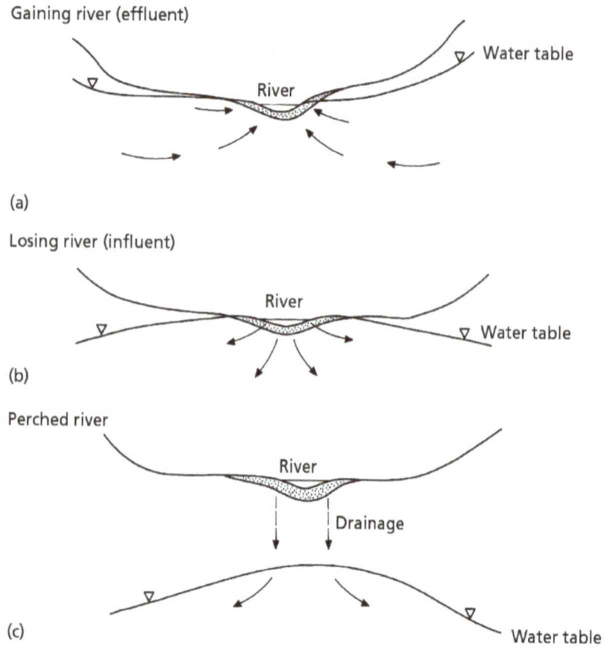

Figure 10.8. *Three hydrogeological situations that represent flow between an aquifer and a river. (a) The water table in the aquifer is above the river stage and there is potential for flow from the aquifer to the gaining river (effluent condition) with the flux generally proportional to the difference between the elevations of the water table and river stage. (b) The water table is below the river stage and the losing river (influent condition) potentially loses water to the aquifer with the flux also generally proportional to the difference between the elevations of the river stage and water table. (c) A common situation is shown in which a partially penetrating river (where the saturated aquifer extends beneath the river) experiences a declining water table below the base of the river. In this situation, water will drain from the perched river under gravity with a unit head gradient. The unit head gradient creates a limiting infiltration rate such that river losses will not increase as the water table falls further. In each case shown, the nature of the river–aquifer interaction will also depend on the properties of the river bed sediments. Sediments with very low permeability can result in a significant resistance to flow.*

Source. https://www.geokniga.org/bookfiles/geokniga-hydrogeologyprinciplesandpractice.pdf.

Approximately 75% of the Chalk's average yearly recharge in the River Ver catchment is devoted to licensed groundwater abstraction, which is currently nearly fully utilized (Owen, 1991). It is effectively lost when about half of the abstracted water is exported to supply areas outside of the catchment. The wastewater treated at sewage treatment plants typically finds its way to the lower reaches of the River Colne valley, where it cannot sustain river flows further upstream in the catchment. This wastewater is then used elsewhere in the catchment.

These springs at the source of the perennial rivers have dried up as a result of the high demand for water, particularly from those boreholes located near the head of the River Colne valley. Regarding the River Ver, the upper ten kilometers of what was once a perennial or sporadic river is now typically dry. This once-typical Chalk stream has undergone significant environmental degradation, with major changes to riparian habitat, the loss of naturally sustained fisheries, the loss of watercress farming, and reductions in the river's overall amenity value (Owen, 1991). Further downstream, the remaining perennial section experiences much reduced flows.

The aforementioned case highlights the necessity of forecasting possible environmental effects and managing catchment water resources with caution. However, it is frequently challenging to assess the effects of abstractions on rivers with any degree of confidence without devoting a substantial amount of time to field research and numerical modeling.

Accurate data availability is a common limitation, especially when it comes to the physical characteristics of the river bed and the sediments that line its banks.

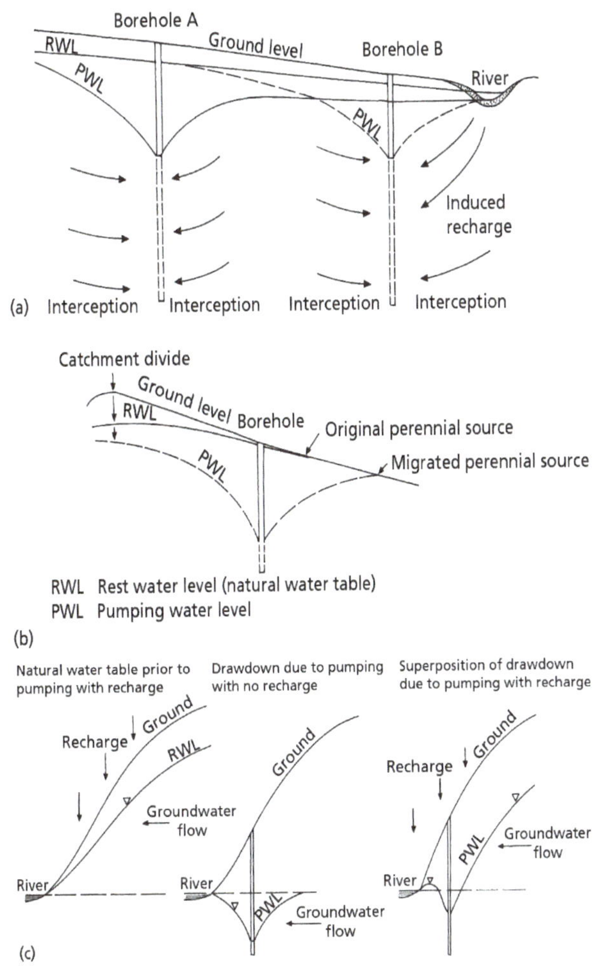

Figure 10.9. *Cross-sections to illustrate the impacts of groundwater abstraction on a groundwater-fed river. In (a) river flow is depleted by the effects of interception and induced recharge and in (b) the perennial source of the river migrates downstream in response to a drawdown in the water table (after Owen, 1991). In (c) the principle of superposition is illustrated as applied to abstraction close to a river.*

Source. https://www.geokniga.org/bookfiles/geokniga-hydrogeologyprinciplesandpractice.pdf.

An additional strategy is to use an analytical solution in the hydrogeological assessment of river flow depletion brought on by groundwater abstraction, as will be covered in the next section (Figure 10.9).

10.2.1. Analytical Solutions of River Flow Depletion

Analytical solutions reduce the complexity of the aquifer's properties by frequently giving the transmissivity and storage parameters a single representative value.

It is also implied that, when the abstraction point is not in play, the aquifer can be described by a single groundwater level value. A cone of depression is formed around the well or borehole as a result of abstraction.

This drawdown is made simpler by using analytical solutions and idealized radial flow assumptions to determine the drawdown zone's shape, which can then be superimposed on the head distribution resulting from the system's natural behavior.

The classic papers by Jenkins (1968, 1970) discuss the application of analytical solutions for the evaluation of stream flow depletion caused by pumping.

Kirk and Herbert's (2002) review of analytical solutions is more recent. Under the following assumptions, Jenkins's (1968) method allows one to estimate the total depletion of stream flow as a function of time caused by nearby abstraction (Figure 10.10).

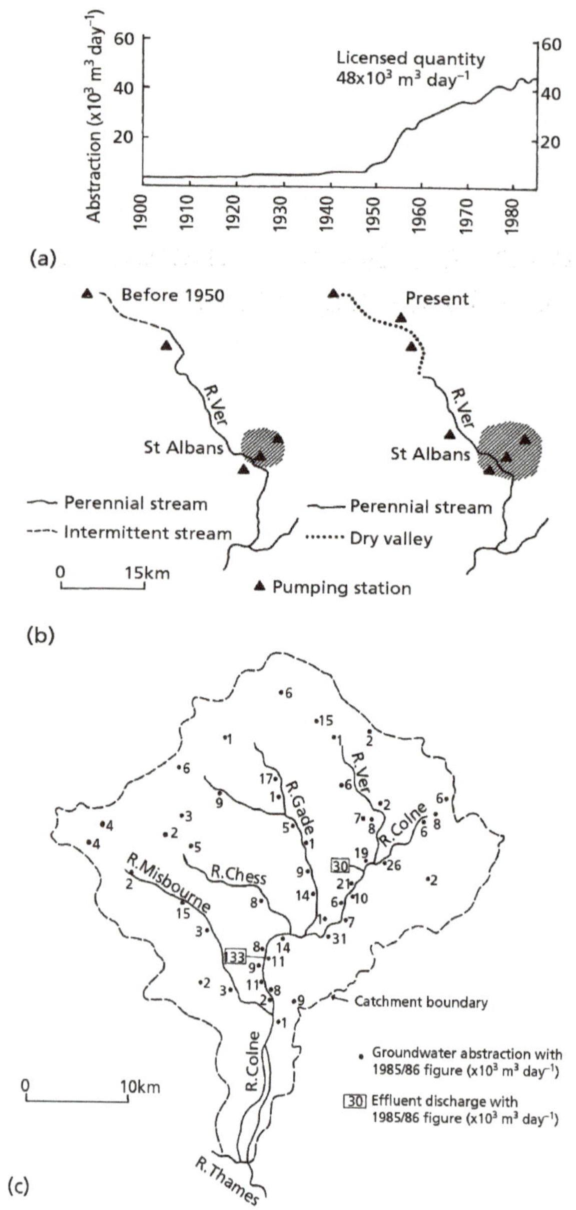

Figure 10.10. *The history of Chalk groundwater abstractions in the River Ver catchment, north London, and the impacts on river flow including: (a) a graph of annual groundwater abstractions; (b) a sketch of the reduction in length of the perennial section of the River Ver; (c) a location map of the River Ver tributary in the Colne catchment.*

Source. https://www.geokniga.org/bookfiles/geokniga-hydrogeologyprinciplesandpractice.pdf.

1. The aquifer is isotropic, homogeneous, semiinfinite in areal extent and bounded by an infinite, straight, fully penetrating stream;
2. Water is released instantaneously from aquifer storage;
3. The borehole or well fully penetrates the aquifer;
4. The pumping rate is steady;
5. The residual effects of previous pumping are negligible.

The model additionally presupposes that the aquifer is contained or, in the event that it is not, that the water-table drawdown is minimal in relation to the thickness of the saturated aquifer (i.e., the transmissivity stays constant). It is believed that the stream's temperature is steady and coincides with the groundwater's temperature.

The mathematical solution to the problem shown in Figure 10.11 gives the rate of stream depletion as a proportion of the groundwater abstraction rate as follows:

$$\frac{q}{Q} = \text{erfc}\left(\frac{1}{2\tau}\right) \tag{3}$$

where τ is a dimensionless length scale for the system given by:

$$\tau = \sqrt{\frac{tT}{L^2 S}} \tag{4}$$

where T is the aquifer transmissivity, S is the aquifer storage coefficient (specific yield for unconfined aquifer approximations), L is the perpendicular distance of the borehole or well to the line of the river, Q is the abstraction rate of the borehole or well, q is the rate of stream flow depletion, t is time and erfc is the complementary error function. Similarly, the volume of stream depletion, v, as a proportion of the groundwater abstraction rate is given by:

$$\frac{v}{Qt} = 4i^2\text{erfc}\left(\sqrt{\frac{1}{2\tau}}\right) \quad (5)$$

where v is the volume of stream depletion during time, t, and i^2 erfc is the second repeated integral of the error function.

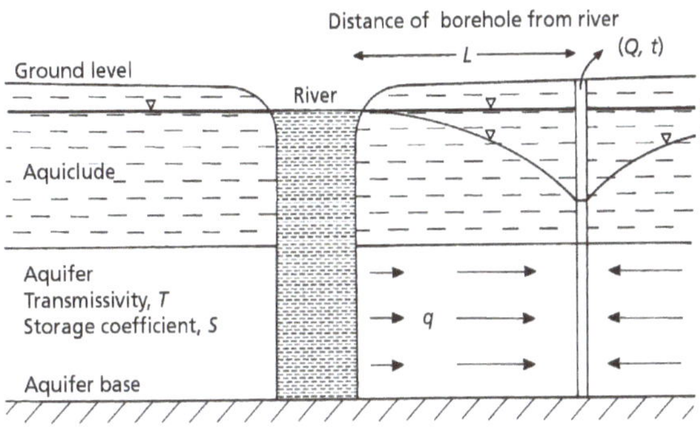

Figure 10.11. *Cross-section, drawn perpendicular to the river, showing the idealized conceptual model of river–aquifer interaction for a confined aquifer. Prior to pumping the initial potentiometric surface (shown as line RWL, the rest water level) is horizontal and equal to the constant water level of the stream. The drawdown due to pumping is shown as line PWL, the pumping water level.*

Source. https://www.geokniga.org/bookfiles/geokniga-hydrogeologyprinciplesandpractice.pdf.

While the solutions to Eqs. (3) and (5) oversimplify the situation, analytical results can yield approximations of the local effects of abstraction on river flow and the durations over which flow depletion takes place. The impact of pumping on stream flow is overestimated, and the amount of time between the start of abstraction and the impact of pumping on stream flow is underestimated, due to the assumption of complete aquifer penetration and the neglect of river bed and bank sediments. However, Wilson (1993, 1994) presented steady-state analytical solutions for two-dimensional, vertically integrated models of induced infiltration from surface water bodies for various combinations of aquifer geometry in the presence of ambient aquifer flow. This suggests that the additional assumption that the initial potentiometric surface is horizontal and equal to the constant water level of the stream prior to pumping is actually not a major limitation as per Kirk and Herbert (2002).

Wallace et al., (1990) expanded on Jenkins' (1968) methodology to demonstrate that pumping impacts could occur over the course of multiple annual cycles. The maximum impact in subsequent years may be greater than the maximum depletion in the first year in situations where the stream depletion impacts develop over long timescales because of the distance between the borehole and river, the type of aquifer properties, or the potential role of river deposits. These delayed effects could be a crucial factor in catchment management to prevent low river flows in the future.

Butler et al., (2001) introduce a novel analytical method to estimate drawdown and stream depletion in natural systems with a finite-width stream of shallow penetration next to a limited lateral extent aquifer. The study challenges the traditional belief of a fully penetrating stream, showing that this assumption can lead to significant overestimation of stream depletion, sometimes exceeding 100% in certain scenarios.

The new solution emphasizes that the stream channel's penetration should be considered negligible compared to the aquifer thickness, though an approximate adjustment allows for accurate estimations in most natural systems with relative penetrations up to 85%. Dimensionless type curves and tables make it easy to calculate stream depletion due to groundwater abstraction, which aids in the practical application of analytical solutions. Jenkins (1968) provided several worked examples, such as calculations of the water volume induced by pumping, the rate of stream depletion for pumping and non-pumping periods, and the effects (both volume and rate of stream depletion) of any chosen pattern of intermittent pumping.

10.2.2. Catchment Resource Modelling of River Flow Depletion

The effects of groundwater abstraction on river flows at the local scale can be examined using the analytical solutions mentioned above or, if data availability allows, numerical models.

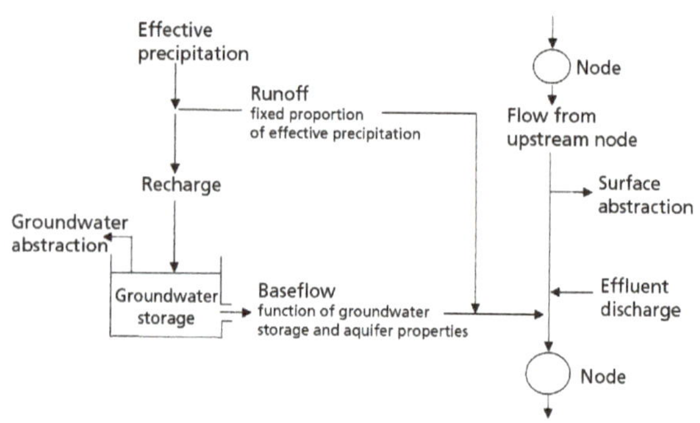

Figure 10.12. *Diagram showing the type of lumped calculations performed for a single river reach within a catchment resource model.*

Source. https://www.geokniga.org/bookfiles/geokniga-hydrogeologyprinciplesandpractice.pdf.

Naturalizing the river flow record is an additional quantitative technique to mitigate the impact of abstractions and discharges from catchments. In order to evaluate the environmental effects of catchment water resources management, it is then convenient to present and compare the derived (naturalized) and actual (historic) river flow records as flow duration curves. (Figure 10.12).

With a basic numerical catchment resource model, one can compute the derivation of a naturalized flow duration curve and simulate the interaction between rivers and aquifers. As an instance, Hiscock et al., (2001) created a resource model to evaluate how groundwater and surface water abstractions affect the flows of lowland rivers in eastern England. The naturalization of measured river flows was made possible by the resource model, which simulated river flows from the fundamental elements of baseflow (aquifer discharge) and surface flows and included the net effect of surface water and groundwater abstractions. Before the river flow record was naturalized, historical flows were used to calibrate the resource model. Using projected mean weekly flows, the model's primary output was the creation of flow duration curves at specific river locations.

One example of a lumped model that relies on adding up all of the inputs into and outputs from specific river reaches inside a catchment or subcatchment is the catchment resource model. The total of surface runoff, baseflow, abstractions, and effluent returns evenly distributed over the reach's length is the inflow to a single reach during a selected time step. The underlying aquifer in each catchment is modeled as a single storage cell that recharges and subtracts abstractions and baseflow from. Hiscock et al., employed their resource model. (2001) used the 95%ile 7-day flows—flows that are equal to or greater than the target 95% of the time—as the calibration target for lowland rivers in eastern England and produced reasonable results. The adoption of a single river reach (without any in-channel storage) and a single aquifer storage cell for the Chalk aquifer beneath it, along with a basic representation of runoff that was not dependent on antecedent soil moisture conditions, naturally led to predictive errors.

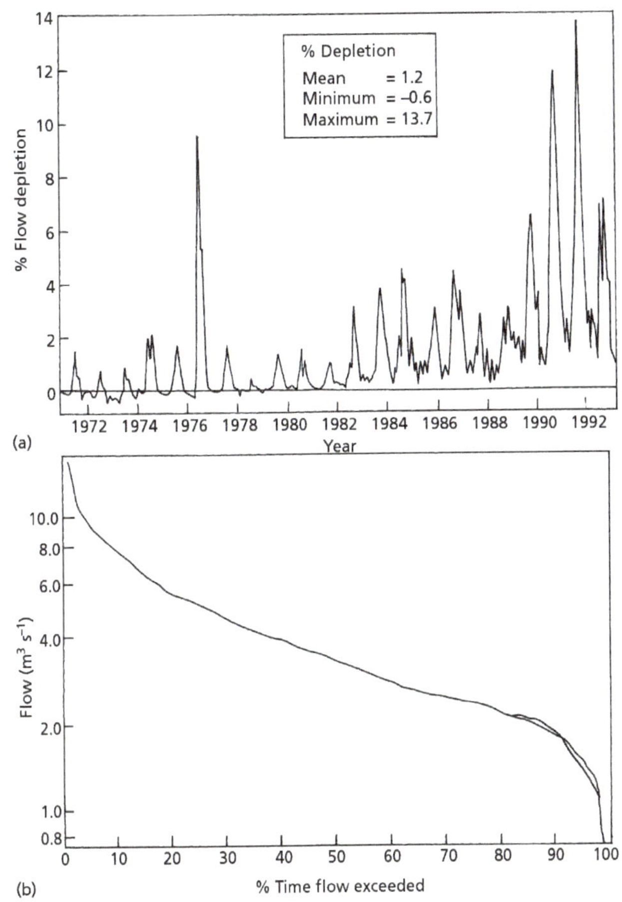

Figure 10.13. *Catchment resource model results illustrating the predicted effects of water resources management on flows in the River Wensum at Costessey Mill in Norfolk, eastern England. (a) The percentage depletion of mean weekly flows are shown due to the net effect of surface water and groundwater abstractions and discharges for the period 1971–1992. A negative percentage depletion indicates a gain in flow. (b) The 7-day flow duration curves are shown for both the naturalized and observed flows for the period 1971–1992. The upper curve shows natural flows, the lower curve observed flows.*

Source. https://www.geokniga.org/bookfiles/geokniga-hydrogeologyprinciplesandpractice.pdf.

The idea of catchment averaging makes modeling aquifer behavior—which assigns the effects of groundwater abstractions to baseflow output predictions—much simpler. Net aquifer storage is determined by this method. Baseflow and storage within the cell are related by the empirical relationship T/L^2 S, which is governed by the aquifer transmissivity, T, storage coefficient, S, and the distance, L, between the river and catchment boundary. The aquifer area is approximated to an equivalent rectangular area. Groundwater abstractions are presumed to be uniformly distributed throughout the entire aquifer area, similar to river abstractions.

River flows can be made naturally by using a catchment resource model and setting all net abstractions to zero. A sequence of net effects of abstractions and discharges are then produced by the arithmetic difference between the historical net abstractions and the nil net abstractions.

A sample is displayed in Figure 10.13a shows the percentage depletion of mean weekly flows for the River Wensum in Norfolk. During the drought of 1989–1992, the mean weekly flows reached their predicted maximum depletion of 14%. Within Figure 10.13b Flow duration curves represent the modeled historic and naturalized surface flow characteristics. Once more, in the case of the River Wensum, net abstractions have minimal effects; rather, the higher percentage of flow depletion happens during extremely low river flows.

The impact of increasing spray irrigation to the total licensed amount from surface water and groundwater sources—an important water use in the primarily agricultural area of eastern England—was also predicted using the catchment resource model. The maximum potential depletion of the River Wensum's long-term (1971–1992) 95%ile 7-day flow is estimated to be 17% under this scenario. The estimated flow depletion is roughly 30% of the mean, as opposed to flows during the drought of 1989–1992, when mean weekly flows in the river dropped to $< 0.8\ m^3\ s^{-1}$. The catchment resource model highlights spray irrigation as a way to show how managing water resources in eastern England will become more difficult in the future due to climate change. The anticipated warmer and drier summers in this region are predicted to put farmers in direct conflict with other water users and the need for water for environmental protection unless they adjust. This will increase the demand for irrigation water.

10.3. WETLAND HYDROGEOLOGY

With an estimated average annual production of 1.125 kg C m^{-2} a^{-1}, wetlands are an incredibly productive part of the landscape when compared to other ecosystems. Their estimated global extent is between 7 and 8 × 10^6 km^2 (Mitsch et al., 1994). Wetlands play a significant role in the landscape's nutrient recycling, species conservation, and plant and animal harvesting due to their comparatively high productivity and biological diversity. Inland water habitats are more varied in their physical and chemical properties, despite being far smaller in area than marine habitats. Wetlands are protected by national and international agreements and laws because of their numerous and highly visible bird species. The Moremi Game Reserve in Botswana's Okavango Delta, the Camargue National Reserve in France, the Keoladeo (Bharatpur) National Park in India, the Doñana National Park in Spain, and the Everglades National Park in the United States are a few notable wetland protected areas (Groombridge and Jenkins, 1990). The Ramsar Convention, also known as the 1975 Convention of Wetlands of International Importance especially as Water-fowl Habitat (Navid, 1989), is an international convention that is specifically dedicated to inland water ecosystems. This makes inland water ecosystems unique. There are two types of inland water habitats: lentic systems, which are lakes and ponds, and lotic systems, which are rivers. According to Groombridge and Jenkins (2000), wetlands are generally diverse habitats with permanent or seasonal shallow water that are dominated by large aquatic plants and divided into a variety of microhabitats that are situated in areas that transition between terrestrial and aquatic habitats. Bogs, fens, marshes, and swamps are the four main types of wetland habitat. Bogs are moist wetlands where organic matter has accumulated over time to produce peat. The only source of water and nutrients is precipitation. Bogs are generally nutrient-poor, acidic environments where Sphagnum moss predominates. Fens are wetlands that produce peat and are usually fed by mineral-rich groundwater. They are also impacted by nutrients from the soil that move through the system. The main vegetation consists of mosses and sedges. Marshes are flooded areas with emergent herbaceous vegetation, usually consisting of reeds, sedges, and grasses. These plants can be seasonal or permanent, and they can be fed by rivers, groundwater, or both. Freshwater wetlands with trees that are covered in water and have little to no peat accumulation are known as swamps. Swamps can be seasonal or permanent, just like marshes.

Wetlands have been classified in a number of ways, and different subdivisions have been identified based on general characteristics like the type of substratum, base status, nutrient status, and water source, as well as water level and successional stage. Wheeler and Proctor (2000) have reviewed the evolution of the primary wetland habitat categories and terms in relation to the primary ecological gradients. Other methods include classifying areas hydrologically and hydrogeologically according to flowpaths, water-level fluctuations, and principal water inflows and outflows (Lloyd et al., Hydromorphological (or hydrotopographical) classification based on the wetland's shape and location in relation to apparent water sources (Goode, 1977; Gilvear and McInnes, 1994). Figure 10.14 depicts a simplified version of the hydrogeological classification. 10.14 to show how topography, geology, and water supply affect the upkeep of wetlands.

A shift in the parameters governing a wetland's water supply, especially one that affects the direction and volume of the flow, may have disastrous effects on the fen community. Widespread destruction of wetland habitat can occur when competing demands for water from wetlands, agriculture, and public supply conflict, as the Florida Everglades graphically illustrate.

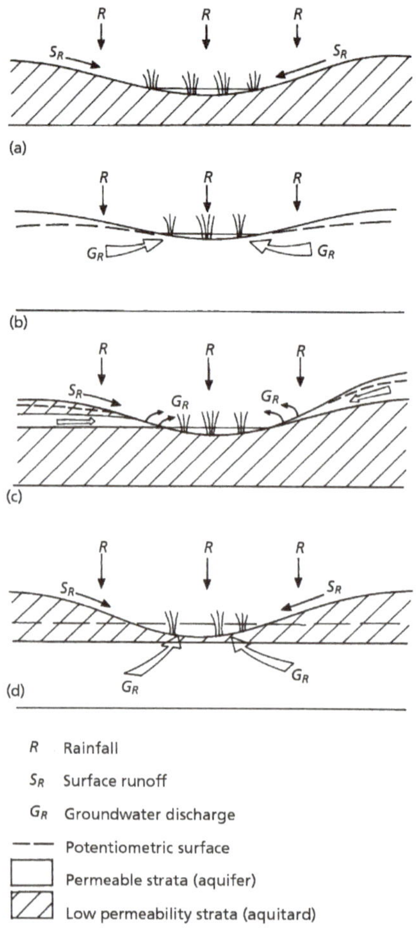

Figure 10.14. *Simple hydrogeological classification of wetland types. (a) Surface runoff is fed by rainfall and collects in a topographic hollow (for example, valley bottom, pingo or kettle hole) underlain by a low permeability layer. (b) Rainfall recharge to an unconfined aquifer supports a wetland in a region of low topography and groundwater discharge. (c) Superficial deposits, both unconfined and semiconfined, and underlain by a low permeability layer, contribute groundwater seepage in addition to surface water runoff. (d) Surface water runoff is in addition to artesian groundwater discharge from a semiconfined aquifer.*

10.4. CLIMATE CHANGE AND GROUNDWATER RESOURCES

Without a doubt, the climate on Earth is shifting. The average global surface air temperature has increased by approximately 0.6°C since the start of the 20th century, according to instrumental records extending back to 1860. Of this warming, approximately 0.4°C has occurred since the 1970s. The warmest year on record was 1998, with 2003 coming in third. According to Hulme et al., the 1990s were the warmest decade on record for the world, and the past 100 years were probably the warmest century of the previous millennium. (Hulme et al., 2002). More intense rainfall events over many mid- to high-latitude land areas in the Northern Hemisphere and a nearly global decline in the extent and ice mass of mountain glaciers are additional indicators of changes in the global climate.

Since 1900, the thermal growing season in central England has gotten one month longer, and over the past 200 years, winters in the United Kingdom have gotten wetter than summers. Furthermore, compared to fifty years ago, a greater percentage of winter precipitation across all regions now occurs on days with high rainfall. The average sea level in the United Kingdom has risen by roughly 10 cm since 1900 after accounting for natural land movements (Hulme et al., 2002).

Both natural and human-caused factors contribute to climate change. The Earth's orbit, interactions between the ocean and atmosphere, variations in incoming solar radiation, and volcanic activity all contribute to the planet's natural climate variations. The primary human cause is most likely the rise in "greenhouse" gas emissions, which include nitrous oxide, carbon dioxide, methane, and chlorofluorocarbons. Approximately 6.5×10^9 t a^{-1} of carbon are released into the atmosphere every year, primarily from the burning of fossil fuels. A further net emission of $1\text{-}2 \times 10^9$ t a^{-1} . is caused by changes in land use, such as clearing tropical rainforests. Over the past 200 years, rising concentrations of "greenhouse" gases in the atmosphere (Table 10.2) have trapped long-wave radiation that would otherwise escape into the lower atmosphere, raising global temperatures. Other atmospheric pollutants, such as sulphate aerosols, which absorb and scatter incoming solar radiation back into space, have opposite effects that work to cool the climate. Countries in the European Union are required by the Kyoto Protocol to reduce their emissions of a "basket" of six gases, including carbon dioxide, methane, and nitrous oxide, by 8% from 1990 levels by 2008–2012. This is in an effort to combat global warming.

Since the hydrological cycle is an essential component of the climate system, it participates in numerous interactions and feedback loops that contribute to the system's complexity (Askew, 1987). Over the course of the next century, climate change is predicted to intensify the global hydrological cycle and have a significant impact on local water supplies (IPCC 1998). Figures 10.15 and 10.16 provides an overview of the probable effects of climate change on natural hydrological systems. Climate change may have detrimental effects on water resources, including a reduction in the length of the precipitation season and an increase in hydrological extremes like floods and droughts.

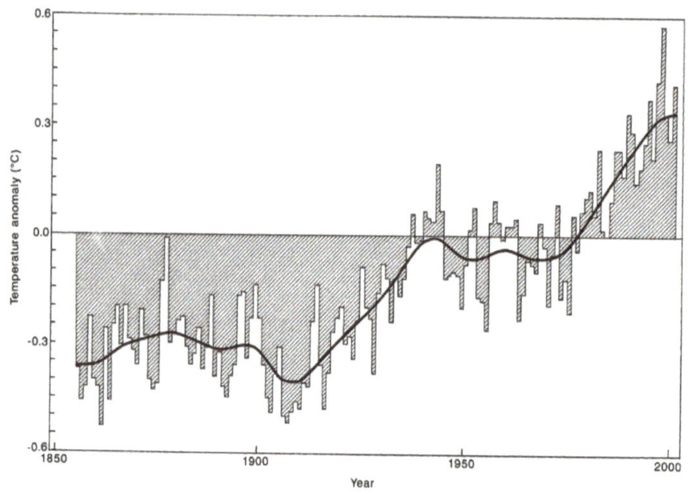

Figure 10.15. *The observed increase in globally averaged surface air temperature. Anomalies are relative to the 1961–1990 average. The smoothed curve is created using a 21-point binomial filter giving neardecadal averages. Two main warming periods are noticeable between about 1910 and 1945 and between 1976 and 2000. The rate of global warming from 1976 to 2000 was about twice as fast (but interannually more variable) than for the period 1910 to 1945.*

Source. https://www.geokniga.org/bookfiles/geokniga-hydrogeologyprinciplesandpractice.pdf.

Table 10.2. *Major Sources, Concentrations and Residence Times of Important 'Greenhouse' Gases in the Atmosphere*

Greenhouse gas	Pre-industrial concentration (1750)	Present concentration (1998)	Residence time	Annual rate of increase*	Major sources
Water vapour	3000 ppm	3000 ppm	10–15 days	n.a.	Oceans
CO_2 (carbon dioxide)	~280 ppm	~365 ppm	5–200 years†	1.5 ppm a^{-1}‡	Combustion of fossil fuels, deforestation
CH_4 (methane)	~700 ppb	1745 ppb	12 years§	7.0 ppb a^{-1}‡	Rice production, cattle rearing, industry
N_2O (nitrous oxide)	~270 ppb	314 ppb	114 years§	0.8 ppb a^{-1}	Agriculture, industry, biomass burning
CFC-11 (chlorofluorocarbon-11)	0	268 ppt	45 years	−1.4 ppt a^{-1}	Aerosols, refrigeration
HFC-23 (hydrofluorocarbon-23)	0	14 ppt	260 years	0.55 ppt a^{-1}	Industrial byproduct
CF_4 (perfluoromethane)	40 ppt	80 ppt	>50,000 years	1 ppt a^{-1}	Aluminium industry

n.a., not applicable. * Rate is calculated over the period 1990 to 1999. † No single lifetime can be defined for CO2 because of the different rates of uptake by different removal processes. ‡ Rate has fluctuated between 0.9 ppm a−1 and 2.8 ppm a−1 for CO2 and between 0 and 13 ppb a−1 for CH4 over the period 1990–1999. § This lifetime has been defined as an 'adjustment' time that takes into account the indirect effect of the gas on its own residence time.

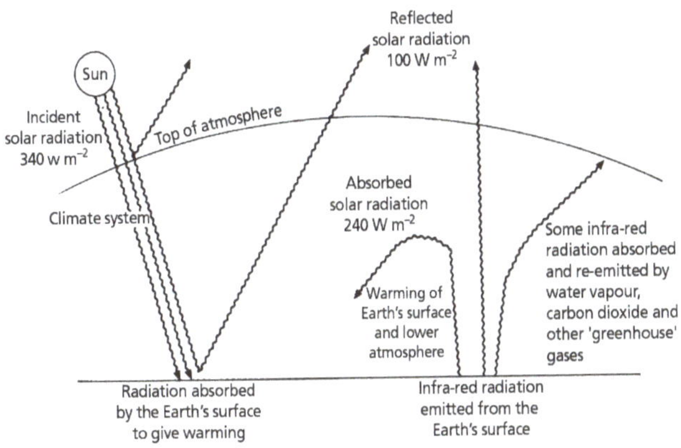

Figure 10.16. *Schematic representation of the global radiation budget. Averaged over the globe, there is 340 W m^{-2} of incident solar radiation at the top of the atmosphere. The climate system absorbs 240 W m^{-2} of solar radiation, so that under equilibrium conditions it must emit 240 W m^{-2} of infra-red radiation. The carbon dioxide radiative forcing for a doubling of carbon dioxide concentrations constitutes a reduction in the emitted infra-red radiation of 4 W m^{-2} producing a heating of the climate system known as global warming. This heating effect acts to increase the emitted radiation in order to re-establish the Earth's radiation balance.*

Source. https://www.geokniga.org/bookfiles/geokniga-hydrogeologyprinciplesandpractice.pdf.

Larger volumes of runoff over shorter time intervals would result from a shorter precipitation season, perhaps combined with heavier precipitation events and a change from snow to rainfall. There would be less runoff from spring snowmelt and more runoff from winter melting. This makes it more difficult to store and manage floodwater in a way that meets water supply goals and protects the environment. Additionally, because there is less chance of groundwater recharge in these circumstances, it might make it more difficult to use surface water and groundwater simultaneously. Water resources in tropical latitudes are unlikely to alter as a result of the anticipated effects of climate change. Changes in temperature, precipitation, and sea level rise also have an impact on the quality of water because they alter the amount of river flow and the extent of saline intrusion (Loáiciga et al., 1996).

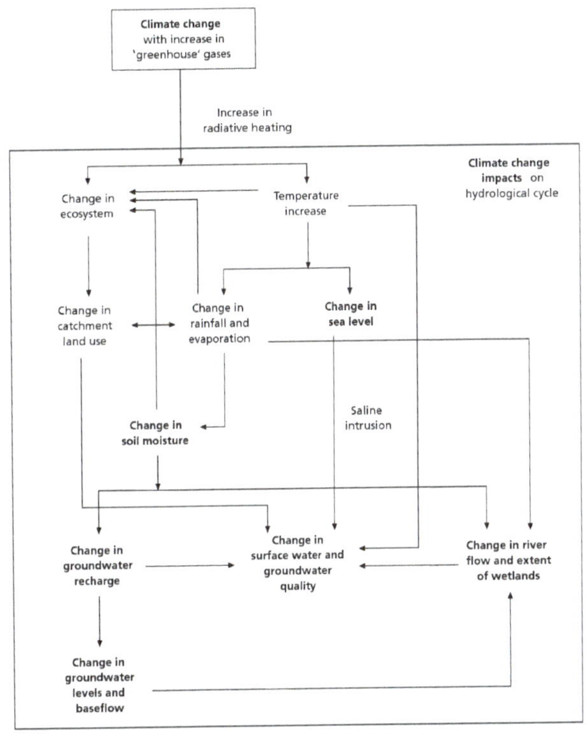

Figure 10.17. *Impacts of increasing 'greenhouse' gas concentrations on the natural hydrological cycle emphasizing changes in hydrogeological conditions.*

Source. https://www.geokniga.org/bookfiles/geokniga-hydrogeologyprinciplesandpractice.pdf.

An objective method for estimating hydrological responses to climate change is to use hydrological models, which come in different levels of complexity to represent current and future climatic conditions. The study of various effects of climate change is made possible by the coupling of physically based hydrological models to output from global circulation models (GCMs) (Conway, 1998). When addressing the scale effects that make it more difficult to predict changes in regional hydrology, caution is needed. At regional scales, runoff and precipitation exhibit significant variability, with 10- to 20-year averages typically varying within ±25% of their long-term means. Given that inherent hydrological variability rises over a shorter time average, the issue is probably worse at shorter timescales (Loáiciga et al., 1996). (Figure 10.17). Vaccaro (1992), who examined the sensitivity of groundwater recharge estimates for a semi-arid basin on the Columbia Plateau, Washington, USA, to historical and projected climatic regimes, provides an early example of hydrological simulation under climate change. Using a daily energy-soil-water balance model, recharge was estimated for predevelopment and current (1980s) land use conditions. Long sequences with parameters estimated from historically significant dry and wet subsets were simulated using a synthetic daily weather generator.

Table 10.3. *Scenarios of Climate and Groundwater Use Change Effects Considered for the Edwards Aquifer, South-Central Texas*

	Scenario			
	I Base	II Climate change effect	III Groundwater use change effect	IV Total effects
Climate (recharge)	$R_{1978-89}$*	$R_{1978-89}(2\times CO_2/1\times CO_2)$†	$R_{1978-89}$*	$R_{1978-89}(2\times CO_2/1\times CO_2)$
Groundwater use	1978–1989 use‡	1978–1989 use	2050 use§	2050 use

Source: https://www.geokniga.org/bookfiles/geokniga-hydrogeologyprinciplesandpractice.pdf.

* *Historical recharge (R) during 1978–1989 (mean = 0.949 × 109 m3 a⁻¹). † Historical recharge scaled to 2 × CO2 climate conditions. ‡ Average groundwater use between 1978 and 1989 = 0.567 × 109 m3 a⁻¹. § Groundwater use forecast for 2050 = 0.784 × 109 m3 a⁻¹.*

It was discovered that the basin's estimated recharge was susceptible to variations in historical precipitation and other climate-related variables. The estimated annual recharge showed less variability when compared to historic data when climate changes were projected by GCMs to double carbon dioxide. Furthermore, the climate scenario's median annual recharge, which was derived from three separate GCM simulation runs, was less than 75% of the historical simulation's median annual recharge. This significant shift illustrates how the climate may have an impact on groundwater recharge.

Loáiciga et al., (2000) and Loáiciga (2003) conducted a recent study to evaluate the impact of aquifer pumping on the Edwards Balcones Fault Zone aquifer system in south-central Texas using climate change scenarios derived from scaling factors from various GCMs. This karst aquifer, located in the Edwards limestone formation, is a key source of agricultural and municipal water in the US. The study identified the Edwards Aquifer as highly vulnerable to climate change, with groundwater recharge primarily coming from stream seepage and feeding major springs supporting a crucial groundwater ecosystem. Multiple pumping scenarios were combined with 2 × CO^2 climate scenarios to assess the vulnerability of these springs to pumping and climate change impacts. The results, detailed in Tables 10.3 and 10.4, show varying effects on spring flows in the Edwards Aquifer. While simulation results indicate an increase in spring flow for the Comal Springs and slight decrease for the San Marcos Springs under certain conditions, there is a substantial depletion in spring flow if both increased groundwater use and climate change are considered. The study concludes that while population growth may pose a greater threat to groundwater use in the Edwards Aquifer than climate change, prolonged droughts under climate change could intensify competition for water resources. It suggests that aquifer strategies must be adjusted to account for climate variability and change in order to protect the Edwards Aquifer's water supply. Since it is difficult to simulate the regional effects of climate change and because future emissions of "greenhouse" gases and aerosols are unknown, there is inherent uncertainty in the development of climate change scenarios. The regional climate changes derived from GCM experiments are referred to as scenarios or projections and cannot be regarded as predictions because of these uncertainties. Carter and La Rovere (2001) provide thorough reviews of climate change scenarios, covering standards and construction. The United Kingdom Climate Change Impacts Programme (UKCIP) methodology, as described by Hulme et al., (2002) is to offer four different climate change scenarios for the UK that cover a fair range of potential future temperatures. The scenarios, which relate to various scenarios of greenhouse gas emissions and changes in global temperature for the 2020s, 2050s, and 2080s, are labeled Low (L), Medium-low (ML), Medium-high (MH), and High (H).

Table 10.4. *Simulated Minimum Spring Flows in the Edwards Aquifer, South-Central Texas, for Climate and Groundwater Use Change Effects Listed in Table 10.3*

Climate change and groundwater use scenario	Edwards Aquifer springs	
	Comal	San Marcos
I	4.84	4.84
II	12.7 (+162%)	5.67 (+17%)
III	0 (−100%)	3.79 (−22%)
IV	1.31 (−73%)	4.79 (−1%)

Source. https://www.geokniga.org/bookfiles/geokniga-hydrogeologyprinciplesandpractice.pdf.

Notes: Spring flows are given in 106 m3 month−1. The numbers in parentheses represent the percentage increase (+) or decrease (−) caused by a scenario relative to the base condition (I).

Yusoff et al., (2002) conducted a study on the effects of climate change on groundwater resources in the Chalk aquifer of eastern England. They created scaling factors for changes in precipitation (P) and potential evapotranspiration (PE), comparing monthly average values for a control run of the Hadley Centre HadCM2 model with future climate change scenarios. The factors, expressed as percentage changes in P and PE, were applied to historical data to predict future groundwater recharge values. These values were used in a numerical groundwater model calibrated against historical records of groundwater levels and river baseflow. One limitation of this approach is that it maintains the same climate variability seen in the historic record in future scenarios, but it still provides a general idea of potential changes in hydrological patterns.

Yusoff et al.'s (2002) climate change impact simulations yielded the most obvious and consistent result was a decrease in groundwater recharge predicted for autumn under all scenarios due to an increase in potential evapotranspiration in the fall and a decrease in summer precipitation. These circumstances result in a 26% decrease in recharge and a 42% increase in the autumn soil moisture deficit for the 2050MH scenario. Eastern England should therefore anticipate longer, drier summers and a postponement of the autumnal and winteral groundwater recharge period. However, a modelled decrease of up to 14% in autumn baseflow volume for the 2050MH scenario suggests that chalk groundwater-fed rivers may show environmental impacts with potential conflicts with other water demands. The drier conditions will have relatively little effect on summer groundwater levels (generally a 1–2% decrease).

Younger et al., (2002) explored how climate change could impact groundwater quality and ocean salinity. They suggested that carbonate aquifers could potentially absorb or release carbon dioxide, affecting the hardness of groundwater. In their simulation, they predicted a slight increase in calcium concentrations in European carbonate aquifers by 2045, which could have minor effects on water resources. However, this research highlights the potential for carbonate aquifers worldwide to play a role in mitigating global warming over extended periods of time.

An estimated 2220 km^3 of groundwater are discharged directly from inland catchments into the world's oceans and seas. A hypothetical 10% increase in global precipitation of 2 mm a^{-1} due to greenhouse warming, according to Zektser and Loaiciga (1993), would result in an additional direct groundwater discharge of about 222 km^3 a^{-1} . Despite the small amount of additional flux, the salt load—which has a dissolved solids content of 585 mg L^{-1}—would add an additional 1.3×10^8 t a^{-1} of salts to the world's oceans and seas. The balance of water inputs to and from the oceans under an intensified hydrological cycle, as well as any volume changes in the oceans due to thermal expansion and ice melting, determine whether this additional salt load would increase the salinity of the oceans. The possibility that these effects of modifications to the direct groundwater flow may only stabilize over periods of hundreds to thousands of years should also be taken into account.

10.5. GROUNDWATER GOVERNANCE

According to the UNDP, governance is the exercise of political, economic, and administrative authority in the management of a country's affairs at all levels. It includes the institutions, procedures, and mechanisms that allow its citizens to express their concerns, settle disputes, and uphold their legal rights and obligations.

As a result, "water governance" refers to the corresponding framework for efficient management of water resources, which includes providing all water services in a way that is socially, environmentally, and economically responsible. It also includes the processes of defining policy options, converting them into goals, setting up institutions, procedures, and means, keeping an eye on things, recording what is happening, and assuming accountability for the results. The relationships, connections, and responsibilities of the relevant public, civil, professional, and private stakeholders are particularly important factors to take into account in this regard. To guarantee the sustainable and effective use of groundwater resources for the benefit of people and dependent ecosystems, "groundwater governance," which is a subset of the aforementioned, is concentrated on the exercise of proper authority and the encouragement of responsible collective action.

But since groundwater is a traditional "common pool resource," it is naturally susceptible to the so-called "tragedy of the commons," in which real and potential stakeholders only act in their own short-term self-interest, ignoring long-term community needs. Therefore, there is always a chance that they will increase (rather than moderate) their extraction and/or relax (rather than intensify) their protection in situations where there is obvious excessive exploitation leading to a falling watertable or where there is progressive quality deterioration due to inadequate protection, simply because they believe that their personal interest in the resource cannot be guaranteed through individual action. A set of guidelines must be put into practice in order to govern "common pool resources.". The characteristics of groundwater also make connections to other land and water resources and environmental governance extremely important. Additionally, as societal demands for better environmental quality and climate change adaptation increase, these demands may have a positive impact on the movement toward better groundwater resource governance.

10.5.1. The Intimate Land-Groundwater Linkage

Groundwater governance cannot be addressed in isolation from consideration of the processes determining or controlling land-use because groundwater resources are highly dependent upon land-use (and changes in land-use) in the main "aquifer recharge areas," which exert a direct influence on both the rates and quality of recharge.

Furthermore, even though a number of legislative frameworks and community standards have been developed to restrict an individual's or a group's access to groundwater, their execution frequently entails significant transaction costs and can be labor-intensive. Therefore, in certain nations, private land ownership still seems to have the greatest direct influence over groundwater access. This contributes to the idea that groundwater is a "private resource," the rights to which are exclusively tied to land ownership.

Land-use classification and control in urban environments are typically the purview of municipal or local government, and a common governance shortcoming is the lack of mechanisms by which water resource agencies can exert influence over the process. Furthermore, laws intended to address unfavorable land-use practices are frequently absent or only loosely enforced in developing countries, and the implementation of groundwater-related controls is heavily reliant on stakeholder participation and awareness.

On the other hand, national agriculture and food policy generally, as well as market incentives for deforestation and pasture land plowing, have a significant impact on rural land-use practices and the intensification of agricultural production (which also have a significant impact on groundwater recharge rates and quality).

10.5.2. Cross-Sector Drivers Impacting Groundwater

The impact of "external drivers" on the use of groundwater resources and pollution pressure must be acknowledged, and in order to improve groundwater governance, mechanisms to influence related macro-level policy decisions will probably be needed.

The more common types of 'external driver' include:

- the process of urbanization, especially given the potential 'coupling' between in-situ sanitation and groundwater, and the frequent inadequacy of utility water supplies
- political (land-jurisdiction) boundaries impeding the rational development of protected urban well fields for major water-supply and influencing the discharge points of potentially-polluting effluents
- planning and development of industrial and mining enterprises
- the development of tourism facilities where this is a major source of income.

And in agricultural areas these drivers often include factors such as:

- the unreliability or absence of alternative surface water-supplies
- highly subsidized or flat-rate electrical energy tariffs for waterwell pumping
- general subsidies on water-well construction, irrigation technology, fertilizers and pesticides, etc
- guarantee prices for certain crop types.

10.5.3. Establishing a Governance Framework for Groundwater Issues

10.5.3.1. A Logical Typology of Groundwater Issues

The experience of GW•MATE indicates that the most appropriate typology for assessing the status and requirements of groundwater governance is based on a practical classification of "groundwater bodies." This is because there is a close relationship between the status and characteristics of groundwater bodies and the issue of why and how governance needs to be strengthened in order to achieve successful implementation of agreed-upon management and protection measures on the ground. The sub-divisions should not be viewed as mutually exclusive because the typologies related to groundwater pollution and quality concerns (1B, 2B, and 3B) can coexist with the other categories of groundwater management problem described. Additionally, many of the aforementioned concerns may be involved in the use and preservation of groundwater resources in and around significant cities.

However, it might be more practical to think of the governance arrangements for urban groundwater as a distinct cross-cutting category given the unique dynamics of the urbanization process and the close relationship between the surrounding environment where groundwater bodies occur and the provision and operation of the water infrastructure. (Table 10.5).

Table 10.5. *Typology of Groundwater Bodies with Situations and Processes Involved*

OVERALL TYPOLOGY OF GROUNDWATER BODY	SUB-DIVISIONS BY TYPE OF SITUATION OR PROCESS INVOLVED #
(1) At Risk of Extensive Quasi-Irreversible Aquifer Degradation and Subject to Potential Conflict Amongst Users ###	(A) Under Intensive Exploitation (provoking land subsidence, saline or polluted water intrusion) ## (B) Vulnerable to Widespread Pollution from Land Surface (depends on aquifer vulnerability and pollutant pressure) ## (C) Undergoing Depletion of Non-Renewable Storage Reserves (normally in aquifers with low contemporary recharge)
(2) Subject to Potential Conflict Amongst Users ### but not at Risk of Quasi-Irreversible Aquifer Degradation	(A) With Growing Large-Scale Abstraction (especially in aquifers with high T/S ratio) (B) Vulnerable to Local Point-Source Pollution (depends on aquifer vulnerability and pollutant pressure) ## (C) With Shared International/Interstate Resources (latter in federal nations with decentralised water management)
(3) Insufficient (or Inadequate Use of) Scientific Knowledge to Guide Development Policy & Process	(A) But Potential to Improve Rural Welfare & Livelihoods (not fulfilling potential role in achieving MDGs) (B) With Presence of Natural Quality Problems (especially with health impacts at low concentrations/eg: As, F) ## (C) But Scope for Large-Scale Planned Conjunctive Use (either for urban water-supply or irrigated agriculture) ##

Source: https://documents1.worldbank.org/curated/zh/603871468323116801/pdf/575550Replacem1ater1Governance12010.pdf.

although covered by this typology it may be preferable in practice to treat urban groundwater situations as a separate cross-cutting category

in all these cases the intrinsic susceptibility or vulnerability to the given type of problem varies widely with aquifer type

users should be taken to include important groundwater-dependent ecosystems.

10.5.3.2. Identification and Prioritization of Needs

Each of the aforementioned concerns calls for a very different governance structure. It is noteworthy to note that only categories 1A–C pertain specifically to safeguarding the "natural resource base" from irreversible degradation (caused by certain members of society engaging in indiscriminate behavior). This must be the most urgent area for the reinforcement of governance provisions. On the other end of the spectrum, typologies 3A-C hold significant importance in terms of development. The primary concern here is whether national and/or local organizations possess the necessary funding and established authority to conduct a groundwater resource survey and subsequently advocate for the effective utilization and development of these resources to meet the UN Millennium Development Goals concerning food security and/or rural water supply.

The starting point to assess the adequacy of existing provisions and to offer specific recommendations for their improvement is careful consideration of:

- the stage of development of the groundwater body
- the typology(ies) into which the groundwater body fall(s).
- the preferred approach in terms of management measures.

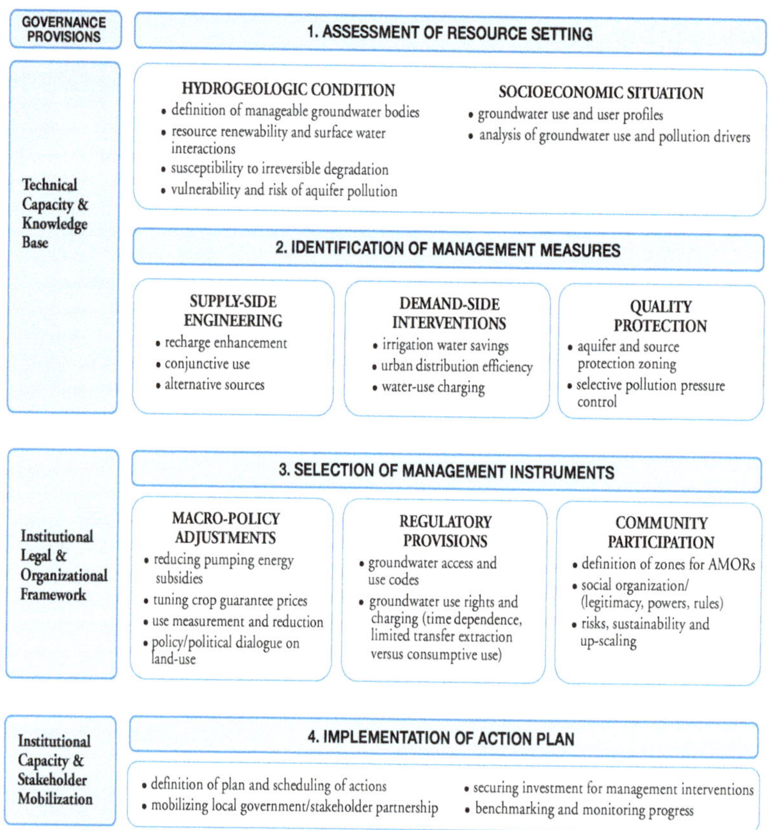

Figure 10.18. *Pragmatic framework for elaboration of an action plan for control of excessive groundwater abstraction or pollution pressure with corresponding governance provisions.*

Source. https://documents1.worldbank.org/curated/zh/603871468323116801/pdf/575550Replacem1ater1Governance12010.pdf.

In relation to specific management measures the following general pointers should prove useful when considering how they can be facilitated through an improved governance framework:

- Groundwater Resource Issues — rationalizing the abstraction from heavily-exploited aquifers, conserving the recharge area of aquifers, and improving the understanding of under-utilized aquifers
- Groundwater Supply Issues (especially for urban and rural domestic water-supply) — local source conservation measures through establishment of protection areas/ perimeters, and adequate quality monitoring for early diagnosis of potential problems.

For particular priority groundwater bodies at risk of irreversible degradation, groundwater management action plans should exist.

These plans should outline the necessary financial resources for supply-side and demand management interventions as well as a distinct institutional structure with regard to authority and responsibility. Figure 10.18 provides a practical framework for developing and implementing such a plan, and it also indicates the types of governance provisions that correspond to it. An important point to note here is that a diagnosis of the resource setting (hydrogeologic condition and socioeconomic situation) should inform the balance of management measures chosen and management instruments employed (explained in more detail below). These factors essentially define the problem and shape its solution.

It may be challenging to transition from disparate measures to a coordinated plan, even in nations where all the components necessary for sustainable groundwater management are present. One potential solution could be to designate "groundwater body management champions" to act as a source of leadership and guarantee better continuity and coordination.

CASE STUDY

GEOPHYSICAL SURVEY & GROUNDWATER RESOURCES ASSESSMENT, SOUTH SUDAN

Solidarités International engaged Groundwater Relief in November 2020 to undertake the study in Fashoda County to assess the available groundwater supply potential in the area for provision of clean drinking water to IDPs and the host communities in the area. The study location is in Upper Nile State in the north east of South Sudan. Previous hydrogeological studies of the area were extremely limited.

The work was completed by Paul Evans (Groundwater Relief) and David Ndago (Rapid Response Team) in December 2020 and comprised a hydrogeological desk study followed by a surface geophysical survey. The survey methodology was based on an integrated approach of electrical horizontal profiling, vertical electrical sounding (VES) and electrical resistivity tomography (ERT). Access to the survey sites was challenging due to flooding and lack of roads; where Landcruiser access was not possible, donkey and boat transport were required to transport the geophysical equipment. The survey was successfully completed across over a period of 17 days. 1D and 2D resistivity data analysis and inversion were undertaken to characterise the ground conditions and to develop a hydrogeological conceptual model for the area. The project findings were used to identify areas where groundwater supply development may be feasible to inform long term project planning. The findings were presented to senior members of the Ministry of Water in Juba in collaboration with Solidarités International.

CLASS ACTIVITY

To understand the importance of groundwater resources, the challenges in their governance and management, and to explore practical solutions through collaborative learning and critical thinking. By engaging in this activity, students will gain a comprehensive understanding of groundwater resources, the complexities of their governance and management, and the importance of sustainable practices. This interactive and collaborative approach will enhance their critical thinking and problem-solving skills in environmental science.

SUMMARY

- Groundwater resources, essential for global water supply, are increasingly vital in sustaining agriculture, industry, and domestic use.
- Groundwater governance involves the frameworks, policies, and institutions that regulate the use and protection of groundwater resources.
- Groundwater management, on the other hand, entails the practical implementation of policies and strategies to monitor, regulate, and preserve groundwater resources.
- The water budget indicates that the decrease in storage would have been worse if 30–40% of the water pumped for irrigation had not infiltrated to the aquifer as irrigation return flows each year.
- Full development of the Great Ouse Groundwater Scheme was envisaged to include almost 350 new boreholes distributed over an area of 2500 km^2 .
- As surface water and groundwater schemes reach full development, the final stage is artificial recharge where water, often treated wastewater, is recharged through basins and returned to the aquifer.
- Diffuse sources of contamination in catchment runoff, especially from agricultural activities, can adversely affect river water quality and therefore bank filtrate.
- The link between groundwater and river flows is fundamental to conserving the riparian environment yet is one of the more difficult hydrogeological situations to predict.
- Local-scale impacts of groundwater abstraction on river flows can be investigated with the above analytical solutions or by numerical models where data availability permits.
- Climate change is influenced by both natural and human causes. The Earth's climate varies naturally as a result of interactions between the ocean and atmosphere, changes in the Earth's orbit, fluctuations in incoming solar radiation and volcanic activity.
- The character of groundwater also means that links with the governance of the environment, and other land and water resources are highly relevant – and since societal demands for improved environmental quality and climate change adaptation are growing they could exert a positive influence on the drive for improved groundwater resource governance.

REVIEW QUESTIONS

1. What are the groundwater resources schemes? Explain each of them.
2. What is catchment resource modelling of river flow depletion?
3. What do you understand by wetland hydrogeology?
4. Discuss about groundwater governance.
5. How to establishing a governance framework for groundwater issues?

REFERENCES

1. Alcamo, J., Moreno, J. M., Nováky, B., et al. (2007). Europe. In M. L. Parry, O. F. Canziani, J. P. Palutikof, et al. (Eds.), *Climate Change 2007: Impacts, Adaptation and Vulnerability. Contribution of Working Group II to the Fourth Assessment Report of the Intergovernmental Panel on Climate Change* (pp. 541–580). Cambridge University Press.
2. Arnell, N. (1996). *Global Warming, River Flows and Water Resources*. John Wiley & Sons, Ltd.
3. Askew, A. J. (1987). Climate change and water resources. *IAHS Publications, 168*, 421–430.
4. Arnell, N. (2002). *Hydrology and Global Environmental Change*. Prentice Hall.
5. Backshall, W. F., Downing, R. A., & Law, F. M. (1972). Great Ouse groundwater study. *Water and Water Engineering, 06*, 3–11.
6. Ballentine, T. M., & Stakhiv, E. Z. (1993). *Climate Change and Water Resources Management: Proceedings of the First National Conference*. US Army Corps of Engineers Institute of Water Resources. IWR Report 93-R-17.
7. Beran, M., & Rodier, J. A. (1985). *Hydrological Aspect of Drought*. Studies and Report in Hydrology 39. UNESCO/WMO.
8. Bertin, C., & Bourg, A. C. M. (1994). Radon-222 and chloride as natural tracers of the infiltration of river water into an alluvial aquifer with significant river/groundwater mixing. *Environmental Science & Technology, 28*, 794–798.
9. Bloomfield, J. P., Gaus, I., & Wade, S. D. (2003). A method for investigating the potential impacts of climate-change scenarios on annual minimum groundwater levels. *Journal of the Chartered Institution of Water and Environmental Management, 17*, 86–91.
10. Bryant, S. J., Arnell, N. W., & Law, F. M. (1992). The long-term context for the current hydrological drought. In *Proceedings of the Institution of Water and Environmental Management Conference on the Management of Scarce Water Resources*.
11. Burgess, D. B. (2002). Groundwater resource management in eastern England: A quest for environmentally sustainable development. In K. M. Hiscock, M. O. Rivett, & R. M. Davison (Eds.), *Sustainable Groundwater Development* (pp. 53–62). Geological Society, London, Special Publications 193.
12. Burke, J. J., & Moench, M. H. (2000). *Groundwater and Society: Resources, Tensions and Opportunities*. United Nations Department of Economic and Social Affairs and Institute for Social and Environmental Transition, United Nations.
13. Carter, T. R., & La Rovere, E. L. (2001). Developing and applying scenarios. In J. J. McCarthy, O. F. Canziani, N. A. Leary, et al. (Eds.), *Climate Change 2001: Impacts, Adaptation and Vulnerability* (pp. 145–190). Cambridge University Press.
14. Commission of the European Communities (CEC). (2007). *Adapting to climate change in Europe–Options for EU Action*. Green Paper from the Commission to the Council, The European Parliament, The European Economic and Social Committee, and the Committee of the Regions. Brussels.

15. Chen, C. X., Wan, J. W., & Zhan, H. B. (2003). Theoretical and experimental studies of coupled seepage-pipe flow to a horizontal well. *Journal of Hydrology, 281*, 159–171.

16. Collins, W. D., Friedlingstein, P., Gaye, A. T., et al. (2007). Global climate projections. In S. Solomon, D. Qin, M. Manning, et al. (Eds.), *Climate Change 2007: The Physical Science Basis. Contribution of Working Group I to the Fourth Assessment Report of the Intergovernmental Panel on Climate Change* (pp. 747–846). Cambridge University Press.

17. Conway, D. (1998). Recent climate variability and future climate change scenarios for Great Britain. *Progress in Physical Geography, 22*, 350–374.

18. Cook, H. F. (1998). *The Protection and Conservation of Water Resources: A British Perspective*. John Wiley.

19. Dai, A., Trenberth, K. E., & Karl, T. (1998). Global variations in droughts and wet spells: 1900–1995. *Geophysical Research Letters, 25*, 3367–3370.

20. Dennehy, K. F., Litke, D. W., & McMahon, P. B. (2002). The High Plains Aquifer, USA: Groundwater development and sustainability. In K. M. Hiscock, M. O. Rivett, & R. M. Davison (Eds.), *Sustainable Groundwater Development* (pp. 99–119). Geological Society, London, Special Publications 193.

21. Dessai, S., & Hulme, M. (2007). Assessing the robustness of adaptation decisions to climate change uncertainties: A case study on water resources management in the East of England. *Global Environmental Change, 17*, 59–72.

22. Doussan, C., Poitevin, G., Ledoux, E., & Detay, M. (1997). River bank filtration: Modelling of the changes in water chemistry with emphasis on nitrogen species. *Journal of Contaminant Hydrology, 25*, 129–156.

23. Downing, R. A. (1993). Groundwater resources, their development and management in the UK: An historical perspective. *Quarterly Journal of Engineering Geology, 26*, 335–358.

24. Downing, R. A., Price, M., & Jones, G. P. (1993). The making of an aquifer. In R. A. Downing, M. Price, & G. P. Jones (Eds.), *The Hydrogeology of the Chalk of North-West Europe* (pp. 1–13). Clarendon Press.

25. Eastwood, J. C., & Stanfield, P. J. (2001). Key success factors in an ASR scheme. *Quarterly Journal of Engineering Geology and Hydrogeology, 34*, 399–409.

26. Foster, S., Morris, B., Lawrence, A., & Chilton, J. (1999). Groundwater impacts and issues in developing cities – An introduction. In J. Chilton (Ed.), *Groundwater in the Urban Environment: Selected City Profiles* (pp. 3–16). A.A. Balkema.

27. Fritts, H. C. (1976). *Tree rings and climate*. Academic Press, Inc.

28. Gilvear, D. J., & McInnes, R. J. (1994). Wetland hydrological vulnerability and the use of classification procedures: A Scottish case study. *Journal of Environmental Management, 42*, 403–414.

29. Goode, D. (1977). Peatlands. In D. A. Ratcliffe (Ed.), *A Nature Conservation Review* (Vol. 1, pp. 249–287). Cambridge University Press.

30. Gringorten, I. I. (1963). A plotting rule for extreme probability paper. *Journal of Geophysical Research, 68*, 813–814.

31. Grischek, T., Hiscock, K. M., Metschies, T., et al. (1998). Factors affecting denitrification during infiltration of river water into a sand aquifer in Saxony, eastern Germany. *Water Research, 32*, 450–460.

32. Grischek, T., Schoenheinz, D., Worch, E., & Hiscock, K. (2002). Bank filtration in Europe – An overview of aquifer conditions and hydraulic controls. In P. J. Dillon (Ed.), *Management of Aquifer Recharge for Sustainability* (pp. 485–488). A.A. Balkema Publishers.

33. Groombridge, B., & Jenkins, M. D. (2000). *Global Biodiversity: Earth's Living Resources in the 21st Century*. UNEP–World Conservation Monitoring Centre, World Conservation Press.

34. Harding, M. (1993). Redgrave and Lopham Fens, East Anglia, England: A case study of change in flora and fauna due to groundwater abstraction. *Biological Conservation, 66*, 35–45.

35. Heberer, T. (2002). Tracking persistent pharmaceutical residues from municipal sewage to drinking water. *Journal of Hydrology, 266*, 175–189.

36. Herrera-Pantoja, M., & Hiscock, K. M. (2007). The effects of climate change on potential groundwater recharge in Great Britain. *Hydrological Processes, 22*, 73–86.

37. Herrera-Pantoja, M., Hiscock, K. M., & Boar, R. R. (2012). The potential impact of climate change on groundwater-fed wetlands in Eastern England. *Ecohydrology, 5*, 401–413.

38. Hiscock, K. M., & Grischek, T. (2002). Attenuation of groundwater pollution by bank filtration. *Journal of Hydrology, 266*, 139–144.

39. Hiscock, K. M., Rivett, M. O., & Davison, R. M. (Eds.). (2002). *Sustainable Groundwater Development*. Geological Society, London, Special Publications 193.

40. Hiscock, K. M., Sparkes, R., & Hodgson, A. (2011). Evaluation of future climate change impacts on European groundwater resources. In T. Holger, J. L. Martin-Bordes, & J. Gurdak (Eds.), *Climate Change Effects on Groundwater Resources: A Global Synthesis of Findings and Recommendations* (pp. 27). CRC Press. IAH International Contributions to Hydrogeology.

41. Holman, I. P. (2006). Climate change impacts on groundwater recharge—uncertainty, shortcomings, and the way forward? *Hydrogeology Journal, 14*, 637–647.

42. Holman, I. P., Nicholls, R. J., Berry, P. M., et al. (2005). A regional, multi-sectoral and integrated assessment of the impacts of climate and socio-economic change in the UK: Part 2 Results. *Climatic Change, 71*, 43–73.

43. Hulme, M., Turnpenny, J., & Jenkins, G. (2002). *Climate Change Scenarios for the United Kingdom. The UKCIP02 Briefing Report*. Tyndall Centre for Climate Change, University of East Anglia, Norwich, and Hadley Centre for Climate Prediction and Research, UK Meteorological Office.

44. IAH. (2006). *Groundwater for Life and Livelihoods–The Framework for Sustainable Use. 4th World Water Forum Invitation and Briefing*. International Association of Hydrogeologists, Kenilworth.

45. Iglesias, A., Garotte, L., Flores, F., & Moneo, M. (2007). Challenges to manage the risk of water scarcity and climate change in the Mediterranean. *Water Resources Management, 21*, 775–788.

46. IPCC. (1990). *Climate Change. The Intergovernmental Panel on Climate Change Scientific Assessment* (Eds. J. T. Houghton, G. J. Jenkins, & J. J. Ephraums). Cambridge University Press.
47. IPCC. (1998). *The Regional Impacts of Climate Change: An Assessment of Vulnerability. A Special Report of the IPCC Working Group II* (Eds. R. T. Watson, M. C. Zinyowera, & R. H. Moss). Cambridge University Press.
48. IPCC. (2001). *Climate Change 2001. The Scientific Basis. A Report of Working Group I of the Intergovernmental Panel on Climate Change*. Cambridge University Press.
49. IPCC. (2001). *Climate Change 2001. The Scientific Basis*. Cambridge University Press. A Report of Working Group I of the Intergovernmental Panel on Climate Change.
50. IPCC. (2007). Summary for Policymakers. In M. L. Parry, O. F. Canziani, J. P. Palutikof, et al. (Eds.), *Climate Change 2007: Impacts, Adaptation and Vulnerability. Contribution of Working Group II to the Fourth Assessment Report of the Intergovernmental Panel on Climate Change* (pp. 7–22). Cambridge University Press.
51. IPCC. (2008). *Technical Paper on Climate Change and Water (Finalized at the 37th Session of the IPCC Bureau)*. Intergovernmental Panel on Climate Change, Geneva.
52. Ireson, A. M., & Butler, A. P. (2011). Controls on preferential recharge to Chalk aquifers. *Journal of Hydrology, 398*, 109–123.
53. Issar, A. S. (2003). *Climate Change during the Holocene and Their Impact on Hydrological Systems*. UNESCO International Hydrological Series. Cambridge University Press.
54. Issar, A. S. (2008). A tale of two cities in ancient Canaan: How the groundwater storage capacity of Arad and Jericho decided their history. In W. Dragoni & B. S. Sukhija (Eds.), *Climate Change and Groundwater* (pp. 137–143). Geological Society, London, Special Publications 288.
55. Kabat, P., van Vierssen, W., Veraart, J., et al. (2005). Climate proofing the Netherlands. *Nature, 438*, 283–284.
56. Keddy, P. A. (2000). *Wetland Ecology: Principles and Conservation*. Cambridge University Press.
57. Krinner, W., Lallana, C., Estrela, T., et al. (1999). *Sustainable Water Use in Europe–Part 1: Sectoral Use of Water*. European Environment Agency, Copenhagen.
58. Lloyd, J. W., Tellam, J. H., Rukin, N., & Lerner, D. N. (1993). Wetland vulnerability in East Anglia: A possible conceptual framework and generalized approach. *Journal of Environmental Management, 37*, 87–102.
59. Loáiciga, H. A. (2003). Climate change and groundwater. *Annals of the Association of American Geographers, 93*, 30–41.
60. Loáiciga, H. A., Maidment, D. R., & Valdes, J. B. (2000). Climate-change impacts in a regional karst aquifer, Texas, USA. *Journal of Hydrology, 227*(1–4), 173–194.
61. Loáiciga, H. A., Valdes, J. B., Vogel, R., et al. (1996). Global warming and the hydrologic cycle. *Journal of Hydrology, 174*(1–2), 83–127.

62. Luckey, R. R., Gutentag, E. D., Heimes, F. J., & Weeks, J. B. (1986). Digital simulation of ground-water flow in the High Plains aquifer in parts of Colorado, Kansas, Nebraska, New Mexico, Oklahoma, South Dakota, Texas, and Wyoming. *United States Geological Survey Professional Paper, 1400–D.*

63. Mawdsley, J., Petts, G., & Walker, S. (1994). *Assessment of Drought Severity*. Institute of Hydrology, Wallingford. British Hydrological Society Occasional Paper No. 3.

64. McKenzie, H. S., & Elsaleh, B. O. (1994). The Libyan Great Man-Made River Project Paper 1. Project overview. *Proceedings of the Institution of Civil Engineers–Water Maritime and Environment, 106*(2), 103–122.

65. McPherson, B. F., & Halley, R. (1997). The South Florida Environment–A Region Under Stress. *United States Geological Survey Circular, 1134*.

66. Mitsch, W. J., & Gosselink, J. G. (2000). *Wetlands* (3rd ed.). John Wiley.

67. Mitsch, W. J., Mitsch, R. H., & Turner, R. E. (1994). Wetlands of the old and new worlds: Ecology and management. In W. J. Mitsch (Ed.), *Global Wetlands: Old World and New* (pp. 3–56). Elsevier Science B.V.

68. Navid, D. (1989). The international law of migratory species: The Ramsar Convention. *Natural Resources Journal, 29*(4), 1001–1016.

69. O'Shea, M. J., & Sage, R. (1999). Aquifer recharge: An operational drought-management strategy in North London. *Journal of the Institution of Water and Environmental Management, 13*(6), 400–405.

70. Ojo, O., Oni, F., & Ogunkunle, O. (2003). Implications of climate variability and climate change on water resources availability and water resources management in West Africa. In S. Franks, S. Bloschl, M. Kumagai, et al. (Eds.), *Water Resources Systems–Water Availability and Global Change* (pp. 37–47). International Association of Hydrological Sciences.

71. Palmer, W. C. (1965). *Meteorological Drought*. Research Paper No. 45. Department of Commerce, Washington, DC.

72. Park, E., & Zhan, H. B. (2003). Hydraulics of horizontal wells in fractured shallow aquifer systems. *Journal of Hydrology, 281*(1–2), 147–158.

73. Pim, R. H., & Binsariti, A. (1994). The Libyan Great Man-Made River Project Paper 2. The water resource. *Proceedings of the Institution of Civil Engineers–Water Maritime and Environment, 106*(2), 123–145.

74. Plummer, C. R., Nelson, J. D., & Zumwalt, G. S. (1997). Horizontal and vertical well comparison for in situ air sparging. *Ground Water Monitoring and Remediation, 17*(4), 91–96.

75. Por, D. F. (2004). The Levantine waterway, riparian archaeology, paleolimnology, and conservation. In I. Goren-Inbar & J. Speth (Eds.), *Human Paleoecology in the Levantine Corridor* (pp. 5–20). Oxbow Books.

76. Price, M. (2002). Who needs sustainability? In K. M. Hiscock, M. O. Rivett, & R. M. Davison (Eds.), *Sustainable Groundwater Development* (pp. 75–81). Geological Society, London, Special Publications 193.

77. Purkey, D. R., Joyce, B., Vicuna, S., et al. (2007). Robust analysis of future climate change impacts on water for agriculture and other sectors: A case study in the Sacramento Valley. *Climatic Change, 87*(Suppl 1), S109–S122.

78. Pyne, R. D. G. (1995). *Groundwater Recharge and Wells: A Guide to Aquifer Storage and Recovery*. CRC Press, Inc.

79. Salgot, M., & Torrens, A. (2008). Impacts of climatic change on water resources: The future of groundwater recharge with reclaimed water in the south of Europe. In W. Dragoni & B. S. Sukhija (Eds.), *Climate Change and Groundwater* (pp. 145–168). Geological Society, London, Special Publications 288.

80. Schubert, J. (2002). Hydraulic aspects of riverbank filtration – field studies. *Journal of Hydrology, 266*(3–4), 145–161.

81. Sheets, R. A., Darner, R. A., & Whitteberry, B. L. (2002). Lag times of bank filtration at a well field, Cincinnati, Ohio, USA. *Journal of Hydrology, 266*(3–4), 162–174.

82. Tanaka, S. K., Zhu, T. J., Lund, J. R., et al. (2006). Climate warming and water management adaptation for California. *Climatic Change, 76*(3–4), 361–387.

83. Trenberth, K., Overpeck, J., & Soloman, S. (2004). Exploring drought and its implications for the future. *Eos, 85*(27), 27.

84. Trenberth, K. E., Dai, A., Rasmussen, R. M., & Parsons, D. B. (2003). The changing character of precipitation. *Bulletin of the American Meteorological Society, 84*(9), 1205–1217.

85. Tufenkji, N., Ryan, J. N., & Elimelech, M. (2002). The promise of bank filtration. *Environmental Science & Technology, 36*(21), 423A–428A.

86. United Nations Environment Programme (UNEP). (2003). *Groundwater and Its Susceptibility to Degradation: A Global Assessment of the Problem and Options for Management*. Early Warning and Assessment Report Series, RS 03-3. Nairobi, Kenya.

87. United States Environmental Protection Agency (EPA). (2006). *National Primary Drinking Water Regulations: Long Term 2 Enhanced Surface Water Treatment Rule*. Environmental Protection Agency, 40 CFR Parts 9, 141 and 142: 654–786. Washington, DC.

88. United States Geological Survey (USGS). (2002). *SOFIA-SFRSF-Hydrology-What is Aquifer Storage and Recovery (ASR)?* Retrieved from: http://sofia.usgs.gov/sfrsf/rooms/hydrology/ASR/index.html (accessed on 18 July 2024).

89. Vaccaro, J. J. (1992). Sensitivity of groundwater recharge estimates to climate variability and change, Columbia Plateau, Washington. *Journal of Geophysical Research, 97*(D3), 2821–2833.

90. Varela-Ortega, C., Blanco-Gutiérrez, I., Swartz, C. H., & Downing, T. E. (2011). Balancing groundwater conservation and rural livelihoods under water and climate uncertainties: An integrated hydro-economic modelling framework. *Global Environmental Change, 21*(2), 604–619.

91. Verstraeten, I. M., Thurman, E. M., Lindsey, M. E., et al. (2002). Changes in concentrations of triazine and acetamide herbicides by bank filtration, ozonation, and chlorination in a public water supply. *Journal of Hydrology, 266*(3–4), 190–208.

92. Wheeler, B. D., & Proctor, M. C. F. (2000). Ecological gradients, subdivisions and terminology of north-west European mires. *Journal of Ecology, 88*(2), 187–203.

93. Wilby, R. L., & Harris, I. (2006). A framework for assessing uncertainties in climate change: Low-flow scenarios for the River Thames, UK. *Water Resources Research, 42*(2), W02419. https://doi.org/10.1029/2005WR004065.

94. Younger, P. L., Teutsch, G., Custodio, E., et al. (2002). Assessments of the sensitivity to climate change of flow and natural water quality in four major carbonate aquifers of Europe. In K. M. Hiscock, M. O. Rivett, & R. M. Davison (Eds.), *Sustainable Groundwater Development* (pp. 303–323). Geological Society, London, Special Publications 193.

95. Yusoff, I., Hiscock, K. M., & Conway, D. (2002). Simulation of the impacts of climate change on groundwater resources in eastern England. In K. M. Hiscock, M. O. Rivett, & R. M. Davison (Eds.), *Sustainable Groundwater Development* (pp. 325–344). Geological Society, London, Special Publications 193.

96. Zektser, I. S., & Loaiciga, H. A. (1993). Groundwater fluxes in the global hydrologic cycle: Past, present and future. *Journal of Hydrology, 144*(1–4), 405–427.

97. Zhan, H. B., & Park, E. (2002). Vapor flow to horizontal wells in unsaturated zones. *Soil Science Society of America Journal, 66*(3), 710–721.

98. Zhang, B. Y., & Lerner, D. N. (2000). Modeling of ground water flow to adits. *Ground Water, 38*(1), 99–105.

Index

H

I

FSC
www.fsc.org
MIX
Papier | Fördert gute Waldnutzung
FSC® C083411

Kolibri 360 GmbH
Ferdinand-Jühlke-Straße 7
99095 Erfurt, Deutschland
produktsicherheit@kolibri360.de